CHILDHOOD NUTRITION

MODERN NUTRITION

Edited by Ira Wolinsky and James F. Hickson, Jr.

Published Titles

Manganese in Health and Disease, Dorothy Klimis-Tavantzis
Nutrition and AIDS: Effects and Treatment, Ronald R. Watson

Forthcoming Titles

Calcium and Phosphorus in Health and Disease, John B. Anderson and Sanford Garner
Nutrition Care for HIV-Positive Persons: A Manual for Individuals and Their Caregivers, Saroj M. Bahl and James F. Hickson, Jr.
Nutrition and Health: An International Perspective, Saroj M. Bahl
Zinc in Health and Disease, Mary E. Mohs

Edited by Ira Wolinsky

Published Titles

Practical Handbook of Nutrition in Clinical Practice, Donald F. Kirby and Stanley J. Dudrick
Advanced Nutrition: Macronutrients, Carolyn D. Berdanier

Forthcoming Titles

Laboratory Tests for the Assessment of Nutritional Status, 2nd Edition, H. E. Sauberlich
Nutrition and Cancer Prevention, Ronald R. Watson and Siraj I. Mufti
Nutrition and Health: Topics and Controversies, Felix Bronner
Nutrition and Hypertension, Michael B. Zemel
Nutrition: Chemistry and Biology, 2nd Edition, Julian E. Spallholz and L. Mallory Boylan
Nutritional Concerns of Women, Ira Wolinsky and Dorothy Klimis-Tavantzis

CHILDHOOD NUTRITION

Edited by

Fima Lifshitz

CRC Press
Boca Raton Ann Arbor London Tokyo

Library of Congress Cataloging-in-Publication Data

Childhood nutrition / editor, Fima Lifshitz.
p. 00 cm.—(Modern nutrition)
Includes bibliographical references and index.
ISBN 0–8493–2764–4
1. Children—Nutrition. 2. Diet therapy for children.
I. Lifshitz, Fima, 1938–0000. II. Series.
RJ206.C5185 1994
618.92'39—dc20

94–14377
CIP

International Standard Book Number 0–8493–2764–4
Library of Congress Card Number 94–14377
Printed in the United States of America 1 2 3 4 5 6 7 8 9 0
Printed on acid-free paper

DEDICATION

To Maimonides: Whose 13th century teachings in nutrition are still inspiring.

APHORISMS OF MOSES BEN MAIMONIDES (1135–1204)

—*The need for food is never ending in health and disease.*

—*Health is a condition all people need. However, not all can follow the rules for this.*

—*Therefore, in medicine, knowledge of dietetics is most helpful.*

—*Immobility is a great detriment for the maintenance of health, as activity is of benefit.*

SERIES PREFACE FOR MODERN NUTRITION

The CRC Series in Modern Nutrition is dedicated to providing the widest possible coverage to topics in nutrition, in all its diversity, both basic and applied.

Published for an advanced or scholarly audience, the volumes of the Series in Modern Nutrition are designed to explain, review, and explore present knowledge and recent trends, developments, and advances across the field. As such, they will also appeal to the educated layman. Volumes in this series will reflect the broad scope of information available as well as the interests of authors. The diversity of topics will appeal to an equally diverse readership from volume to volume. The format for the series will vary with the needs of the author and the topic, including but not limited to, edited volumes, monographs, handbooks, and texts.

Contributors from any bona fide area of nutrition, including the controversial, are welcome.

James F. Hickson, Jr., Ph.D., R.D.
Ira Wolinsky, Ph.D.
Series Editors

PREFACE

As we approach the 21st century, pediatricians with nutrition expertise and nutritional scientists ponder the most important nutritional concerns for children. In this book, we have gathered an outstanding group of specialists who have written extensively in their fields of expertise. *Childhood Nutrition* provides the reader with accurate facts on current concepts in pediatric nutrition with regards to the pathophysiology of disease and developing nutritional health care plans. It fulfills the needs of students, clinicians, and professors involved in children's nutrition. Each chapter is thoroughly researched and referenced, and the book should become a valuable resource for physicians, nutritionists, dieticians, and anyone else interested in the field. Health promotion through disease prevention with particular attention to nutrition has now attracted the attention of politicians and has become prominent enough to be the backbone of the future healthcare policy for the United States.

Fima Lifshitz, M.D.

FOREWORD

Many times, in attending International Pediatric Congresses, we have spoken with fellow participants from different parts of the world, and we have all agreed that, for obvious reasons, the field of nutrition has not received sufficient attention. On the other hand, in International Nutrition Congresses, we have observed the same situation concerning pediatrics. These facts were discussed with interested colleagues from various countries and precipitated the creation of the International Society of Pediatric Nutrition (ISPEN). This society has formed to gather professionals (medical doctors, nutritionists, psychologists, biologists, and others) involved in the study of nutrition in children, adolescents, and pregnant women. It will also promote the dissemination of scientific research projects and papers and organize the international congress and other meetings on Pediatric Nutrition. Foremost, this society will stimulate the investigation of nutrition concerns for children.

The society has already promoted the First World Congress of Nutrition in Pediatrics, in March 1990, in the city of São Paulo, Brazil, and the Second International Congress of Pediatric Nutrition in Lisbon, Portugal, in March 1994. Additional meetings included the Latin America meeting held in September 1991, coordinated by myself, the Asian regional meeting held in December 1992 with Dr. Perla Santos Ocampo as coordinator, and the African meeting held in December 1993 with Dr. Mamdouh Gabr as coordinator. Regional preparatory meetings for the nutritional congress were also organized under the auspices of ISPEN, specifically for the discussion of "Main Nutrition Problems of the World: Risk Groups and Proposals for Intervention Programs." Regional encounters with local specialists were organized to prepare reports by world experts considering the different nutritional concerns of their continents. The North America Meeting resulted in the publication of this book. This regional meeting and the book were organized and edited by Dr. Fima Lifshitz.

Finally, I would like to thank the Kellogg Company, which, through Dr. Victor Fulgoni, has contributed greatly to the realization of the North American Meeting. I would also congratulate Dr. Fima Lifshitz who organized and presided over the meeting and through his editorial efforts, has also facilitated publication of this book. This volume contains the most current information on the nutritional concerns of North America for this decade and issues for the future century. The contributors to the book are all recognized authorities in their fields. This book will surely be of help as a resource and reference for physicians, nutritional scientists, dietitians, and all those involved in childhood nutrition as we approach the 21st century.

Fernando Nobrega, M.D.

THE EDITOR

Fima Lifshitz, M.D., is Professor of Pediatrics and Chairman of the Department of Pediatrics at Maimonides Medical Center—State University of New York Health Science Center at Brooklyn, New York. Dr. Lifshitz is the author of more than 300 journal articles, reviews, and book chapters. He is the editor of 16 books, including *Clinical Disorders in Pediatric Gastroenterology and Nutrition, Carbohydrate Intolerance in Infancy, Pediatric Nutrition: Infant Feeding-Deficiencies-Diseases, Common Pediatric Disorders, Nutrition for Special Needs in Infancy, Food Allergy, Children's Nutrition,* and *Pediatric Endocrinology,* which is now in its 3rd edition. He is also editor of the book series *Clinical Pediatrics* and *Clinical Disorders in Pediatric Nutrition* (Marcel Dekker, Inc.). His latest accomplishment is the editorship of the book *Childhood Nutrition,* and becoming the Editor-in-Chief of the *Journal of the American College of Nutrition* in 1993. Dr. Lifshitz is a member of numerous professional societies, including the American Pediatric Society, Society for Pediatric Research, Society of Experimental Biology and Medicine, American Endocrine Society, American Diabetes Association, and the Lawson Wilkins Pediatric Endocrine Society.

CONTRIBUTORS

Marvin E. Ament, M.D.
Professor of Pediatrics
University of California
Los Angeles School of Medicine
Los Angeles, California, and
Chief
Division of Pediatric Gastroenterology and Nutrition
UCLA Medical Center
Los Angeles, California

Ronald Bainbridge, M.B.B.S.
Assistant Professor of Pediatrics
SUNY Health Science Center at Brooklyn, and
Associate Attending Neonatologist
Department of Pediatrics
Maimonides Medical Center
Brooklyn, New York

Ranjit Kumar Chandra, M.D.
Director of Immunology and Pediatrics
Memorial University of Newfoundland
St. John's, Newfoundland, Canada

Lloyd J. Filer, Jr., M.D., Ph.D.
Professor Emeritus
Department of Pediatrics
College of Medicine
University of Iowa
Iowa City, Iowa

Laurence Finberg, M.D.
Professor and Chairman of Pediatrics
Children's Medical Center
SUNY Health Science Center at Brooklyn
Brooklyn, New York

Arturo R. Hervada, M.D.
Professor of Pediatrics
Jefferson Medical College
Thomas Jefferson University
Philadelphia, Pennsylvania

Maria Hervada-Page, M.S.S.
Department of Family Medicine
Jefferson Medical College
Thomas Jefferson University
Philadelphia, Pennsylvania

Robert Karp, M.D.
Associate Professor of Clinical Pediatrics
Children's Medical Center
SUNY Health Science Center at Brooklyn, and
Director
Pediatric Resource Center
Kings County Hospital Center
Brooklyn, New York

Ronald E. Kleinman, M.D.
Associate Professor of Pediatrics
Harvard Medical School, and
Chief
Division of Pediatric Gastroenterology and Nutrition
Massachusetts General Hospital
Boston, Massachusetts

Donald P. Kotler, M.D.
Associate Professor of Medicine
Columbia University College of Physicians and Surgeons
Gastrointestinal Division
Department of Medicine
St. Luke's Roosevelt Hospital Center
New York, New York

Norman Kretchmer, M.D., Ph.D.
Professor of Nutritional Sciences, Pediatrics, and Obstetrics
University of California
Berkeley, California

Emanuel Lebenthal, M.D.
Professor of Pediatrics
Director, International Institute for Infant Nutrition and Gastrointestinal Disease
Hahnemann University
Philadelphia, Pennsylvania

Fima Lifshitz, M.D.
Professor of Pediatrics
SUNY Health Science Center at Brooklyn, and
Chairman
Department of Pediatrics
Maimonides Medical Center
Brooklyn, New York

Andrea Maggioni, M.D.
Research Fellow
Division of Pediatric Gastroenterology and Nutrition
Department of Pediatrics
Maimonides Medical Center
Brooklyn, New York

Reuben Matalon, M.D., Ph.D.
Chief
Genetics and Metabolism, and
Director
The Research Institute
Miami Children's Hospital
Miami, Florida

Kimberlee Michals-Matalon, Ph.D., R.D.
Miami Children's Hospital
Miami, Florida

Adib A. Moukarzel, M.D., Ph.D.
Assistant Professor of Pediatrics
SUNY Health Science Center at Brooklyn, and
Director
Division of Pediatric Gastroenterology and Nutrition
Department of Pediatrics
Maimonides Medical Center
Brooklyn, New York

Thomas B. Newman, M.D., M.P.H.
Associate Professor of Pediatrics
University of California Medical Center
San Francisco, California

Fernando José de Nóbrega, M.D.
Executive Director
International Society of Pediatric Nutrition
São Paulo, Brazil

Russell Rising, Ph.D.
Chief
Body Composition and Nutrition Laboratory
Department of Pediatrics
Maimonides Medical Center
Brooklyn, New York

Audrey Rosner, M.A., R.N., C.P.N.P.
Schulman Medical Associates
Brooklyn, New York

Susan K. Schulman, M.D.
Attending Pediatrician
Department of Pediatrics
Maimonides Medical Center
Brooklyn, New York

Mildred S. Seelig, M.D., M.P.H.
Adjunct Professor
Department of Nutrition
School of Public Health
North Carolina University Medical Center
Chapel Hill, North Carolina

Melanie M. Smith, M.N.S., R.D.
Instructor of Pediatrics
SUNY Health Science Center at Brooklyn, and
Chief Pediatric Nutritionist
Department of Pediatrics
Maimonides Medical Center
Brooklyn, New York

Omer Tarim, M.D.
Fellow
Pediatric Endocrinology
Department of Pediatrics
Maimonides Medical Center
Brooklyn, New York

Reginald Tsang, M.B.B.S.
Professor of Pediatrics, Obstetrics, and Gynecology
University of Cincinnati College of Medicine
Cincinnati, Ohio

Myron Winick, M.D.
Williams Professor of Nutrition (Emeritus)
Columbia University College of Physicians & Surgeons
New York, New York

Michael B. Zimmermann, M.D.
Fellow
Department of Nutritional Sciences
University of California
Berkeley, California

TABLE OF CONTENTS

Chapter 1
Historical Overview and Future Perspectives on Pediatric Nutritional Science 1
Laurence Finberg

Chapter 2
Controversies in Childhood Nutrition 5
Ronald E. Kleinman

Chapter 3
Cholesterol Screening and Dietary Intervention in Childhood for Prevention of Adult Onset Cardiovascular Disease 13
Omer Tarim, Thomas B. Newman, and Fima Lifshitz

Chapter 4
A Clinician's Approach to Initiating Breastfeeding 21
Susan K. Schulman and Audrey Rosner

Chapter 5
Optimal Nutrition in Low Birth Weight Infants 33
Ronald Bainbridge and Reginald Tsang

Chapter 6
Infant Nutrition: The First Two Years 43
Arturo R. Hervada and Maria Hervada-Page

Chapter 7
Iron Deficiency 53
Lloyd J. Filer, Jr.

Chapter 8
Food Allergy and Atopic Disease 61
Ranjit Kumar Chandra

Chapter 9
Nutrition Therapy of Inborn Errors of Metabolism 71
Kimberlee Michals-Matalon and Reuben Matalon

Chapter 10
The Pathophysiology of Childhood Obesity 81
Russell Rising

Chapter 11
Nutrition for Special Needs—In Pediatric Gastrointestinal Diseases 99
Emanuel Lebenthal

Chapter 12
Diarrhea and Malnutrition 107
Andrea Maggioni and Fima Lifshitz

Chapter 13
Nutritional Complications of HIV Infection 137
Donald P. Kotler

Chapter 14
Nutrition and Growth 143
Fima Lifshitz and Melanie M. Smith

Chapter 15
Trace Elements in Parenteral Nutrition 159
Adib A. Moukarzel and Marvin E. Ament

Chapter 16
Home Parenteral Nutrition 183
Adib A. Moukarzel and Marvin E. Ament

Chapter 17
Prenatal and Genetic Magnesium Deficiency in Cardiomyopathy: Possible Vitamin and Trace Mineral Interactions 197
Mildred S. Seelig

Chapter 18
Nonnutritive Dietary Supplements in Pediatrics 225
Michael B. Zimmermann and Norman Kretchmer

Chapter 19
The Ecology of Poverty, Undernutrition, and Learning Failure 239
Robert J. Karp

Chapter 20
Nutrition Education in Medical Schools and Residency Training 251
Myron Winick

Index 255

Chapter 1

HISTORICAL OVERVIEW AND FUTURE PERSPECTIVES ON PEDIATRIC NUTRITIONAL SCIENCE

Laurence Finberg

The history of pediatric contributions to nutritional science is much too rich to relate in a brief presentation. I shall therefore be selective and choose to use some seminal events as examples of an illustrious past.

I shall begin by retelling events that are of particular interest to me. This first event I believe marks the start of the scientific era for nutrition. During the cholera epidemic in Great Britain in 1831, a young man by the name of William Brooke O'Shaughnessy could not practice medicine because he had graduated from a Scots medical school. Licensing was controlled by the physicians of London; graduates of London schools would not grant him a license. He therefore began to study cholera patients. Although he was not the first to analyze human blood for its constituents, he was one of the early ones to do so. He performed chemical analyses of both the blood and stools of cholera patients. As a chemist he only had inorganic chemistry techniques to use for his analyses; therefore what he analyzed were the electrolytes. He noticed that the composition for the stool water in cholera victims was very similar to that of their plasma. These values were close to those of normal controls, except that the patients had markedly reduced water content. He deduced they had lost large amounts of water and salts. He suggested from these data that replacing water and salt would be beneficial to them.

The history that has been handed down to us shows that he first made known his findings by publishing his observations in a very brief letter to the *Lancet* in December, 1831.[1] By May 1832, scarcely five months later, Thomas Latta, a practitioner in Leith, Scotland, through a surrogate, published in the *Lancet* a report successfully carrying out O'Shaughnessy's suggestion.[2] Mind you, he had no needles, syringes, or tubing and no one to make the solutions for him. He had to make all the equipment and solutions himself while carrying on a busy practice. In five months he had treated 16 patients, 8 of them successfully.

This landmark in medicine constitutes the first example of someone solving a problem by gathering data on human subjects and initiating the therapeutic process. Interestingly enough, of course, physicians being conservative, the general implementation of this therapy in medical practice was delayed for 80 years. It was not until roughly 1910, when technology became available and more people had demonstrated that O'Shaughnessy and Latta were right, that fluid therapy became a staple in medical practice. Fluid therapy thus begins scientific nutrition.

0-8493-2764-4/95

Referring again to the 19th century, when the industrial revolution was well under way, there were a number of social changes relevant to nutritional events. Among them were women entering the work place; mothers and, perhaps in some way more important, women who had served as wet nurses now found that they could seek more remunerative employment. The availability of breast feeding for infants had been the only source of feeding available until this time. Alternative infant feeding began to appear as an item for public and medical concern. Liebig in Germany developed a food to feed infants he considered to be perfect. The food was made of a mixture of wheat flour, malt, and cow milk. When cooked, it could be fed to babies. The mixture began to be used for the feeding of infants; however, the mortality rate for babies not breastfed was enormous, just as it continues to be today in underdeveloped parts of the world.

Looking back, we can be fairly sure that the principal problem was contamination and subsequent infection; at the time, physicians worried most about composition. Different feedings continued to appear both in Europe and North America. Science again emerged, primarily around Ruebner's department in Berlin, where he and his associates began to look at nutrition as a science. First, they showed the interchangeability of fat and carbohydrates in metabolic systems. They then went on to develop calorimetry so that they could assess food in the production of energy. This constitutes another landmark.

A number of pediatric names surface at the turn of the century—Czerny and Finkelstein from Germany, the latter emigrating to Chile. Thomas Morgan Rotch—the Pediatric Professor of Harvard—built his career on infant nutrition using "percentage feeding," which primarily kept women out of the work place because they had to figure out in his complicated system what to feed the baby each day. In this same exciting period, the concept of vitamins as essential substances which had to be ingested in order for metabolism to proceed came along. Mellanby, Alfred Hess, and Edwards Park began the Vitamin D story in the 1920s.

Gerstenberg, striving to find the ideal infant food, developed a product in which he modified cow milk. In order to make it resemble human milk, he took out the butter fat and replaced it with beef and vegetable fat.[3] He began to promote this product—the beginning of the infant formula industry in the United States. A number of companies at this time were promoting additives, some famous ones such as Horlicks malted milk, Ovaltine, and dextrimaltose, among them.

Meanwhile, the electrolyte physiology story advanced through the work of two famous pediatricians, James Gamble and Daniel Darrow. Grover Powers, a clinician who did not do research himself but had enormous appreciation for the value of research and an intuitive understanding for the translation of the research accomplishment into clinical action, published a paper in 1925 pointing out something that remains, for me, the basis for the teaching of nutrition to physicians.[4] He stated simply, "Energy is the currency of the analysis of nutrition." He understood that of the various nutrient components of a diet, both the sources of energy and non-energy requirements are best expressed using energy as the denominator. This is something we do not teach well even today. Unfortunately, many authorities often talk in terms of requirements for protein in grams per kilogram, overlooking the fact that differently aged and sized people expend different energy differently.

By 1930, as we began to move more and more toward the "artificial" feeding of infants, the standard eventually became the use of evaporated milk modified by the addition of some sort of carbohydrate. By that time, the purpose of adding carbohydrate was to reduce renal solute load, not to reduce the fat portion of the energy as a preceding generation had postulated. In the 1950s and 1960s, manufactured infant formula took over and for a while it reduced breast feeding in the United States to a minority of infants. The "return to nature" movement of the late 1960s and 1970s brought a welcome return to breast feeding which, unfortunately, again appears to be eroding.

The most recent trend in nutritional science application to feeding infants and children concerns attempts to prevent adult "degenerative diseases" through dietary practices in childhood. In the chapters to follow, this topic will emerge more fully.

REFERENCES

1. O'Shaughnessy, W.B., Experiments on the Blood in Cholera, *Lancet,* 1, 490, 1931–32.
2. Latta, T., Letter to Secretary of Central Board of Health, London. Affording view of rationale and results of his practice in treatment of cholera and saline injections, *Lancet,* 2, 274, 1831–32.
3. Gerstenberg, H.J., Haskins, H.D., McGregor, H.H., and Roh, H.O., Studies in the adaptation of an artificial food to human milk, *Am. J. Dis. Child.,* 10, 249, 1915.
4. Powers, G.F., Comparison and interpretation on a caloric basis of milk mixture used in infant feeding, *Amer. J. Dis. Child.,* 30, 453–475, 1925.

Chapter 2

CONTROVERSIES IN CHILDHOOD NUTRITION

Ronald E. Kleinman

TABLE OF CONTENTS

I. Introduction 5

II. Goals 5

III. Commandments 8

IV. Guidelines 11

References 12

I. INTRODUCTION

Over the past fifty years there has been an interesting shift in our expectations of what our diets will provide for us. This is particularly true with regard to the diets of our children. Whereas at one time we expected only that the diet would promote optimal growth, we now take that for granted and instead expect that the diet will promote optimal long-term health and minimize the risk of chronic illness due to heart disease, hypertension, and obesity. This has led to the classification of some foods as good and healthy and others as bad and potentially responsible for causing illness. Cartoons now feature vignettes in which mothers scold their children for drinking milk or eating meat. In the following discussion, I will review some of the recent recommendations with regard to children's diets and attempt to place them in the context of the science, or lack thereof, that underpins them.

II. GOALS

Recommendations for what children should be eating seem to fall into three categories: goals, commandments, and guidelines. For the first half of the twentieth century, the principal goal of nutritional science was to identify and quantify the nutrients which are necessary for optimal growth and development in infants, children, and adolescents. This goal has been realized, for the most part, for healthy infants and children, but remains an area of fertile research for those with unique nutritional needs, such as premature infants or children with chronic illnesses. There has therefore been a shift in current goals away from identification and quantification of nutrients to ensuring that all children have access to optimal nutrition, to identifying the effects of childhood nutrition on long-term health, and most recently, to identifying the ways in which genes and nutrients interact.

0-8493-2764-4/95

In spite of the fact that we have quantified the daily requirements for nutrients such as iron, vitamin A, iodine, calcium, and vitamin K, to name just a few, controversy still surrounds these and other nutrients and their roles in children's diets. Iron deficiency remains the most prevalent single nutrient deficiency among children in the United States. It is particularly prevalent among low-income infants and children, and this has been well documented and recognized.[1] Yet for fifty years between 1920 and 1970, unfortified or low iron formulas were recommended by a majority of pediatricians, largely because of anecdotal experiences that suggested that iron in infant formulas was responsible for significant intolerance to those feedings.[2] It was also felt that iron could be obtained later in infancy or early childhood in sufficient amounts from solid foods in the diet to make fortification of infant formulas with more than 1.5 mg/l unnecessary. The American Academy of Pediatrics, however, has recommended the exclusive use of high iron-supplemented formulas, except for those very few infants who suffer from iron overload conditions such as hemochromatosis.[3] The adoption of this recommendation by the WIC program has led to a very significant decrease in the prevalence of iron deficiency anemia among this group of high-risk infants.

The requirement for vitamin A has also been well established at approximately 1300 IU/d. As is the case for iron, the role of vitamin A in growth and development has been well established, and it should therefore be an unlikely candidate for controversy. Yet recently we have come to understand that vitamin A has significant immune adjuvant effects, and recent work strongly suggests that serum retinol levels correlate well with the risk of mortality from measles.[4,5] This has led to controversial recommendations that all children—the well-nourished, those at high risk for malnutrition, and the malnourished—receive supplemental vitamin A. However, the benefits of providing a high dose of vitamin A to all children with measles remain uncertain. Efforts have now been directed towards providing sufficient vitamin A to children at high risk of under-nutrition and supplemental vitamin A to those who acquire measles and who are at high risk for developing complicated illness.[6]

Seventy years ago, it was recognized that simple goiter could easily be prevented by ensuring adequate amounts of iodine in the diet, and it was expected that goiter would be excluded from the list of human diseases within a short period of time as soon as society determined to make the effort to do so. Yet even today, iodine deficiency remains a common problem in some areas of the world,[7] where one in ten infants has an elevated serum thyrotropin level, compared to one in fourteen hundred in areas where iodine is present in sufficient quantities in the diet. We also continue to argue over what the optimal iodine supplement should be, with recommendations varying tenfold from 48 ug/d to as high as 480 ug/d.

In the absence of vitamin K administration at birth, classical hemorrhagic disease of the newborn will occur in between .25% and 1.7% of infants. Similarly, late hemorrhagic disease of the newborn has an incidence of 4.4 to 72/100,000 births in Europe and Asia. A single oral dose of vitamin K will reduce this incidence to approximately 1.4 to 6.4/100,000 births. Multiple oral doses or a single intramuscular dose of vitamin K completely eliminates the incidence of late hemorrhagic disease of the newborn in these populations.

Recently, it was reported that there was a doubled incidence of leukemia among children less than ten years of age who had received vitamin K intramuscularly as prophylaxis in the neonatal period. There also was an increase in other forms of cancer, although this was not statistically significant.[8] It was, therefore, recommended that infants no longer receive intramuscular vitamin K. This obviously caused significant anxiety among parents whose children already received this vitamin by the intramuscular route and caused consternation among healthcare providers who recognize that intramuscular vitamin K is the most effective way to prevent both classical and late hemorrhagic disease of the newborn. Careful scrutiny of the data show that in the United States between 1947 and the present time, there has been no increase in childhood leukemia.[9] As well, tests in bacteria and rodent lymphocytes and examination of human chromosomes demonstrate no evidence for mutagenicity of vitamin K. Therefore, vitamin K should continue to be administered intramuscularly to all infants following birth in a dose of 0.5 to 1 mg.[10]

Finally, several recent studies have challenged our concept of the optimal daily calcium intake necessary to achieve peak bone mass during childhood and have suggested that the recommended daily amounts of 800 to 1200 per day may not, in fact, achieve this goal. A recent study conducted in seventy pairs of identical twins between the ages of six and fourteen years examined the effect of increased daily calcium intake on bone mineralization over a period of three years.[11] One twin in each pair received approximately 900 mg of calcium per day with the sibling receiving approximately 1600 mg per day. The radius, hip, and spine were examined serially by absorptiometry. Prepubertal twins who received the higher calcium intake were found to have increased bone mass in the mid-shaft of the radius, in the distal radius, and at the lumbar spine. There were no significant differences in bone mass of other sites in prepubertal children and no significant differences at any site when supplementation occurred during or after puberty.

We recognize that most children today meet their daily requirement for calcium at the current recommended daily allowance. Recommendations to increase the recommended daily allowance and to fortify foods such as fruit juices and cereals with calcium are at this time premature. Issues such as whether the increased bone mass will be maintained following supplementation in childhood, whether supplementation later in life for those most at risk for osteoporosis is as effective as supplementation in childhood, and the adverse effects of calcium supplementation overall must be answered first.

Thus, many nutrients whose role and quantity in the diet have been well established remain subjects of controversy even today. It is probably well, therefore, that we continue to subscribe to the nutrition goals articulated at the World Summit for Children sponsored by the United Nations in 1990. By the year 2000, we should have reduced severe and moderate malnutrition among under five-year-old children by half of the 1990 levels. There should be a reduction of iron deficiency anemia in women by one-third of the 1990 level, and iodine deficiency should be eliminated by the year 2000. Vitamin A deficiency should be eliminated along with its consequences, including blindness. Of critical importance is the goal of empowering all women to breast feed their children exclusively for at least four to six months and to continue breast feeding with complementary food into the second year of life.

III. COMMANDMENTS

Having reviewed controversies surrounding nutrition goals—some of which have been achieved over fifty years ago and some which remain as goals—it is now appropriate to turn to examining recommendations which speak to the effects of childhood nutrition on long-term health. These goals are unfortunately often better viewed as "commandments" because they lack any basis in science and therefore must be taken on faith. There have been repeated calls for significant reductions in the amount of salt in children's diets in an effort to reduce the risk of hypertension later in life. There is no credible evidence which demonstrates that limiting salt in the diet of young children will effect the development of hypertension in adult years. Salt intake should be matched to sodium requirements, and there is clearly no need for the high levels of sodium intake found in some children's diets. Nevertheless, homeostatic mechanisms control sodium levels in the blood very well in children and permit a wide range of sodium intakes without any significant consequences for blood pressure or long-term health.

An often-heard recommendation is that we should eliminate, as much as possible, all simple sugars from the diet in order to reduce diabetes, tooth decay, and in particular, hyperactivity in children. There is no scientific evidence to suggest that simple sugars are by themselves responsible for any of these conditions. Diabetes in children is a genetically restricted disorder, and children with diabetes must match their diets and their insulin injections to produce normal serum glucose levels. There is nothing to suggest, however, that non-diabetic children are put at risk for diabetes from eating sugar. The development of tooth decay depends upon the presence of bacteria and substrate in the oral cavity. The prevention of tooth decay can best be achieved through good oral hygiene and the presence of fluoride in the diet or supplementation in the water or in toothpaste. Simple sugars do not appear to be any more cariogenic than complex sugars in promoting tooth decay. Finally, hyperactivity is unrelated to sugar or carbohydrates in the diet. Those children with true attention deficit disorder often require pharmacologic agents for adequate treatment. Running and jumping around is normal behavior during childhood and may be intensified at times when children are gathered with their friends and eat sweetened foods such as cookies and cake. Some children in fact may be so delighted in getting a piece of cookie or finally getting their parents to provide them with a piece of cake that they go a bit overboard.

Among the most frightening commandments that have recently been handed down is the one which tells us to eliminate all dairy products from the diet of infants and children to eliminate the risk of diabetes, food allergies, colic, and heart disease. Headlines in newspapers have trumpeted the potential hazards of cow's milk and questioned whether children should drink milk or instead eat lots of broccoli. The recommendations to eliminate cow's milk from the diet were inaccurately based in large measure on guidelines from the American Academy of Pediatrics concerning the use of whole cow's milk in infancy[12] and a recent study relating insulin-dependent diabetes mellitus and a bovine albumin peptide.[13] The Committee on Nutrition of the American Academy of Pediatrics recommended delaying the introduction of whole cow's milk until one year of age because whole cow's milk is not an optimal source of nutrition for rapidly growing infants when compared to human milk or infant formula. For most infants over the age of six months, whole cow's milk can be introduced into the diet without any significant danger to the infant's health, growth, or development, provided that the diet is supplemented to provide adequate amounts of iron, trace minerals, and vitamins. The Academy's recommendation was framed in the context of the most effective way to provide optimal nutrition support to vulnerable rapidly growing infants. The relationship between dairy products and diabetes remains highly speculative.[14,15] Animal studies conducted in rats and mice show that in a minority of these animals diabetes can be triggered by exposure to a cow's milk protein. However, this response can be blocked by other cow's milk proteins, and human case control breast feeding studies remain inconclusive with regard to the relationship between cow's milk proteins and diabetes. Among the Finnish children reported in a recent study of the relationship between diabetes and cow's milk protein, all had antibodies to cow milk proteins; the diabetic children had higher levels of antibodies than the healthy children.[13] Dairy products are a concentrated and familiar source of minerals, protein, and energy for American children. There is no data to support the contention that cow's milk in the diet causes diabetes and insufficient evidence at present to even suggest that infants who are genetically at risk for diabetes be restricted in their intake of dairy products.

Allergies to food proteins occur in approximately 2% of infants and children. A milk feeding, and often cow's milk-based formula, is the only food that most children under the age of four months experience, and therefore it stands to reason that cow's milk would account for most cases of food allergy at this age. Cow's milk, along with eggs, wheat, soy, and nuts are all common foods in infancy or childhood and are those foods most commonly associated with allergy. Ninety-eight percent of children or more tolerate these foods without any adverse reactions. Therefore, to suggest we eliminate milk or all of these foods from the diets of infants and young children would be to deprive them of a significant portion of their nutrient intake. While children may become intolerant to milk sugar for a brief period of time during or following an acute episode of gastroenteritis, this too is a generally insignificant problem for healthy, well-nourished children and should not be taken as an excuse or a reason to exclude these foods permanently from children's diets.

Heart disease is a major public health problem in the United States, and the relationship between saturated fat in the diet, blood cholesterol, and the risk of heart disease in adults has now been well established in studies involving hundreds of thousands of study subjects. This has led to recommendations that adults limit the amount of fat in their diet to 30% of the total calories consumed and to reduce the saturated fat to no more than 10% of total calories.[16] Whether these same findings and recommendations should be strictly applied to children remains an area of controversy, and therefore recommendations such as those to eliminate dairy products (in fact, all animal products) from the diet of children again fall into the realm of commandments (Chapter 3).

Recommendations for early intervention in childhood with regard to diet and blood cholesterol depend upon an agreement that the measurement of blood cholesterol in childhood allows for the identification of children who are at increased risk of coronary disease as adults.[17] In addition, these recommendations are credible only if intervention to reduce the future risk of coronary heart disease is proven to be more effective if it is begun in childhood rather than during adult years. And finally, intervention with regard to diet and cholesterol screening can only be recommended if it can be accomplished with a minimum of risk to the child. We know that the antecedents of atherosclerosis can be found in major blood vessels of the peripheral circulation and of the heart during childhood, and an increasing percentage of children and adolescents demonstrate fatty streaks in these vessels along with minimal amounts of atherosclerosis.[18] In addition, there is a well-identified clustering of premature coronary heart disease as well as abnormal lipid profiles among family members. Yet dietary and environmental influences on the actual progression of these fatty streaks and early atherosclerotic lesions have not been fully quantified. The major influence appears to be genetic, at least among younger individuals.[19] The uncertainties with regard to which children will go on to develop early or definite coronary heart disease is further supported by the tracking data of the Muscatene study, in which only 32% of children with total cholesterol values greater than the 75th percentile for age went on to qualify for intervention as adults, and only 52% percent of those children with total cholesterol values which fell at the 90th percentile or higher qualified for intervention as adults.[19]

Because of these uncertainties, current recommendations from the American Academy of Pediatrics and the National Cholesterol Education program are for selective screening of children and adolescents whose parents and grandparents at age 55 years or less suffered a documented myocardial infarction, angina pectoris, peripheral vascular disease, cerebral vascular disease or sudden cardiac death.[17,20] In addition, children of parents who have high blood cholesterol should be screened and children and adolescents whose parental or grandparental history is unobtainable, particularly those with other risk factors for coronary heart disease, should be screened at the discretion of the health care providers. Infants and young children should have no limitations placed upon their fat intake. A transition should occur during toddler years so that the fat content of the diet gradually moves down over a period of several years to *approximately* 30% of total calories, with less than ten percent coming from saturated fat.[21] Cholesterol in the diet should be limited to less than 300 mg/d. These dietary recommendations are intended to apply over a period of several days and do not suggest that any food that contains more than 30% fat is unhealthy or should be avoided.

As with heart disease, obesity is another condition where the influences of genetics, diet, and environment have not been clearly defined. While the goal should be to match energy intake with energy requirements, "commandments" often focus solely on eliminating fat from foods in the diet. As with all the other examples of commandments cited, this too has very little basis in scientific fact. The major influence on the development of body fatness appears, for the most part, to be genetic[22], and as with heart disease, the tracking of obesity from childhood to adult years is imperfect. Not all fat children become fat adults, and many studies even show that the majority of fat children do not become fat adults. Genetic influences also program the distribution of body fat, a major consideration in the development of complications of obesity. The role of exercise, smoking, and other psychological factors on the development of obesity may be equally as important as diet.[23]

IV. GUIDELINES

In summary then, we have largely achieved our goals of identifying and quantifying the nutrients necessary for growth and development of young children. Commandments should be recognized for what they are, and it is incumbent on health professionals and the media to support recommendations with an adequate scientific foundation. A number of guidelines with regard to children's diets seem quite reasonable. Dietary choices should be regarded as a necessary and important part of a prudent lifestyle. The diet should be balanced to meet the needs for growth, activity, and development of infants, children, and adolescents. Children should be encouraged to eat a diet which is varied and which is similar to that which the whole family is eating. These diets should emphasize fruits, vegetables, cereals and grains, lean meats, poultry and fish, and low-fat dairy products and fish, and individual foods within the diet should not be classified as good or bad. Finally, any population recommendations with regard to diet should recognize that a significant and increasing proportion of American children live in poverty and are at risk for under-nutrition. The threat to their growth and development from poor or under-nutrition and violence is far greater than largely unsubstantiated relationships between childhood diets and long-term chronic illness at this time.

REFERENCES

1. Yip, R., Binkin, N. G., Fleshood, L., et al., Declining prevalence of anemia among low income children in the United States, *JAMA,* 258, 1619–1623, 1987.
2. Reeves, J. D. and Yip, R., Lack of adverse side effects of oral ferrous sulfate therapy in one year olds, *Pediatrics,* 75(2), 352–355, 1985.
3. Committee on Nutrition, American Academy of Pediatrics, Iron-fortified infant formulas (RE9169), *Pediatrics,* 84(6), 1114–1115, 1989.
4. Frieden, T. R., Sowell, A. L., Henning, K. J., Huff, D. L., and Gunn, R. A., Vitamin A levels and severity of measles New York City, *AJDC,* 146, 182–186, 1992.
5. Arrieta, A. C., Zaleska, M., Stutman, H. R., and Marks, M. I., Vitamin A levels in children with measles in Long Beach, CA, *J. Pediatr.,* 121, 75–78, 1992.
6. Committee on Infectious Disease, American Academy of Pediatrics, Vitamin A treatment of measles, *Pediatrics,* 91(5), 1014–1015, 1993.
7. Dunn, J. T., Iodine deficiency—the next target for elimination? *New Eng. J. Med.,* 326(4), 267–268, 1992.
8. Golding, J., Greenwood, R., Birmingham, K., and Mott, M., Childhood cancer, intramuscular Vitamin K, and pethidine given during labor, *Br. Med. J.,* 305, 341–346, 1992.
9. Klebanoff, M. A., Read, J. S., Mills, J. L., and Shiono, P. H., The risk of childhood cancer after neonatal exposure to Vitamin K, *New Eng. J. Med.,* 329, 905–908, 1993.
10. American Academy of Pediatrics, Vitamin K Ad Hoc Task Force, Controversies concerning Vitamin K in the newborn, *Pediatrics,* 91(5), 1001–1003, 1993.
11. Johnston, C. C., Miller, J. Z., Slemenda, C. W., Reister, T. K., Hui, S., Christian, J. C., and Peacock, M., Calcium supplementation and increases in bone mineral density in children, *New Eng. J. Med.,* 327, 82–87, 1992.
12. Committee on Nutrition, American Academy of Pediatrics, The use of whole cow's milk in infancy, *Pediatrics,* 89, 1105–1109, 1992.
13. Karjalainen, J., Martin, J. M., Knip, M., et al., A bovine albumin peptide as a possible trigger of insulin dependent diabetes mellitus, *New Eng. J. Med.,* 327, 302–307, 1992.
14. Scott, F. W. and Marliss, E. B., Conference summary: diet as an environmental factor in development of IDDM, *Can. J. Physiol./Pharmacol.,* 69, 311–319, 1991.
15. Sheard, N. F., Cow's milk, diabetes and infant feeding, *Nutr. Rev.,* 51(3), 79–89, 1993.
16. National Cholesterol Education Program, Report on the Expert Panel on Population Strategies for Blood Cholesterol Reduction, Washington, D.C.: National Heart, Lung and Blood Institute Report No. 90-3046, 1990.
17. National Cholesterol Education Program, Report of the Expert Panel on Blood Cholesterol Levels in Children and Adolescents, Rockville, MD: U.S. Department of Health, Education, and Welfare, 1991, National Institutes of Health publication N4LB1.
18. Pathobiological Determinants of Atherosclerosis in Youth (PDAY) Group, Natural history of aortic and coronary atherosclerotic lesions in youth, *Arterioscler. Thromb.,* 13, 1291–8, 1993.
19. Heller et al., Contributions of genetics and environment to lipid profiles, *New Eng. J. Med.,* 1150, 1993.
20. Lauer, R. M. and Clarke, W. R., Use of cholesterol measurements in childhood for the prediction of adult hypercholesterolemia: The Muscatene Study, *JAMA,* 264, 3039–3043, 1990.
21. Committee on Nutrition, American Academy of Pediatrics, Statement on Cholesterol in the diets of infants and children, *Pediatrics,* 90(3), 469–473, 1992.
22. Stunkard, A. J., Sørensen, T. I. A., Hanis, C., et al., An adoption study of human obesity, *New Eng. J. Med.,* 314, 193–198, 1986.
23. Rosenbaum, M. and Liebel, R. L., Obesity in childhood, *Pediatr. Rev.,* 11(2), 43–55, 1989.

Chapter 3

CHOLESTEROL SCREENING AND DIETARY INTERVENTION IN CHILDHOOD FOR PREVENTION OF ADULT ONSET CARDIOVASCULAR DISEASE

Omer Tarim, Thomas B. Newman, and Fima Lifshitz

TABLE OF CONTENTS

I. Introduction 13

II. Recommendations and Policies for Cholesterol Screening in Childhood 13

III. Problems Associated with Cholesterol Screening 15

IV. Trials of Dietary Treatment of High Blood Cholesterol 16

V. Conclusion 17

References 18

I. INTRODUCTION

It is well known that there is an increased risk of death from coronary heart disease (CHD) at higher serum cholesterol levels.[1] This positive association has led to U.S. policies directed at lowering high blood cholesterol.[2] However, recent studies indicate there is also a higher risk of death from non-CHD causes at lower serum cholesterol levels. In addition, randomized trials of cholesterol lowering have failed to demonstrate a reduction in mortality. These findings have received inadequate attention from health policy makers.[3,4] In this review, we will present implications of these findings for cholesterol screening in childhood and our reservations about recommendations for childhood nutrition aiming to prevent atherosclerosis later in life.

II. RECOMMENDATIONS AND POLICIES FOR CHOLESTEROL SCREENING IN CHILDHOOD

National Cholesterol Education Program (NCEP) Expert Panel on Blood Cholesterol Levels in Children and Adolescents[5] and the American Academy of Pediatrics Committee on Nutrition[6] have recommended that children should be

0-8493-2764-4/95

screened selectively, based on a family history of high blood cholesterol or premature cardiovascular disease.

They estimate that at least one-fourth of the children in the U.S. will meet their criteria for screening. These criteria are defined as a parent or grandparent with cardiovascular disease before age 55 or a parent with a blood cholesterol >240 mg/dl. However, selective screening misses at least half the children with hypercholesterolemia.[7–9] Therefore, some studies have raised questions about screening only children with a family history of cardiovascular disease,[10] and it was suggested that thorough identification of children with elevated low-density lipoprotein (LDL) cholesterol would be most effectively accomplished by an inclusive, whole population approach.[11] In the last few years, many authorities have proposed that all children be screened for high blood cholesterol levels.[12–15]

On the other hand, in order to accept the policy of universal or selective cholesterol screening in childhood, the benefits of this approach must be documented. First, measurement of blood cholesterol levels should allow the identification of children at substantially increased risk of CHD. Second, interventions to reduce the risk must be shown to be more effective if started in childhood than later in life. Third, and most important, cholesterol screening and risk intervention trials in childhood should be accomplished in a cost-efficient manner with little risk to the child.[16]

To date, the evidence for the positive predictive value of hypercholesterolemia in childhood regarding the future development of CHD is indirect. CHD usually occurs several decades after childhood, and there are no long-term prospective studies directly linking hypercholesterolemia in childhood with CHD. The relationship of the fatty streaks present in the arteries of teenagers with high blood cholesterol levels with later CHD is unclear.[17]

Although an elevated blood cholesterol level in childhood is a statistically significant predictor of an elevated cholesterol level one or two decades later,[13,18,19] it is at best "a risk factor for a risk factor" for CHD.[16] Considering the inaccuracy of cholesterol measurements in a community or office setting, the association between the measured childhood cholesterol level and future CHD may be further weakened.[16]

The cumulative death rates from CHD before age 65 years in individuals studied in accordance with their serum cholesterol quintile showed that the risk for CHD death was relatively small, even among males in the highest cholesterol quintile. CHD deaths before age 65 years were uncommon.[16] Newman et al. found that even under very unrealistically optimistic assumptions, about 100 to 200 boys in the top quintile of serum cholesterol levels would need to be treated with a cholesterol-lowering diet and/or medication for 50 years to prevent one premature CHD death.[16] Since CHD death rates for women before age 65 are about one-third of those for men, the number of girls who would need to follow dietary treatment to prevent one premature death would be about 300 to 600.[16] For children with serum cholesterol levels in the lower to normal quintiles, the benefit of any intervention which attempts to lower the cholesterol levels would be expected to be less, and the number of individuals treated to show a positive effect would be significantly greater.[16]

Finally, cholesterol screening and intervention do not appear to be cost-effective. Cholesterol testing, clinic visits, repeat tests, and possible long-term intervention and treatment for those with high serum levels, all contribute to the increased expense of cholesterol screening. For example, the cost of such a program would probably be between $1 million and $10 million per year of life prolonged for women aged 25 to 44 years or for men aged 25 to 34 years.[20] The cost would be even higher if the population involved were younger. The cost per year of life prolonged of cholesterol screening and intervention in childhood is exponentially increased above that of young adults. These figures are more than 10 times the cost of other preventive interventions or treatments such as those directed at smoking or those effective high-technology treatments such as left main coronary by-pass surgery.[20–22]

III. PROBLEMS ASSOCIATED WITH CHOLESTEROL SCREENING

In addition to the frequent laboratory inaccuracies in an office setting and the cost, another potential adverse effect of cholesterol screening is suboptimal nutrition and nutritional deficits which could follow the restrictive diets recommended for those patients found to have high cholesterol levels. There is reported evidence of growth failure in children referred for dietary treatment of high blood cholesterol levels detected on routine screening.[23,24] The nutritional growth retardation was due to unsupervised application of low-fat, low-cholesterol diets recommended by their pediatricians.[23,24] Additionally, there may be iron deficiency and other nutrient alterations when there is dietary restriction and low-fat, low-cholesterol diets are followed, particularly when this is done with inadequate supervision.[24] Children ingesting less than 30% of calories from fat consumed inadequate diets with increased carbohydrate and sucrose and decreased vitamin and minerals such as folic acid and iron.[25]

Another adverse effect is the problem of labelling as a ''sick child'' and of the possibility of resulting family conflicts. There may be significant long-term psychosocial consequences of separating individual children for special medical diagnoses or treatment. Treatments prescribed may seem unfair to the children on restricted diets, and the families often find it difficult to force their child to adhere to an unappealing diet.[16]

The final adverse effect is perhaps the most worrisome: the possibility that reduction of cholesterol levels may cause an increased risk in non-cardiovascular mortality.[26] Although the mechanisms for unexpected and highly significant increase in both injury and cancer deaths among individuals with low serum cholesterol levels are not understood, this causal inference has been supported by recent reports from randomized trials.[26–29] Cumulative data from pooled cohort studies revealed that men with total cholesterol levels of less than 160 mg/dl had a 20% higher age-adjusted rate of cancer deaths and 40% higher rate of non-cardiovascular non-cancer deaths compared to those with cholesterol levels between 160 and 199 mg/dl. The latter included increased rates of injury deaths (by 35%), respiratory system deaths (by 15%), digestive system deaths (by 50%), and "other" causes of death (by 70%). Among women, the pattern of non-cardiovascular deaths were similar to that in men, except that the excess in cancer mortality was smaller (about 5%).[30–32] A recently reported meta-analysis of randomized primary prevention trials of cholesterol reduction also showed that subjects who received cholesterol-lowering interventions (diet or drugs) were 76% more likely to die of accidents, violence, or suicide than subjects in the control groups ($p = 0.004$).[26] A recent report of a multifactorial study also found a striking increase in injury deaths ($p = 0.002$).[28] This risk is not to be taken lightly when children are involved, since trauma is the major cause of death in this age group.

Animal studies provide additional evidence for potential adverse effects of cholesterol-lowering intervention trials. Monkeys fed a prudent diet, low in fat and cholesterol, exhibited significantly more aggressive behavior.[33] A twofold to fivefold increase in mortality was observed in monkeys given clofibrate.[34] Additionally, associations with cancer were noted in the product information for lovastatin, pravastatin, cholestyramine, clofibrate, and gemfibrozil.[34] Although the mechanisms are not clear, several authors have speculated that lowering blood cholesterol levels may alter cellular function and increase the propensity to injuries or cancer.[35–37]

IV. TRIALS OF DIETARY TREATMENT OF HIGH BLOOD CHOLESTEROL

In order to reduce the risk for chronic disease, dietary modifications have been recommended by the American Cancer Society, the U.S. Department of Agriculture (USDA), Health and Human Services, the American Heart Association, and the American Academy of Pediatrics.[38–41] Currently, the expert panels recommend avoidance of obesity and reduction of the intake of fat and cholesterol with diets providing <30% of calories from fat.[42] These recommendations have also been extended for children over two years of age by the Expert Panel on Blood Cholesterol Levels in Children and Adolescents, American Academy of Pediatrics.[43]

However, there are few studies of cholesterol-lowering interventions by diet in children. These are unfortunately of short duration and showed that reductions of cholesterol levels achieved were 5% or less.[13,44,45] In a recent prospective study,[46] ninety-seven adults underwent four consecutive nine-week periods of treatment according to a randomized, balanced design: a high-fat diet-placebo period, a low-fat diet-placebo period, a high-fat diet-lovastatin period, and a low-fat diet-lovastatin period. The dietary regimens were Step 2 diets. Serum low-density lipoprotein (LDL) cholesterol levels on average were 5% lower during the very low-fat diet (Step 2) than during the high-fat diet. The reduction in serum

LDL cholesterol with lovastatin treatment was 27% in patients who were on high-fat diet and 32% in patients on low-fat diet. Thus, drug therapy was far more effective than diet in reducing total cholesterol levels. In addition, the level of high density lipoprotein (HDL) cholesterol was also reduced by 6%, compared to a drop of 5% in LDL cholesterol, during the dietary treatment, thus off-setting the possible beneficial effect of reducing LDL cholesterol.[46]

A recent meta-analysis of long-term studies also showed that Step 1 diet with a total fat intake of <30% of calories given as individual intervention for >6 months to a total of 21,342 healthy adults resulted in a net reduction in serum cholesterol of only about 2% (range 0–4%).[47] Over 5–10 years, population education programs combined with Step 1 diet caused only minimal reductions of 0.6–2% in serum cholesterol levels. Diets more intensive than Step 1 diets reduced serum cholesterol concentrations by 13% over 5 years in selected high-risk men, by 6–15% in hospital outpatients, and by only 12–15% in institutionalized patients who followed very restrictive diets.[47]

Considering that the effect of cholesterol-lowering diets on serum cholesterol is so small, and not even clearly favorable, it appears prudent not to enforce marked reductions in dietary fat intake in childhood.[48] Restricting dietary intake of fat and cholesterol to <30% of calories is not necessary for children and may be harmful. Children are not "little adults". Guidelines should be based on moderation, not arbitrary numerical targets.[48]

However, information gathered through medical research in adults should not be ignored.[1,26,49–51] It would be inappropriate to justify a very high intake of saturated fat and cholesterol in children. The principle of moderation should be followed to avoid both extremes, strict restriction resulting in nutritional deficiencies on one end, and indulgence in food consumption leading to obesity on the other.

V. CONCLUSION

Universal cholesterol screening in children is difficult to justify considering the lack of evidence that hypercholesterolemia is more easily treatable in childhood, the evidence that lowering cholesterol levels later in life will be nearly as effective at reducing CHD, the psychosocial problems associated with cholesterol screening in children, and the possibility that reduction of cholesterol levels in this age group might increase total mortality.[52] Since blood cholesterol is only one of many risk factors for CHD, and since only one-fifth of all CHD deaths can be attributed to having an adult cholesterol level in the top quintile,[1,16] cholesterol screening in selected groups is unjustified. In addition, when a high cholesterol level is detected, it is questionable whether these children should be subjected to an intervention.

Selective approaches carry similar risks, and most of the arguments against universal screening and intervention apply to more selective screening and intervention as well. Excluding the rare (less than 1%) children at high risk of premature CHD because of a parent with familial hypercholesterolemia, the risk of cholesterol screening in children—even those with a positive family history—are likely to exceed the benefits.[52]

When, if ever, a safe and effective method of lowering cholesterol levels is identified, most of the benefit can be achieved by measuring the cholesterol levels of adults and treating high levels as they approach the age when CHD begins to occur. Reduction in CHD mortality from treating high blood cholesterol levels in patients with coronary disease is more likely to outweigh any adverse effects of the treatment.[53] Moreover, there is no evidence that dietary restrictions cause lasting decrease in blood cholesterol levels in children, or that the changes persist into adulthood, or decrease in CHD incidence. It has to be kept in mind that hypercholesterolemia in childhood can only be a risk factor for a risk factor for CHD. During the first two decades of life, growth and psychosocial adaptation should not be compromised in an attempt to prevent CHD.

Therefore, the dietary guidelines for children should differ from those recommended for adults. Over- or under-feeding should be avoided by monitoring weight and height progression in any child who is being given dietary recommendations. Infants <2 years of age should not be given dietary restrictions. The risk-benefit ratio of general recommendations should be carefully assessed before suggesting adult diets for children. At present, there are no data suggesting that marked reduction in fat intake is needed for children or that this would be healthful. Moderation and ingesting a large variety of foods is the best advice that can and should be given for young infants and children.

REFERENCES

1. Martin, M.J., Hulley, S.B., Browner, W.S., Kuller, L.H., and Wentworth, D., Serum cholesterol blood pressure and mortality: implications from a cohort of 361,622 men, *Lancet,* 2, 933–936, 1986.
2. NCEP Expert Panel, Report of the National Cholesterol Education Program Expert Panel on Detection, Evaluation and Treatment of High Blood Cholesterol in Adults, *Arch. Intern. Med.,* 148, 36–69, 1988.
3. Hlatky, M.A. and Hulley, S.B., Plasma cholesterol: can it be too low? *Arch. Intern. Med.,* 141, 1132, 1981.
4. Frank, J.W., Reed, D.M., Grove, J.S., and Benfante, R., Will lowering population levels of serum cholesterol affect total mortality? *J. Clin. Epidemiol.,* 45, 333–346, 1992.
5. NCEP Expert Panel on Blood Cholesterol Levels in Children and Adolescents, National Cholesterol Education Program (NCEP): highlights of the report of the Expert Panel on blood cholesterol levels in children and adolescents, *Pediatrics,* 89, 3, 495–501, 1992.
6. American Academy of Pediatrics Committee on Nutrition, Indications for cholesterol testing in children, *Pediatrics,* 83, 141–142, 1989.
7. Forster, J.H., Elevated lipids: whom to screen and how to treat, *Contemp. Pediatr.,* 22–32, 1985.

8. Cresanta, J.L., Burke, G.L., Downey, A.M., Freedman, D.S., and Berenson, G.S., Prevention of atherosclerosis in childhood, *Pediatr. Clin. North Am.*, 33, 835–838, 1986.
9. Long, T.J., Routine screening tests for the adolescent, *Prim. Care,* 14, 41–47, 1987.
10. Dennison, B.A., Kikuchi, D.A., Srinivasan, S.R., Webber, L.S., and Berenson, G.S., Parental history of cardiovascular disease as an indication for screening for lipoprotein abnormalities in children, *J. Pediatr.,* 115, 186–194, 1989.
11. Griffin, T.C., Christoffel, K.K., Binns, H.J., and McGuire, P.R., Pediatric Practice Research Group, Family history as a predictive screen for childhood hyper-cholesterolemia, *Pediatrics,* 84, 365–373, 1989.
12. Garcia, R.E. and Moody, D.S., Routine cholesterol screening in childhood, *Pediatrics,* 84, 751–755, 1989.
13. Wynder, E.L., Perenson, G.S., Strong, W.B., and Williams, C., eds., Coronary artery disease prevention, cholesterol, a pediatric perspective, *Prev. Med.,* 18, 323–409, 1989.
14. Strong, W., You are a preventive cardiologist: the scope of pediatric preventive cardiology, *AJDC,* 143, 1145, 1989.
15. Strong, W.B., Point-counterpoint: to test or not to test in pediatric practice, *Pediatr. Rev,* 12, 36–38, 1990.
16. Newman, T., Browner, W., and Hulley, S., The case against childhood cholesterol screening, *JAMA,* 264, 3039–3043, 1990.
17. Newman, W.P., Freedman, D.S., Voors, A.W., et al., Relationship of serum lipoprotein levels and systolic blood pressure to early atherosclerosis: the Bogalusa Heart Study, *N. Engl. J. Med.,* 314, 138–143, 1986.
18. Lauer, R.M., Lee, J., and Clarke, W.R., Factors affecting the relationship between childhood and adult cholesterol levels: the Muscatine Study, *Pediatrics,* 82, 309–318, 1988.
19. Lauer, R.M. and Clarke, W.R., Use of cholesterol measurements in childhood for the prediction of adult hypercholesterolemia: the Muscatine Study, *JAMA,* 264, 3034–3038, 1990.
20. Hulley, S.B., Newman, T.B., Grady, D., Garber, A.M., Baron, R.B., and Brower, W.S., Should we be measuring blood cholesterol levels in young adults? *JAMA,* 269, II, 1416–1419, 1993.
21. Weinstein, M.C., Economic assessment of medical practices and technologies, *Med Decision Making,* 1, 309–330, 1981.
22. Toronto Working Group, Asymptomatic hyper-cholesterolemia: a clinical policy review, *J. Clin. Epidemiol.,* 43, 1029–1121, 1990.
23. Pugliese, M.T., Weyman-Daum, M., Moses, N., and Lifshitz, F., Parental health beliefs as a cause of nonorganic failure to thrive, *Pediatrics,* 80, 175–182, 1987.
24. Lifshitz, F. and Moses, N., Growth failure: a complication of dietary treatment of hypercholesterolemia, *AJDC,* 143, 537–542, 1989.
25. Nicklas, T.A., Webber, L.S., Koschak, M.L., and Berenson, G.S., Nutrient adequacy of low fat intakes for children: the Bogalusa Heart Study, *Pediatrics,* 89, 2, 221–228, 1992.
26. Muldoon, M.F., Manuck, S.B., and Matthews, K.A., Lowering cholesterol concentration and mortality: a quantitative review of primary prevention trials, *BMJ,* 301, 309–314, 1990.
27. Oliver, M., Might treatment of hypercholesterolemia increase non-cardiac mortality? *Lancet,* 337, 1529–1531, 1991.
28. Strandberg, T., Saloman, V., Naukkarinen, V., Vanhanen, H., Sarna, S., and Miettinen, T., Long-term mortality after 5-year multifactoral primary prevention of cardio-vascular diseases in middle-aged men, *JAMA,* 266, 1225–1229, 1991.
29. Smith, G.D. and Pekkanen, J., Should there be a moratorium on the use of cholesterol lowering drugs? *Br. Med. J.,* 304, 431–434, 1992.
30. Hulley, S.B., Walsh, J.M.B., and Newman, T.B., Health policy on blood cholesterol: time to change directions, *Circulation,* 86, 3, 1026–1029, 1992.
31. Jacobs, D., Blackburn, H., Higgins, M., et al., Report of the Conference on Low Blood Cholesterol: mortality associations, *Circulation,* 86, 1046–1060, 1992.
32. Neaton, J., Blackburn, H., Jacobs, D., et al., Serum cholesterol level and mortality: findings for men screened in the MRFIT, *Arch. Intern. Med.,* 152, 1490–1500, 1992.

33. Kaplan, J., Manuck, S., and Shively, C., The effects of fat and cholesterol on social behavior in monkeys, *Psychosom. Med.*, 53, 634–642, 1991.
34. Duffy, M.A., *Physicians' Desk Reference,* 47th Ed., Montvale, N.J.: Medical Economics Data, 1993, 2547–2548, 1556–1560, 1863–1865, 732–734, 1787–1790.
35. Mason, R.P., Herbette, L.G., and Silverman, D.I., Can altering serum cholesterol affect neurologic function? *J. Mol. Cell. Cardiol.*, 23, 1339–1342, 1991.
36. Engelberg, H., Low serum cholesterol and suicide, *Lancet,* 339, 727–728, 1992.
37. Jacobs, D.R., Why is low blood cholesterol associated with risk of nonatherosclerotic disease death? *Annu. Rev. Public Health.*, 14, 95–114, 1993.
38. American Cancer Society, *Nutrition and Cancer: Cause and Prevention,* New York: American Cancer Society, 1984.
39. U.S. Department of Agriculture, U.S. Department of Health and Human Services, Nutrition and your health: dietary guidelines for Americans, *Home and Garden Bulletin,* No. 232, Washington, D.C.: Government Printing Office, 1985.
40. American Heart Association, Nutrition Committee, Dietary guidelines for healthy American adults, *Circulation,* 74, 1465A–1468A, 1986.
41. American Academy of Pediatrics, Committee on Nutrition, Prudent life style for children: dietary fat and cholesterol, *Pediatrics,* 78, 521–525, 1986.
42. Expert Panel, Report on the national cholesterol education program expert panel on detection, evaluation, and treatment of high blood cholesterol in adults, *Arch. Int. Med.*, 148, 36–69, 1988.
43. Expert Panel, National cholesterol education program, Report of the expert panel on blood cholesterol levels in children and adolescents, *Pediatrics,* 89, 525–584, 1992.
44. Puska, P., Vartiainen, E., Pallonen, U., et al., The North Karelia Youth Project: evaluation of two years of intervention on health behavior and CVD risk factors among 13–15 year-old children, *Prev. Med.*, 11, 550–570, 1982.
45. Walter, H.J., Hofman, A., Vaughan, R.D., and Wynder, E.L., Modification of risk factors for coronary heart disease, five year results of a school-based intervention trial, *N. Engl. J. Med.*, 318, 1093–1100, 1988.
46. Hunninghake, D.B., Stein E.A., Dujovne, C.A., et al., The efficacy of intensive dietary therapy alone or combined with lovastatin in our patients with hypercholesterolemia, *N. Eng. J. Med.*, 328, 17, 1213–1219, 1993.
47. Ramsay, L., Yeo, W.W., and Jackson, P.R., Dietary reduction of serum cholesterol concentrations: time to think again, *Br. Med. J.*, 303, 953–957, 1991.
48. Lifshitz, F., Children on adult diets: is it harmful? Is it healthful? *J. Am. Col. Nutr.*, 11S, 84–90S, 1992.
49. National Academy of Sciences, *Diet and health, implications for reducing chronic disease risk,* Washington, D.C.: National Academy Press, 1989.
50. Dorr, A.E., Gundersen, D., Schneider, J.C., et al., Colestipol hydrochloride in hypercholesterolemic patients: effect on serum cholesterol and mortality, *J. Chronic Dis.*, 31, 5–14, 1978.
51. Dayton, S., Pearce, M.L., Hashmoto, S., et al., A controlled clinical trial of a diet high in unsaturated fat in preventing complications of atherosclerosis, *Circulation,* 39–40, (Suppl II), 1–63, 1969.
52. Newman, T.B., Browner, W.S., and Hulley, S.B., Childhood cholesterol screening: contraindicated, *JAMA,* 267–1, 100–101, 1992.
53. Criqui, M.H., Cholesterol, primary and secondary prevention, and all-cause mortality, *Ann. Intern. Med.*, 115, 973–976, 1991.

Chapter 4

A CLINICIAN'S APPROACH TO INITIATING BREASTFEEDING

Susan K. Schulman and Audrey Rosner

TABLE OF CONTENTS

I. Introduction 21

II. Physiology 22
  A. The Breast 22
  B. Milk Production 22
  C. Let-down Reflex 23
  D. Lactational Infertility 23

III. Maternal-Infant Diad—The Vital Connection 23
  A. Proper Nursing Technique 24
  B. Timing of Feedings 24
  C. Engorgement 25

IV. Common Maternal Problems and Their Solutions 26

V. Common Problems—The Infant 28

VI. Conclusion 30

References 30

I. INTRODUCTION

Although most physicians favor breast milk as the most advantageous infant nutrition, relatively few possess the knowledge to provide effective counseling if problems arise. Many infants are deprived of this wonderful milk because the mother encountered one of numerous difficulties with her nipples, her milk supply, the baby's sucking grasp, fever, jaundice, or other factors which result in cessation of breastfeeding.

With basic understanding of the normal processes of milk production, milk ejection, and the care of the maternal-infant diad, a physician or other health

0-8493-2764-4/95

professional should be able to initiate breastfeeding and deal with these problems, thereby helping the breastfeeding to continue. The following pages contain a brief summary of the breastfeeding process with emphasis on useful clinical information.

II. PHYSIOLOGY

A. THE BREAST

Anatomically, the breast is made up of alveolar milk-producing cells arranged around alveoli. These are connected by ducts to a collecting system which then connects to distal ducts leading to the nipple pores. Around each alveolus is a band of myoepithelial cells which contract on hormonal stimulus and "let down" the milk into the collecting system. The glandular breast tissue is supported by variable amounts of fatty tissue and extensive vascular and lymphatic networks. The skin over the aureola and nipple is highly innervated and has smooth muscle fibers which allow it to become erect upon tactile stimulation. There is extensive innervation of these areas beginning in the nipple and aureola and travelling via spinothalamic tracts to the hypothalamus and the pituitary.[1]

There are also Montgomery glands, a type of sebaceous gland, opening onto the surface of the aureola that produce lubricants. The breast tissue which matures with puberty is stimulated to hypertrophy under the influence of high levels of prolactin, estrogen, progesterone, and placental hormones during pregnancy.

With a normal amount of breast tissue, there is a substantial increase in breast size during pregnancy. This is due to increase in vascularity, increase in gland size, fluid retention, and in the second half of the pregnancy, the presence of colostrum. The colostrum remains in the breast until after birth when milk production begins.[1]

B. MILK PRODUCTION

The milk production begins after parturition, any time from 18 h to 3 d after birth. From the time of birth until the milk production starts, the breast contains colostrum. This clear fluid, rich in immune substances, has nutrient content but is low in volume compared to milk.[2] Initially, the milk is produced in response to a fall in estrogen which blocks the effect of the high levels of prolactin and placental lactogen on milk production during pregnancy. In order for milk production to continue, prolactin must be present. In the first hours after the birth, prolactin levels also drop, but rise again dramatically in response to suckling.[3] The more frequent the early suckling, the earlier the milk supply increases.[4] The prolactin acts directly on the alveolar cells to induce the production of milk, a mixture of milk protein, milk sugar (lactose), and milk fat.

The primary stimulation for prolactin release is suckling. The nerve fibers in the aureola and nipple carry stimulation via the hypothalamus which then suppresses a prolactin-inhibiting factor. This then allows the pituitary to secrete prolactin. Milk pumping and expression of milk by hand also provide some stimulation but much less than suckling.[2]

C. LET-DOWN REFLEX

Another hormone, oxitocin, is responsible for the "let-down" reflex which makes the milk accessible to the baby by causing contractions of the muscle cells surrounding the alveoli, ejecting the milk into the peripheral ducts and even sometimes out of the nipple pores. Like prolactin, oxitocin is also released from the pituitary in response to suckling. This hormone is also released in response to thinking about the baby, hearing the baby cry, and other similar cognitive stimuli.[5]

D. LACTATIONAL INFERTILITY

The complex hormonal interactions during lactation usually cause prolonged amennorhea and infertility. This effect is variable, and the timing of the first ovulation during lactation is not predictable. It has been shown that exclusively breastfeeding mothers have an average of ten months of additional birth interval when compared to non-breastfeeding mothers when neither group is using birth control.[6] It is for this reason that family planning organizations in Third World countries, with the strong encouragement of the World Health Organization, have begun a massive campaign to promote breastfeeding.[7]

Useful facts in review:

1. When there is adequate breast tissue present, there is marked enlargement of the breasts during pregnancy.
2. Prolactin is the main hormone responsible for milk production.
3. The most potent stimulus for prolactin release is the nerve stimulation during suckling.
4. The hormone responsible for the "let-down" reflex which ejects the milk from the alveoli into the collecting system and out through the nipple pores is oxitocin.
5. The most potent stimulus for oxitocin release is suckling. Other cognitive stimuli such as hearing the baby cry also contribute to oxitocin production.
6. During lactation mothers usually experience amennorhea and infertility.

III. MATERNAL-INFANT DIAD— THE VITAL CONNECTION

The most important single task of the mother and baby in establishing breastfeeding is getting the baby to grasp the nipple properly. Although it might seem simple in theory, the actual deed is far from easy to accomplish.

To be successful, the mother must have a breast that is usable to the infant. Her breast size should have increased during the pregnancy. Her nipples must be soft and easily stretched. The papilla must evert and not invert when the aureola is compressed. Her nipples and aureola should have darkened during the pregnancy to provide thicker, less fragile skin to withstand the trauma of suckling.

The baby, too, must meet certain minimal requirements in order to be able to grasp the nipple properly. The term infant must be alert and not drugged after the birth. He must demonstrate a rooting reflex which causes him to open his mouth wide enough to take in the nipple papilla as well as a significant amount of aureola. He then must clamp down with his gums on the outer rim of the aureola in order to compress the milk from the lacuna (collecting areas) through the nipple ducts and into the mouth. With his tongue he must stroke the papilla and compress it against the hard palate to "milk" the nipple.[8]

If the infant has a poor grasp and clamps down only on the papilla itself, the mother experiences pain, and the baby gets very little milk. It is common for this to occur if the baby is given bottle feeds during the first few days of life. The baby experiences "nipple confusion" and tries to use the same grasp on the human nipple that works so well on the rubber one.[9]

A. PROPER NURSING TECHNIQUE

It is vital that a mother be taught immediately how to nurse properly. Ideally, this first lesson should take place in the delivery room right after the birth.[10] As stated previously, the earlier the feedings are initiated, the sooner an adequate milk supply is established. It is very helpful to show the mother what a proper grasp looks and feels like before bad habits are ingrained.

There are several nursing positions which can be learned that help to avoid sore nipples and encourage complete drainage of the milk from the breast.

The Cradle Position—This is the way one naturally holds an infant while feeding with the baby's head in the crook of the arm and the baby's body resting across the mother's lap. In this position, the baby grasps the nipple on an angle with his nose pointing to the mother's axilla.[11]

The Football Hold—In this position, the mother holds the baby's head in her hand and places the baby's body under her arm, resting on her hip. The baby grasps the nipple on the same side, and his nose will be pointing to the mother's sternum.

The Lying Down Position—In this position, the baby lies on the bed next to the mother as she lies on her side with her arm under a pillow under her head. In this position, the baby grasps the nipple in a straight up and down hold with his nose pointed at the top of the breast.

Since the strongest emptying occurs in the area directly under the baby's clamping gums, these changes in position help to empty different areas more efficiently and prevent nipple soreness by allowing different papillar areas to be exposed to friction each time.[11]

B. TIMING OF FEEDINGS

A lot of opinions have been stated regarding the timing of early feedings. In the United States, the nursery feeding schedules have been geared to formula feeding. The traditional four-hour wait from one feeding to the next is totally unacceptable to the breastfeeding mother and baby. Since breast milk production requires frequent vigorous stimulation by suckling, early frequent feeds allow the whole process to proceed naturally. There is good reason for the observation that breast milk production starts three to five days after birth in the United States and only 18 h after birth in primitive cultures where the baby remains at the mother's breast during the first few days of life. These babies are allowed to suckle whenever they are awake, and their efforts bring in the mother's milk very quickly.[12]

The ideal arrangement for the immediate post-partum period is "rooming-in," with baby in the mother's room and a nurse available to care for both of their needs while encouraging the mother by teaching her the proper methods of breastfeeding. The baby can be fed whenever hungry and is not allowed to become frantic by delaying the feedings to meet nursery schedules.[10]

In the beginning, the feedings should be limited to a few minutes on each breast with a different nursing position used each time. After the first day, the feedings can be lengthened, but care must be taken to prevent the nipples from getting too sore. Meticulous attention must be paid to the grasp of the baby's mouth on the nipple. The aureola should be gathered and pushed into the baby's mouth before he clamps down. The deeper the grasp, the more comfortable the nipples will remain. When releasing the grasp, the mother should allow the baby to bite on her finger as she slips it into the mouth, and then she can withdraw her nipple without resistance.[7]

C. ENGORGEMENT

When the milk first comes in, some mothers experience engorgement. This painful, hard swelling is a result of edema of breast tissue combined with milk overproduction. The breast can become so indurated that there may be little or no flow of milk. The swelling extends to the aureola and the nipple causing a flattening of the papilla, making it extremely difficult for the baby to nurse. The treatment for engorgement is to empty the breast by use of a breast pump to start the flow and bring out the papilla and then allowing the baby to suck. At the end of the feeding, the breast should be further emptied by pumping until the breast feels soft. Frequent feedings will prevent the further buildup of milk and decrease edema. Mothers are often cautioned to limit or delay feedings to avoid excess milk production, but in reality this often makes the engorgement worse. The fact is that mothers who do not allow more than three hours to pass between feedings rarely get engorged in the first place.[13]

Useful facts in review:

1. The baby must have a deep grasp of the nipple including the aureola for proper suckling and to prevent nipple soreness.
2. The mother should use at least three different nursing positions to prevent soreness and allow for proper drainage.
3. The best way to prevent engorgement is to nurse frequently in different positions when the milk first comes in. The best way to treat engorgement is to empty the breast as often as possible.

IV. COMMON MATERNAL PROBLEMS AND THEIR SOLUTIONS

It is important to note that the single most important factor in getting a mother to start and continue breastfeeding is family and professional support. A reassuring nurse or doctor who encourages the mother and tells her what a wonderful job she is doing can overcome almost any problem. Conversely, a single negative word will cause a mother to run for a formula bottle.

Nipples that are not flexible and do not protrude into the baby's mouth during suckling—Flat or inverted nipples should be treated prior to starting breastfeeding. The goal is to stretch the adhesions that bind the papilla down so that the infant will be able to "latch-on" easily. The technique called "Hoffman's Technique" is done by placing both thumbs on the aureola next to the nipple on opposite sides and pulling the thumbs away from the nipple, stretching the skin and the adhesions beneath it. This process is repeated in a circular fashion around the entire nipple several times a day. It is also helpful to grasp the papilla and gently tug on it to pull it out. Turning the pulled nipple gently in both directions further stretches the adhesions.[14]

If these methods are not adequate, it is advisable to use milk cups worn inside the bra to evert the nipple. When it is still too difficult for the baby to latch on, it is sometimes necessary to use soft silastic nipple shields which cover the aureola and nipple and give the baby a papilla that can be utilized. During the suckling, the natural papilla is pulled into the shield, and eventually this also helps it to evert. After a short time, the shield can be removed, and nursing can proceed unaided.[15]

Sore nipples—Nipple soreness often develops if the baby is not grasping the nipple properly. It is important to observe the nursing infant to determine if this is the source of the problem. Some women, particularly Caucasians who have very light skin, are prone to soreness because of the lack of melanin on the skin of the nipples. In most cases, it is not advisable to stop nursing even temporarily to help the nipples heal. The best treatment is to air dry the nipples by leaving the bra flaps open between feedings. Sometimes a lanolin-based breast cream gives some relief, although experts disagree as to its use. Changing nursing positions with each feeding is essential until the nipples heal. The baby should always be fed when he seems hungry to prevent him from becoming frantic by delaying a feeding too long. The suck of a frantically hungry infant is extremely strong, and the infant might not take the time to latch on properly, which can cause more trauma to the nipple.[16]

Breast engorgement—(see above).

Blocked milk ducts—Occasionally, a particular milk drainage duct will become blocked causing a cord-like swelling in one area of the breast which becomes tender and firm. Close observation will sometimes reveal a plug of dried milk in one of the pores on the papilla. When this plug is found, it can be gently removed with a sterile needle to relieve the obstruction. When the blockage is more internal to the breast, hot compresses as well as gentle massage and using the baby or a pump to empty the breast usually relieves the blockage.[7]

Mastitis—Blocked ducts can become infected causing a tender hot area in the breast with a concomitant fever and flu-like aches and malaise. Since the organisms causing mastitis are usually staphylococcus, the best choice of antibiotics would be cephalexin or an anti-staphylococcal penicillin. In most cases, nursing should continue as well as breast massage, warm compresses, and further emptying with a pump after feedings. If the mother is unable to feed the baby from that side, a pump must be used to keep the breast as empty as possible. If despite medication an abscess forms, it is sometimes necessary to incise and drain it surgically.[17]

Maternal fever and infections—When the mother is ill, even if she is febrile, there are only a few instances when breastfeeding should be discontinued to protect the infant. Since most antibiotics are not contraindicated during lactation, bacterial infections such as urinary infections, pneumonia, and breast infections can be treated while the mother continues breastfeeding.[18] Many studies have shown that breast milk protects the infant from various infections such as gastroenteritis[19] and even decreases the incidence of otitis media.[20] Respiratory viruses can infect the newborn infant but are not spread through the milk. The nursing mother should take precautions such as washing her hands before handling the baby and her nipples and avoid coughing on the infant.[7] The HIV virus passes through breast milk. Therefore, in the United States where clean formula is available, HIV-positive mothers are told not to nurse.[21] Mothers with active pulmonary tuberculosis and active pertussis should not breastfeed until their infections are under control. Mothers who develop Varicella within three days of the birth must be isolated from their infants until they are no longer contagious. Mothers with herpes simplex infections must take care to wash their hands but can nurse if the lesion is not on the breast itself.[22] Aside from these rare exceptions, mothers with fever should be allowed to breastfeed their newborn infants.

Cesearean section—Mothers who have had a Cesearean section can breastfeed successfully if they are given special support. Having a nurse or helper lift the baby to the breast and using nursing positions that do not stress the incision site are very important in the first few days after the birth. A pillow on the abdomen of the seated mother protects her from strain on the incision when the baby is placed in her lap. The football hold, also with the help of a pillow for support, is very useful.[23]

Maternal medications—Many drugs, such as analgesics and even mild narcotics, are acceptable during breastfeeding since they have no clinical effect on the infant. Most antibiotics are also acceptable. It is important to use references such as *Drugs in Pregnancy and Lactation* to check the acceptability of any medication given to the mother. Frequently, if a contraindicated drug is prescribed, an acceptable alternative is often available.[18]

Maternal illness—If the mother has a medical illness such as diabetes, autoimmune disease, inflammatory bowel disease, or chronic pulmonary disease, it is important to communicate with her own internist to coordinate the best interests of the baby and the mother. With special attention to diet, medications, and the general health of the mother, breastfeeding can be successful in many instances.

Maternal nutrition—Breastfeeding mothers have to eat well to maintain a good milk supply and to feel well during lactation. The societal emphasis on thinness can influence a mother to undernourish herself during the first few months after the birth. It is up to her professional advisor to guide her on proper nutrition during lactation. Very low-fat and low-calorie diets are not acceptable as they adversely effect the quality of the milk.[24]

Useful facts in review:

1. A strongly supportive professional advisor is critical for a new breastfeeding mother.
2. Nipples that are retractile or flat and inflexible should be treated with stretching exercises, the Hoffman Technique, milk cups, and occasionally silastic nipple shields to make the breast usable for the baby.
3. Sore nipples can be treated with air drying, creams, and meticulous attention to the baby's grasp.
4. Blocked milk ducts can be relieved by removing the visible plug, heat, massage, and thorough emptying of the breast.
5. Mastitis can be treated with antibiotics and heat and massage while continuing to breastfeed the infant.
6. Mothers who have had cesearean sections can nurse with proper help and pain medications.
7. Medications should be checked in special references for acceptability during lactation.
8. If the mother has a chronic medical problem, there should be communication between her breastfeeding advisor and her internist to maintain proper care of both the mother and her infant.
9. Breastfeeding mothers need good nutrition to support a baby's growth and maintain their own health and well-being.

V. COMMON PROBLEMS—THE INFANT

Prematurity—Because of the size and strength of the baby, it is difficult to breastfeed a premature infant under thirty-four weeks gestation. The smaller, less mature infant can be fed breast milk (with fortifiers, if necessary) by tube or other means until he is able to suck. Many mothers of very small premature infants have successfully breastfed their babies starting with pumped feedings and later switching over to suckling. This practice is strongly encouraged by neonatologists because the expressed breast milk has been shown to reduce the risk of necrotizing enterocolitis in these babies.[25]

Since suckling requires a lot of energy, the premature infant usually requires short frequent feedings until the weight approximates 2500 grams. At that point, the feedings will naturally become longer, and the baby will consume more so the interval between feedings will lengthen.[26]

When placing a premature infant on the breast, it is very important to make sure the baby latches on properly so that the effort of sucking produces the most intake. This takes patience and practice. It also requires a strongly supportive breast feeding advisor who will encourage the mother and reassure her that she is doing what is best for the baby and, most important of all, that she is doing the baby no harm.

Illness—If the infant is ill, has congenital neurologic or neuromuscular deficiencies, or other neonatal problems, the breastfeeding may have to be delayed or forgone. Each individual case should be evaluated by the breastfeeding advisor and the pediatrician, and decisions should be made for the benefit of the baby and the mother. Downs Syndrome infants usually suck well. When this diagnosis is made, the mother should be encouraged to nurse since this may improve maternal infant bonding.[27]

Jaundice—More babies have been taken off of breast feedings for jaundice than for any other reason. In the first three days of life, physiological jaundice is common, and in breastfed infants it is sometimes more exaggerated. This phenomenon has nothing to do with true breast milk jaundice which persists after the first week of life.[28] In the vast majority of infants, cessation of breastfeeding does nothing to improve their health which had not been threatened by this physiologic process. Simply by increasing the number and frequency of feedings, bilirubin levels will usually fall. The practice of giving water by nipple to these infants is of no medical use, and it can cause nipple confusion. Increasing breast feedings causes more milk to come in, increases the number of bowel movements (thus decreasing enterohepatic circulation of bilirubin), and increases hydration and urinary output.[29]

In babies with hemolytic jaundice requiring phototherapy, the baby can continue to breastfeed as long as weight and hydration are monitored, and the baby is vigorous with a strong suck.[30]

Infants with true breast milk jaundice, which starts after the fourth day and persists for several weeks, are usually not threatened by the level of bilirubin. Once it has been determined that the problem is not obstructive jaundice (as in biliary atresia), it is not necessary to intervene. Stopping breastfeeding for one to two days will cause a dramatic fall in the bilirubin level, but the benefit to the baby is strictly cosmetic.[7]

Useful facts in review:

1. Premature infants can be breastfed when they reach about thirty-four weeks gestation. If they are too immature to suck, they can be fed expressed breast milk until they are more developed.
2. Breast milk helps prevent necrotising enterocolitis in premature infants.
3. Sick or weak newborns need individual evaluation regarding their ability to nurse.

4. Physiologic jaundice can be improved by increasing the frequency of breastfeedings.
5. Hemolytic jaundice requiring phototherapy is not a contraindication to breastfeeding.
6. True breast milk jaundice does not cause illness in the baby, so stopping breastfeeding temporarily is necessary only if the baby's appearance is a problem.

VI. CONCLUSION

Everyone agrees that breast milk is the best milk for all infants. Knowledge of the basic process of initiating breastfeeding is essential for anyone dealing with the mother and infant in the immediate post-partum period. Every effort must be made to get the baby onto the breast properly and to encourage the mother in the first few (most difficult) days after the birth. There is no more gratifying and appreciated task.

REFERENCES

1. Vorherr, H., *The Breast,* New York: Academic Press, 1974.
2. McClelland, D. B. L., McGrath, J., and Samson, R., Antimicrobial factors in human milk, *Acta. Paediatr. Scand,* Suppl. 271, 1978.
3. Noel, G. L., Suh, H. K. et al., Prolactin release during nursing and breast pump stimulation in postpartum and nonpostpartum subjects, *J. Clin. Endocrinol. Metab.,* 38, 413, 1974.
4. Egli, G. E., Egli, N. S., and Newton, H., Influence of the number of breastfeedings on milk production, *Pediatrics,* 27, 314, 1961.
5. Jelliffe, D. B. and Jelliffe, E. F. P., *Human Milk in the Modern World,* Oxford: Oxford University Press, 1978.
6. Rosner, A. E. and Schulman, S. K. Birth interval among breast-feeding women not using contraceptives, *Pediatrics,* 86(5), 747, 1990.
7. Goldfarb, J. and Tibbits, E., *Breastfeeding Handbook,* New Jersey: Enslow, 1989.
8. Woolridge, M. W., The "anatomy" of infant sucking, *Midwifery,* 2, 164, 1976.
9. Frantz, K. B., *Managing nipple problems,* LaLeche League International, Reprint No. 11, 1982.
10. Salariya, E. M., Easton, P. M., and Cater, J. I., Duration of breastfeeding after early initiation and frequent feeding, *Lancet,* 1141, 1978.
11. Frantz, F. B., Techniques for successfully managing nipple problems and the reluctant nurser in the early postpartum period, Infrier, S., ed., *Human Milk, Its Biological and Social Value,* Excerpta Medica, Amsterdam, 1980.
12. Steven, H., Unlimited suckling time improves breastfeeding (letter), *Lancet,* 393, 1981.
13. Blake, B., et al., *A Guide to Handling Breastfeeding Problems,* Philadelphia: CEA of Greater Philadelphia, Spring 1977.
14. Alexander, J. M., Randomized controlled trial of breast shells and Hoffman's exercises for inverted and non-projectile nipples, *Br. Med. J.,* 304(6833), 1030, 1992.
15. Countryman, B. A., *How the Maternity Nurse Can Help the Breastfeeding Mother,* Illinois: LaLeche League International, 1977.

16. Worthington-Roberts, B. and Rodwell Williams, S., *Nutrition in Pregnancy and Lactation,* St. Louis: Mosby Year Book, 1989.
17. Marshall, B. R. et al., Sporadic puerperal mastitis, *JAMA,* 223, 1975.
18. Briggs, G. G., Freeman, R. K., and Yaffe, S. J., *Drugs in Pregnancy and Lactation,* Baltimore: Williams and Wilkins, 1990.
19. Kanaaneh, H., The relationship of bottle feeding to malnutrition and gastro-enteritis in a pre-industrial setting, *J. Trop. Pediat. Env. Chld. Hlth.,* 18, 302, 1972.
20. Schaefer, O., Otitis media and bottle feeding, *Can. J. Publ. Hlth.,* 62, 478, 1971.
21. Oxtoby, M. G., Human immunodeficiency virus and other viruses in human milk: placing the issues in a broader perspective, *Pediatr. Infec. Dis.,* 7, 825, 1988.
22. Sullivan-Bolyai, J. Z., Fife, K. H., Jacobs, R. F., et al., Disseminated neonatal herpes simplex virus type I from a maternal breast lesion, *Pediatrics,* 71, 455, 1983.
23. Frantz, K. B. and Kalmen, B. A., Breastfeeding works for caesareans too, *RN,* 42, 39, 1979.
24. Jelliffe, E. F. P., Maternal nutrition and lactation, *CIBA Foundation Symposium,* 45, 119, 1976.
25. Wharton, B., *Nutrition and Feeding of Preterm Infants,* Oxford, U.K.: Blackwell Scientific Publications, 1987.
26. Pearce, J. L. and Buchanan, L. F., Breast milk and breastfeeding in very low-birth weight infants, *Arch. Dis. Child.,* 54, 897, 1979.
27. *Nursing the Down's Syndrome Baby,* Illinois: Avery Publishing, 1983.
28. Osburn, L. M., Reiff, M. I., and Bolus, R., Jaundice in the full-term neonate, *Pediatr.,* 73, 520, 1984.
29. Nevill, N. C. and Neifert, M. R., *Lactation,* New York: Plenum Press, 1983.
30. Lauwers, J. and Woessner, C., *CEA of Greater Philadelphia: Counselling the Nursing Mother—The Reference Handbook for Health Care Providers and Lay Counselors,* Wayne, N.J.: Avery Publishing, 1983.

Chapter 5

OPTIMAL NUTRITION IN LOW BIRTH WEIGHT INFANTS

Ronald Bainbridge and Reginald Tsang

TABLE OF CONTENTS

I. Introduction 33

II. Calcium, Phosphorus, and Vitamin D 35
  A. Recommendations 37

III. Vitamin E 37
  A. Recommendations 38

IV. Vitamin A 38
  A. Recommendations 39

V. Acknowledgment 40

References 40

I. INTRODUCTION

Optimal nutrition of the low birth weight (LBW) infant requires consideration of multiple factors. Some of the factors contributing to the definition of optimal nutrition are gestational and chronologic age, birth weight, route of nutrient intake, type of feed, the absolute and relative amounts of different nutrients, and the clinical status of the infant. In our attempts therefore, to come to a better understanding of what constitutes optimal nutrition in the low birth weight infant, it is customary to consider certain "gold standards" by which to measure the appropriateness of the various nutritional manipulations we make in these infants: intrauterine growth rate and accretion of different nutrients; and the growth rate of term breastfed infants and their serum concentrations of different nutrients and enzymes.

Low birth weight is arbitrarily defined as a birth weight of less than 2500 g. Clearly, however, this birth weight criterion defines a wide spectrum of infants with different nutritional needs. Birth weight is a reflection of intrauterine growth, and infants of similar birth weights may be of vastly different gestational ages. The small for gestational age (SGA) LBW infant, with relatively more developed

0-8493-2764-4/95

organ functions than the appropriate for gestation age (AGA) LBW infant, will likely have a quicker transition to full enteral feeds. Because of the relative maturity of their organs, and the catch-up growth that SGA infants often exhibit, they may have higher energy needs and may tolerate greater solute loads than their AGA counterparts. Within the group of LBW infants, nutritional requirements will vary depending on whether they are very low birth weight (<1500g) or extremely low birth weight (<1000g). As a rule, the nutritional requirements relative to body weight varies inversely with birth weight and gestational age. The maturity of the gastrointestinal (GI) system is gestational age dependent.[1] There is an orderly progression in the appearance of different GI functions through intrauterine life. Disturbed GI motility resulting in gastric stasis, delayed emptying, distension, and reflux limits enteral nutrition in smaller and more premature infants. Delayed enteric function is compounded by the maturational delay in the appearance of various GI enzymes and secretions necessary for digestion. Chronologic age affects the timing and volume of enteral feeds. During the transitional period from intrauterine to extrauterine life, which may extend up to 14 days for the most premature infants, emphasis is placed on meeting the fluid, electrolyte, acid-base, and respiratory needs of the LBW infant. There is no single criterion on which the decision to initiate enteral feedings is based. The decision is essentially a clinical one and varies widely among practitioners.

Human milk feedings generally are inadequate for optimal nutrition of the LBW preterm infant. For initiation of enteral feeds, however, it has advantages over infant formulas of being better tolerated, producing less GI disturbance and having a lower incidence of necrotizing enterocolitis.[2] Fortification of breast milk is necessary to provide the increased minerals, protein, and calories required by the preterm infant to achieve growth that more closely approximates that occurring in utero. Preterm infant formulas have been formulated to contain higher concentrations of protein, calcium, phosphorus, and several vitamins. Preterm infants fed these formulas demonstrate improved weight gain and may have improved neurodevelopmental outcome compared to infants fed standard formulas.[3]

What constitutes optimal nutrient intake depends on both the absolute and relative amounts of different nutrients and the interaction of such nutrients when ingested. Probably the most often discussed and understood interactions between nutrients are those between calcium (Ca) and phosphorus (P). A calcium to phosphorus ratio of approximately 2:1 appears to provide the balance required for optimal absorption and utilization of both minerals and for optimal bone mineralization. Any mismatch of Ca and P intake (e.g., relatively high P intake) resulting in lower serum calcium concentrations may cause stimulation of parathyroid hormone (PTH) secretion. PTH secretion promotes hyperphosphaturia lowering serum phosphorus while mobilizing bone Ca to increase serum Ca concentration. The end results of imbalances in Ca and P intake are altered Ca and P and bone metabolism.

For practitioners, nutrition of the low birth weight infant is often dictated by clinicopathologic factors over which they have variable control; neonatal asphyxia, respiratory distress, the placement of central vascular catheters, neonatal sepsis, the presence of a patent ductus arteriosus and electrolyte disturbance are some of the factors that contribute to a delay in the initiation of enteral and parenteral feeds. A select number of nutrients (Ca, P, vitamins D, A, and E) will be discussed in the rest of this manuscript as illustrations of optimal nutritional intake in low birth weight infants. Nutrients will be discussed in relation to their known developmental physiology, how they are delivered and utilized, and current recommendations for their use.

II. CALCIUM, PHOSPHORUS, AND VITAMIN D

If we accept that in utero growth and nutrient accretion rates in large part define optimum nutrient requirements, then we could attempt to achieve third trimester calcium and phosphorus accretion rates in preterm infants. The third trimester is the period of maximal calcium and phosphorus accretion. Between 24 weeks gestation and term, the fetus accretes calcium and phosphorus at rates of approximately 92 to 119 mg/d and 59 to 74 mg/d respectively, reaching a peak during the last 5 weeks of pregnancy.[4,5] Calcium and phosphorus accretion in utero occurs in a relatively restricted ratio of approximately 1.7:1 for the whole body and 2:1 if only the skeleton is considered. This ratio of Ca to P (1.7 to 2:1) appears to be important for maximal accretion of both and is the ratio of these minerals in human milk.[6,7] Intestinal Ca absorption and retention varies directly with Ca intake. Ca absorption is maximal in the proximal small intestine: its active absorption is facilitated predominantly by the actions of 1,25 Dihydroxy vitamin D ($1,25(OH)_2$ D) the active metabolite of vitamin D.[8,9] However, in preterm infants, a large fraction of Ca absorption appears to be non-vitamin D-dependent, probably by diffusion mechanisms. Individual variability in Ca absorption and endogenous secretion has been clearly established in preterm infants.[10,11,13] High P intake may inhibit intestinal Ca absorption presumably by forming insoluble poorly digestible intraluminal complexes. However, at levels of low Ca intake, Ca absorption may be enhanced by increasing P intake.[14]

Fat malabsorption may result in increased fecal Ca loss and decreased intestinal Ca absorption. This is of particular importance as up to 20% of breast milk Ca is in the milk fat, and fat absorption in premature infants is less efficient than in term infants. The precise relationship between milk lactose content and intestinal Ca absorption is not settled. However, lactose appears to enhance Ca absorption.[15] The multiple interactions of dietary magnesium (Mg), Ca, and P have been examined in LBW and VLBW infants.[16] The data from this study suggest that increased dietary Mg promoted P absorption and retention in VLBW infants from 20 d onward, but had no effect on Ca balance in either VLBW or LBW infants. How dietary cereals with their high phytate content affect Ca absorption in the low birth weight infant remains largely unexplored. The use of dual tracer stable isotope studies (^{44}Ca orally and ^{46}Ca intravenously) has allowed for more precise assessment of Ca absorption, net retention, and fecal and urinary excretion in

LBW infants.[10–12] The data from these studies suggest that net Ca retention in LBW infants fed high Ca- and P-containing formulas is primarily dependent on total dietary Ca absorbed, rather than endogenous gastrointestinal secretion. These studies also confirm that LBW infants fed such formulas can achieve in utero rates of Ca retention. Normally only 1 to 2% of filtered renal Ca is excreted. This will be increased with the use of medications such as furosemide and theophylline, by glucosuria and high sodium intake, and by feeding high mineral containing preterm formulas to LBW infants.[12,17]

Phosphorus absorption occurs maximally in the jejunum by both active transport and simple diffusion.[18] The precise mechanisms of absorption have not been fully elucidated, but they involve the interactions of vitamin D, Na^+–K^+ ATPase, and thyroxine.[18,19] Phosphorus absorption increases directly with P intake.[20,21] However, the percent of P absorbed decreases with increased intake. High Ca intake may inhibit P absorption by forming insoluble complexes as mentioned earlier. Net P retention is affected by renal P excretion, which increases with increased P intake and with factors resulting in increased parathyroid hormone secretion. Dietary phytate may bind P, inhibiting its absorption.[22]

Vitamin D requires intact hepatic 25 hydroxylation and renal 1 alpha hydroxylation systems to form its most active metabolite 1,25 $(OH)_2D$. These systems appear functional in most preterm babies.[23] 1,25 $(OH)_2D$ enhances small intestinal absorption of both Ca and P. If breastfed full term infants are used as the reference population, then vitamin D requirement may in fact be very low in preterm infants. Human milk vitamin D content is relatively low but may be increased by sunlight exposure, vitamin D intake and low skin pigmentation.[24] The adequacy of vitamin D supplementation has been evaluated by assessing serum 25 hydroxy vitamin D (25 (OH)D) concentrations. However, relating 25 (OH)D to vitamin D status in preterm infants is an inexact science, as it may be affected by Ca and P intake. At low levels of Ca and P intake there may be increased clearance of serum 25 (OH)D which may be related to increased levels of 1,25 $(OH)_2$ D.[26] However, even with high serum 25 (OH)D concentrations following vitamin D supplementation, radiologically detected osteopenia and rickets still occur.[27] The recommendations for vitamin D intake in premature babies should therefore be accepted in the context of an incomplete understanding of what constitutes normality in such babies. What constitutes optimal requirements of Ca, P, and vitamin D in the low birth weight infant receiving total parenteral nutrition (TPN) is still being decided. The requirement for vitamin D appears to be minimal and may be related to the fact that its site of action, the GI system is being by-passed. There are insufficient data that authoritatively define vitamin D requirements of infants on TPN. However, 160 i.u./Kg/d up to a maximum of 400 i.u./d of vitamin D appears acceptable.[28,29] The optimal TPN content of Ca and P depends on the Ca:P ratio, the absolute concentrations of both, and the amino acid concentration. As mentioned earlier, the Ca:P ratio appears to be very important in maximizing accretion of both. In TPN solutions, Ca:P ratios of 1.3:1 to 1.7:1 appear at present to promote optimal accretion. Parenteral solutions with Ca and P concentrations of up to 600 and 450mg/L respectively have been well tolerated by preterm infants and have resulted in very high retention of both minerals.[29,30] The greater the amino acid concentration of TPN, the more Ca and P will remain in solution and be delivered to the infant.

A. RECOMMENDATIONS

To assess the adequacy of Ca, P, and vitamin D intake, several measurements may be helpful. Serum Ca, P, and PTH may reflect adequacy of Ca and P intake, but may change acutely with clinical conditions. Low serum P concentrations (<4mg/dl) are a harbinger of rickets in preterm infants; possibly associated with increased serum Ca concentrations. Serum 25 (OH)D may be affected by factors mentioned earlier, while alkaline phosphatase measures are widely variable in this population. Radiologic evaluation of the wrists especially; for osteopenia, fractures, and rickets may provide a measure of gross bone disturbance. Despite the limitations of these assays and measurements, together they provide a clinical means to monitor optimal Ca, P, and vitamin D intakes. Measurements of bone mineral content (BMC) may provide useful information regarding bone mineralization, but remain predominantly a research tool. Although human milk does not provide adequate Ca and P intake for most preterm babies, it offers advantages in early feeds of less GI disturbance. Once full feeds are established, however, the high Ca- and P-containing preterm formulas or human milk supplemented to increase its mineral content should be used to promote mineral accretion and bone growth. Vitamin D of 400 i.u./d appears sufficient for most enterally fed preterm infants. In parenterally nourished preterm infants the Ca and P content of the TPN solutions should be gradually increased over several days while monitoring serum concentrations of Ca and P. When the child is stable, higher Ca- and P-containing solutions as mentioned earlier can then be used. Similar to enterally fed infants 400 i.u./d of vitamin D appears adequate in parenterally nourished infants.

III. VITAMIN E

''Poor vitamin E. Every time it appears to gain some respectability, it gets involved in some medical misadventure,'' Frank Oskie, 1984.[31] This statement may still reflect our slowly evolving knowledge of the role of vitamin E in the nutrition of low birth weight infants.

Vitamin E functions biologically as a scavenger of free radicals and as an antioxidant, while dietary iron and polyunsaturated fatty acids (PUFA) promote lipid peroxidation. There is a steady increase in fetal whole body vitamin E content with increasing gestational age which is proportional to percent body fat.[32] It is still unclear what serum vitamin E concentration constitutes normality for preterm infants. Vitamin E status has been defined relative to serum lipids and red blood cell hemolysis.[33] It is logical to define vitamin E status in this latter manner, as one of its primary functions is to prevent peroxidation of cell membrane lipids.[32]

Recently interest in vitamin E has increased in part because of its potential for the amelioration or prevention of intraventricular hemorrhage, retinopathy of prematurity (ROP), and bronchopulmonary dysplasia (BPD). The value of vitamin E in prevention of anemia of prematurity has long been established.[34] Vitamin E given within the first hours of life appears to reduce the incidence of types 2 and 3 intraventricular hemorrhages, especially in infants less than 750 g birth weight.[35,36] The data supporting the use of vitamin E in prevention of BPD and ROP is less convincing.[37] Fat soluble vitamin E requires normal pancreatic and biliary systems for optimum GI absorption. Absorption is by passive diffusion and increases with gestational age, breast milk feeding, and amount ingested.[38] It should be noted that there are higher concentrations of vitamin E in milk of mothers of preterm infants vs. mothers of term infants which persist for several weeks.[38] In this regard feeding preterm infants their own mother's milk may be advantageous. However, because of their reduced fat stores and relatively reduced fat absorption, preterm infants may require higher vitamin E intake. Optimum parenteral vitamin E intake is yet to be determined. The recommended parenteral dose of vitamin E is based on very limited data.[39] Caution should be exercised in the use of vitamin E especially at pharmacologic doses. An increased incidence of necrotizing enterocolitis and sepsis has been reported at higher doses.[40] The vitamin E supplement E-Ferol was also associated with hepatic and renal failure, coagulopathy, and pulmonary insufficiency;[41] the medication vehicle may have been the toxic agent in that situation.

A. RECOMMENDATIONS

The recommendations for vitamin E intake in the preterm infant are based on serum concentrations of vitamin E and are relative to the PUFA content of their feed. Preterm infants should receive 1 i.u. of vitamin E per gram of linoleic acid in the diet, or 0.7 i.u. vitamin E/100 Kcal. This may be provided by supplementing infant feeds with 5 to 25 i.u. vitamin E/d. Whether preterm infants fed their own mother's milk require supplementation is unclear. For infants on parenteral nutrition, 3 to 5 i.u. per day is probably sufficient.

IV. VITAMIN A

Similar to vitamin E, the mechanisms of human fetal vitamin A accretion are unknown. Several studies support the hypothesis of active placental transport of vitamin A.[42,43] Fetal vitamin A status appears to be well maintained irrespective of maternal vitamin A status. Vitamin A is a lipid soluble vitamin requiring an intact biliary/pancreatic system for optimum digestion. Ingested plant and animal vitamin A sources are hydrolyzed by pancreatic secretions releasing retinol which is absorbed in the small intestine. The mode of absorption is unclear but may be a passive process. After absorption, vitamin A is stored predominantly in the liver from where it is transported to other tissues bound to retinol binding protein (RBP).[44] Preterm infants are at increased risk for vitamin A and fat malabsorption secondary to the immaturity of their GI enzymes.[1] However, human milk feeding may facilitate vitamin A absorption in these infants.[45] The precise serum vitamin A concentration which defines a deficiency state in the preterm infant is unclear. Measurements of serum vitamin A, RBP and vitamin A to RBP molar ratio have all been used to assess vitamin A status.[43,46] Serum vitamin A concentrations below 10 mcg/dl probably represent vitamin A deficiency.[28]

Vitamin A is essential for the support of growth, differentiation of epithelial cells, and formation of rhodopsin. Vitamin A deficiency affects multiple organ systems resulting in growth failure, poor vision, and a skin rash. Of particular importance in the preterm infant, however, is that vitamin A deficiency may produce lung pathology similar to that observed in BPD.[47] This has prompted some to propose the use of pharmacologic doses of vitamin A to prevent the development of BPD.[48]

Human milk vitamin A content varies widely and is generally highest at birth, decreasing gradually over several months.[49,50] Similar to vitamin E, vitamin A content of preterm human milk is higher than that of term milk.[49,50] Human milk-fed full term infants receiving an estimated 90 mcg vitamin A/100Kcal of milk display no signs of vitamin A deficiency[51] (1mcg vitamin A = 3.33 i.u.). Despite the higher vitamin A content of preterm milk, preterm infants fed own mothers milk may be at risk for vitamin A deficiency. Preterm infants require much higher vitamin A intake than provided in preterm milk to achieve serum vitamin A and RBP concentrations within normal limits. Providing vitamin A to the infant on parenteral nutrition is problematic. Vitamin A undergoes significant photodegradation and is readily absorbed by intravenous tubing.[52] More than 80% of the vitamin A supplied in parenteral solutions may be lost by these two mechanisms. Administering vitamin A in intravenous lipids, however, may improve its delivery.[28]

A. RECOMMENDATIONS

Enterally fed preterm infants require over 700 i.u. vitamin A/kg/d (210 mcg/kg/d) to maintain normal serum vitamin A concentrations.[28] The precise upper limits of vitamin A requirements for stable growing preterm infants is yet to be determined. Up to 2800 i.u./kg/d appear to be well tolerated in these infants. When administering the higher doses of vitamin A, periodic measurements of serum vitamin A and RBP should be implemented to guard against vitamin A toxicity. There is limited data available to support the use of vitamin A in the prevention of BPD. No firm recommendations are yet available for intravenous vitamin A. The suggested intake of 500 mcg/kg/d is a useful starting point but may be changed as more data become available.

ACKNOWLEDGMENT

The authors would like to acknowledge the secretarial assistance of Ms. Nilda Barbieri in the preparation of this manuscript.

REFERENCES

1. Bucuvalas, J. and Balistreri W., The Neonatal Gastrointestinal Tract, in *Neonatal Perinatal Medicine: Disease of the Fetus and Infant,* Fanaroff, A. and Martin, R., eds., Mosby Year Book, 1992, 1010–1024.
2. Lucas, A. and Cole, T., Breast milk and neonatal necrotizing enterocolitis, *Lancet,* 336, 1519–1523, 1990.
3. Lucas A., Morley R., Cole, T., et al., Early diet in preterm babies and developmental status at 18 months, *Lancet,* 335, 1477–1481, 1992.
4. Ziegler, E., O'Donnell A., Nelson S., et al., Body composition of the reference fetus, *Growth,* 40, 329–341, 1976.
5. Sparks, J., Human intrauterine growth and nutrient accretion, *Semin. Perinatal.,* 8, 74–93, 1984.
6. Lonnerdal, B., Smith, C., and Keen, C., Analysis of breast milk: current methodologies and future needs, *J. Ped. Gastroent. Nutr.,* 3, 290–295, 1984.
7. Picciano, M., What constitutes a representative human milk sample, *J. Ped. Gastroent. Nutr.,* 3, 280–283, 1984.
8. Reichel, H., Koeffler, H., and Norman, A., The role of the vitamin D endocrine system in health and disease, *N. Eng. J. Med.,* 320, 980–990, 1989.
9. Habener, J., Rosenblatt, M., and Potts, J., Parathyroid hormones: biochemical aspects of biosynthesis, secretion, action, and metabolism, *Physiol. Rev.,* 64, 985–1053, 1984.
10. Hillman, L., Tack, E., Covell, D., et al., Measurement of true calcium absorption in premature infants using intravenous ^{46}Ca and oral ^{44}Ca, *Pediatr. Res.,* 23, 589–594, 1988.
11. Abrams, S., Esteban, N., Vieira, N., et al., Dual tracer stable isotopic assessment of calcium absorption and endogeneous fecal excretion in low birth weight infants, *Pediatr. Res.,* 29, 615–618, 1991.
12. Abrams, S., Yergey, A., Schanler, R., et al., Hypercalciuria in premature infants receiving high mineral containing diets, *J. Ped. Gastroent. Nutr.,* 18, 20–24, 1994.
13. Barltrop, D., Mole, R., and Sutton, A., Absorption and endogenous fecal excretion of calcium by low birth weight infants on feeds with varying contents of calcium and phosphate, *Arch. Dis. Child.,* 52, 41–49, 1977.
14. Senterre, J., Putet, G., Salle, B., et al., Effects of vitamin D and phosphorus supplementation on calcium retention in preterm infants fed banked human milk, *J. Ped.,* 103, 305–107, 1983.
15. Ziegler, E. and Fomon, S., Lactose enhances mineral absorption in infancy, *J. Pediat. Gastroent. Nutr.,* 2, 288–294, 1983.
16. Giles, M., Laing, I., Elton, R., et al., Magnesium metabolism in preterm infants: effects of calcium, magnesium, and phosphorus, and of postnatal and gestational age, *J. Pediatr.,* 117, 147–154, 1990.
17. Senterre, J. and Salle, B., Renal aspects of calcium and phosphorus metabolism in preterm infants, *Biol. Neonate.,* 53, 220–229, 1988.
18. Cross, H., Debiec, H., and Peterlik, M., Mechanism and regulation of intestinal phosphate absorption, *Miner. Electrol. Metab.* 16, 115–124, 1990.
19. Taylor, A., In vitro phosphate transport in chick Ileum: effect of cholecalciferol, calcium, sodium, and metabolic inhibitors, *J. Nutr.,* 104, 489–494, 1974.

20. Rowe, J., Goetz, C., Carey, D., et al., Achievement of in utero retention of calcium and phosphorus accompanied by high calcium excretion in very low birth weight infants fed a fortified formula, *J. Pediatr.*, 110, 581–585, 1987.
21. Shenai, J., Reynolds, J., and Babson, S., Nutritional balance studies in very low birth weight infants: enhanced nutrient retention rates by an experimental formula, *Pediatrics*, 66, 233–238, 1980.
22. Shenai, J., Jhaveri, B., Reynolds, J., et al., Nutritional balance studies in very low birth weight infants: role of soy formula, *Pediatrics*, 67, 631–637, 1981.
23. Salle, B., Glorieux, F., Delvin, E., et al., Vitamin D metabolism in preterm infants: serial serum calcitriol values during the first four days of life, *ACTA Pediatr. Scand.*, F2, 203–206, 1983.
24. Specker, B., Tsang, R., and Hollis, B., Effect of race and diet in human milk vitamin D and 25-hydroxy vitamin D, *Am. J. Dis. Child*, 139, 1134–1137, 1985.
25. Clements, M., Johnson, L., and Fraser, D., A new mechanism for induced vitamin D deficiency in calcium deprivation, *Nature*, 325, 62–65, 1987.
26. Halloran, B., Bikle, D., Levens, M., et al., Chronic 1,25 dihydroxy vitamin D_3 administration in the rat reduces the serum concentration of 25-hydroxy vitamin D by increasing metabolic clearance rate, *J. Clin. Invest.*, 78, 622–628, 1986.
27. Evans, J., Allen, A., Stinson, D., et al., Effect of high dose vitamin D supplementation on radiographically detectable bone disease of very low birth weight infants, *J. Pediatr.*, 115, 779–786, 1989.
28. The American Society for Clinical Nutrition, Guidelines for the use of vitamins, trace elements, calcium, magnesium, and phosphorus in infants and children receiving total parenteral nutrition, *Am. J. Clin. Nutr.*, 48, 1324–1342, 1988.
29. Koo, W., Tsang, R., Succop, P., et al., Minimal vitamin D and high calcium and phosphorus needs of preterm infants receiving parenteral nutrition, *J. Pediatr. Gastroent. Nutr.*, 8, 225–233, 1989.
30. Pelegano, J., Rowe, J., Carey, D., et al., Simultaneous infusion of calcium and phosphorus in parenteral nutrition for premature infants use of physiologic calcium-phosphorus ratio, *J. Pediatr.*, 114, 115–119, 1989.
31. Yearbook of Pediatrics Chicago, Oski, F. and Stockman, J., Eds., Yearbook Publishers, Inc., 1986, 72.
32. Karp, W. and Robertson, A., Vitamin in neonatology, *Adv. Pediatr.*, 33, 127–147, 1986.
33. Horwitt, M., Harvey, C., Dahm, C., et al., Relationship between tocopherol and serum lipid levels for determination of nutrition adequacy, *Ann. N.Y. Acad. Sci.*, 203, 223–236, 1972.
34. Oski, F. and Barness, L., Vitamin E deficiency: a previously unrecognized cause of hemolytic anemia in the premature infant, *J. Pediatr.*, 70, 211–220, 1967.
35. Fish, W., Cohen, M., Franzek, D., et al., Effect of intramuscular vitamin E on mortality and intracranial hemorrhage in neonates of 1000 grams or less, *Pediatrics*, 85, 578–584, 1990.
36. Chiswick, M., Gladman, G., Sinhas, et al., Vitamin E supplementation and periventricular hemorrhage in the newborn, *Am. J. Clin. Nutr.*, 53, 370S–372S, 1991.
37. Roberts, R. and Knight, M., Pharmacology of Vitamin E in the newborn, *Clin. Perinatal.*, 14, 843, 1987.
38. Gross, S. and Gabriel, E., Vitamin E status in preterm infants fed human milk or infant formula, *J. Pediatr.*, 106, 635–639, 1985.
39. Phillips, B., Franck, L., and Greene, H., Vitamin E levels in premature infants during and after intravenous multivitamin supplementation, *Pediatrics*, 80, 680–683, 1987.
40. Johnson, L., Bowen, F., Abbasi, S., et al., Relationship of prolonged pharmacologic serum levels of vitamin E to incidence of sepsis and necrotizing enterocolitis in infants with birth weight 1500 g or less, *Pediatrics*, 75, 619–638, 1985.
41. Martone, W., Williams, W., Mortensen, M., et al., Illness with fatalities in premature infants: association with an intravenous vitamin E preparation E-Ferol, *Pediatrics*, 78, 591–600, 1986.
42. Vobecky, J. S., Vobecky, J., Shapcott, D., et al., Biochemical indices of nutritional status in maternal, cord, and early neonatal blood, *Am. J. Clin. Nutr.*, 36, 630–642, 1982.

43. Jansson, L. and Nilsson, B., Serum retinal and retinal binding protein in mothers and infants at delivery, *Biol. Neonate,* 43, 269–271, 1983.
44. Goodman, D., Vitamin and retinoids in health and disease, *N. Eng. J. Med.,* 310, 1023–1031, 1984.
45. Fredrikzon, B., Hernell, O., Blackberg, L., et al., Bile salt stimulated lipase in human milk: evidence of activity in vivo of a role in the digestion of milk retinol esters, *Pediatr. Res.,* 12, 1048–1052, 1978.
46. Shenai, J., Chytil, F., Jhaveri, A., et al., Plasma vitamin A and retinol binding protein in premature and term neonates, *J. Pediatr.,* 99, 302–305, 1981.
47. Shenai, J., Chytil, F., and Stahlman, M., Vitamin A status of neonates with bronchopulmonary dysplasia, *Pediatr. Res.,* 19, 185–188, 1985.
48. Shenai, J., Kennedy, K., Chytil, F., et al., Clinical trial of vitamin A supplementation in infants susceptible to bronchopulmonary dysplasia, *J. Pediatr.,* 111, 269–277, 1987.
49. Vaisman, N., Mogilner, B., and Sklan, D., Vitamin A and E content of preterm and term milk, *Nutr. Res.,* 5, 931–935, 1985.
50. Chappell, J., Francis, T., and Clandinin, M., Vitamin A and E content of human milk at early *Hum. Dev.,* 11, 157–167, 1985.
51. Olson, J., Recommended dietary intakes (RDI) of Vitamin A in humans, *Am. J. Clin Nutr.,* 45, 704, 1987.
52. Greene, H. et al., Persistently low blood retinol levels during and after parenteral feeding of very low birth weight infants: examination of losses into intravenous administration sets and a method of prevention by addition to a lipid emulsion, *Pediatrics,* 79, 894, 1987.

Chapter 6

INFANT NUTRITION: THE FIRST TWO YEARS

Arturo R. Hervada and Maria Hervada-Page

TABLE OF CONTENTS

I. Introduction 43

II. Follow-on Formulas 44

III. When to Start Weaning Foods 44
A. Recommended Dietary Allowances 46

IV. The Toddler's Diet 46

V. Brain Development and the Critical Period of Brain Growth 49

VI. Non-Hematologic Complications of Iron Deficiency Anemia in Infants 50

References 52

No doctor seems able to advise you what to eat any better than his grandmother or the nearest quack.

—G. B. Shaw
The Doctor's Dilemma

I. INTRODUCTION

In the United States, the nutritional needs of infants in the first six months of life have been satisfactorily solved. Breastfeeding should be the first choice; human milk is the best food for babies. Infants receiving human milk should receive 400 IU of vitamin D daily. Unfortunately, there is no oral vitamin preparation that contains only vitamin D in the correct dose. A supplement containing vitamins A, D, and C is the only choice. At four months of age, breast-fed infants should also receive a daily prophylactic dose of iron; otherwise, they are at risk of becoming anemic.[1,2] For non-breast-fed infants, an adapted, low-osmolality, iron-supplemented formula is sufficient to provide optimal growth.

The feeding of infants from 6 to 24 months of age poses the real nutritional challenge for the 1990s. Pediatricians, family physicians, dietitians, and pediatric

0-8493-2764-4/95

nurse practitioners should concentrate their efforts on the second half of the first year and the entire second year, for two main reasons. First, this age is the critical period of human brain growth, the so-called "brain growth spurt". During this time, the right diet, one containing not only carbohydrates but also protein and fat, will provide the optimal substrate for brain development. The right nutrition may play a crucial role in the child's maximum intellectual and genetic potential. Second, the diet must provide enough iron to avoid iron depletion and iron deficiency anemia. Studies from the last decade clearly suggest that iron deficiency anemia in infancy, even after it is treated, may cause developmental and cognitive defects that are diagnosed five years later.

II. FOLLOW-ON FORMULAS

A relatively new type of infant formula, called follow-on or second age formula, have been in use in Europe for over a decade. Infant formulas are defined by the Codex Alimentarius, a Joint Commission of the World Health Organization (WHO) and the Food Agricultural Organization (FAO),[3] as "food intended as a liquid part of the weaning diet for the infant from the sixth month on and for young children". The Commission further regulates how these formulas are to be made and specifies the nutrient content, including a minimum and a maximum for most nutrients.

In Europe such formulas are used as a better alternative to cow's milk for 40% of babies aged 6 months and 75% aged 9 months. These formulas have advantages over cow's milk in that they contain less protein, are iron enriched, contain all necessary vitamins, and do not produce microscopic rectal bleeding.

In the United States, the Committee on Nutrition of the American Academy of Pediatrics addressed the use of these formulas in 1989[4]: "Recently 'follow-up feedings' (formula) have been marketed in the United States as they have been in Europe. Such food should provide no more than 50 to 65% of the total calorie intake. The nutrient content is satisfactory in available follow-on formulas." The statement concludes, "Thus, follow-up formulas, although nutritionally adequate, offer no clear advantage for infants receiving sufficient amounts of iron and vitamin-containing solid foods." At present, only one follow-on formula, Carnation's "Follow-Up", is available in the United States.

Since follow-on formulas are being used extensively in Europe, however, detailed guidelines on how to make them have been provided by the Committee on Nutrition of the European Society for Pediatric Gastroenterology and Nutrition (ESPGAN)[5] and more recently by the European Community "Commission Directive" (May 14, 1991).[6]

III. WHEN TO START WEANING FOODS

In the past decade, North American pediatricians and possibly mothers—though there are no good data on this—have been delaying the introduction of

TABLE 1
Dr. Sackett's Feeding Schedule

Age of baby	Type of food
2–3 days	Oatmeal or barley cereal twice a day
10 days	Strained vegetables (peas, beans, and carrots to start)
14 days	Concentrated cod liver oil and drops
17 days	Strained foods (the midnight feeding should be stopped at this age)

From Sackett, J. R., *South. Med. J.*, 1953, 46:358–363. With permission.

weaning foods. In 1980 the Committee on Nutrition of the American Academy of Pediatrics stated: "On the basis of present knowledge, no nutritional advantage results with the introduction of supplemental foods prior to the fourth to sixth month of age."[7] Based on present knowledge, no nutritional advantages result from the introduction of supplemental foods prior to that age. Indeed, most infants are not interested in solid foods until about six months of age. In the past, especially during the 1950s and 1960s, infants were fed these foods earlier and earlier without any demonstrable clinical or nutritional benefits. This trend reached its peak with the introduction of solid foods right in the nursery. Such a feeding schedule is shown in Table 1.

There are four developmental reasons why the introduction of solid and semi-solid foods should not start until the fifth or sixth month of life. First, the *extrusion reflex,* this defensive reflex in which infants use their tongue in a back and forth motion to reject foods or foreign objects placed in their mouth, is present in infants until the fourth or fifth month of age. Second, food spitting and regurgitation due to *gastroesophageal reflux* increase the possible risk of aspiration. These activities are present in over half of normal newborn infants. Gryboski and Walter[8] describe this phenomenon as "a physiologic gastrointestinal reflux". Gastrointestinal reflux is due to a developmental delay of the closing of the esophageal sphincter when liquid reaches the stomach and starts to contract. This, by the way, is the main reason why mothers have been placing bibs on babies for centuries, well before infant formulas were invented.

Infants do not have *head control* until about four months of age. Before this, they sit up with help, with the head flexed over the chest. This is an awkward position, to say the least, in which to swallow semi-solid foods. That babies can do it is an example of their remarkable versatility.

Finally, the ability to produce *pancreatic amylase,* an important enzyme in the digestion of complex carbohydrates, is not fully developed until nine months of life.[9] Therefore, starches do not appear to be appropriate as the first food to initiate. Solid foods like fruits, with their simpler sugars, would seem to be more physiologic. The same can be said regarding the development of gastric pepsin.

A. RECOMMENDED DIETARY ALLOWANCES

The recommended daily allowances (RDA) are defined by the Food and Nutrition Board of the National Research Council,[10] as "the levels of intake of essential nutrients that, on the basis of scientific knowledge, are judged by the Food and Nutrition Board to be adequate to meet known nutrient needs of practically all healthy persons". The Committee further explains that "RDAs are neither minimal requirements nor necessarily optimal levels of intake . . . rather, RDAs are safe and adequate levels . . . reflecting the state of knowledge concerning a nutrient, its bioavailability, and variations among U.S. population."

The RDA for protein intake is 1.38 g/kg of body weight per day for the first six months of life and 1.21 g/kg per day for the second six months. Interestingly enough, there are no RDAs for the other two macronutrients, fat and carbohydrates.

RDAs for calcium in infants and children are 400, 600, and 800 mg, for the first six months, the second six months, and one to three years of age, respectively. RDAs for calcium are only possibly achieved with a diet rich in dairy products.

IV. THE TODDLER'S DIET

By the time infants reach 12 months of age, the speed of their growth clearly decelerates. In the first year, weight increases from 3.5 kg at birth to 10 kg at 12 months. Weight gain in the second year, about 2.5 kg, will be only one-third of the previous year's gain. Now the infant's world has enlarged; he or she develops great curiosity. As the infant crawls and soon walks, food is no longer the only attraction and source of entertainment. The toddler is now busy discovering the world, and meals become an imposition. He or she must stop whatever he or she was doing, be placed in a high chair, and be fed by teaspoon a food that needs to be chewed. The child's appetite reaches a plateau, and choice of foods becomes erratic and capricious. One day he loves carrots, so the parent buys dozens of jars. The next day, the child does not like them anymore. This is the period when the mother, in a voice filled with desperation, tells the pediatrician, "But, doctor, he doesn't eat anymore," and she is right. In the first year, babies eat more and more every month. In the second year, a child's appetite is the same.

In preparation for this, the parent should plan a diet that is varied and contains foods rich in fats, proteins, and heme iron, namely, meats, chicken, and fish. The overuse and abuse of the plastic baby bottle become evident for children at this age. What a toddler with an erratic appetite does not need is several bottles filled with milk or juices every day. For a child of this age, a plastic bottle is no longer a vehicle to provide alimentation; it has become a pacifier, a source of entertainment and distraction. A restless toddler who starts to cry, especially in a public place, and is quieted with a bottle is learning to eat without being hungry.

Baby juices come in a variety of pleasant, sweet flavors; they provide calories and a few vitamins but nothing else. Intakes of formula over 600 to 850 ml of milk in the second year of life are excessive. In general, an infant who drinks four

TABLE 2
How to Successfully Feed a Toddler

Be engaging, but not overwhelming.
Talk in a quiet and encouraging manner.
Be careful not to overwhelm with talking or behavior.
Seat infant so that he/she is straight up and facing forward.
Wait for infant to pay attention and open up before feeding.
Feed when infant wants to eat but involve structure.
Let infant touch his/her food and feed him/herself with his/her fingers.
Let the infant decide how fast he/she wants to eat.
Respect the infant's food preferences.
Respect the infant's caution; all children eventually learn to eat.
Let the infant eat as much as desired.
Stop when the infant indicates that he/she is full.

From Arena, J., *Pediatr. News,* 1977, 11:18. With permission.

to five 8-oz. bottles of milk and one or two 4- to 6-oz. bottles of juice has no appetite left for anything else.

Apple juice is frequently recommended for the therapy of acute gastroenteritis in infants and children. Apples and apple juice have had a long empiric reputation as antidiarrheals since Binberg many years ago recommended raw apples for the treatment of diarrhea.[11]

Recently Vega Franco[12] has demystified the questionable therapeutic properties of the apple. In truth, apple juice, especially when drunk in large amounts, can produce diarrhea in infants and toddlers.[13,14] This effect was assumed to be due to the sorbitol (3 g/L) present in apple juice. Recent work using breath H_2 tests has shown, however, that diarrhea is caused by the high content of fructose in apple juice (64 g/L).[15] On the other hand, it now appears that sorbitol may be the carbohydrate responsible for the laxative effects of prunes, owing to the large amount (23 g/L) present in this fruit.[15]

A noted pediatrician, J. Arena, some time ago made recommendations on how to feed a toddler who was a finicky eater. These are given in Table 2. In Table 3, we list foods that are taboo for adult dieters but which, by their high caloric and nutritious content, may appeal to a finicky toddler with a poor appetite.

The content of the weaning diet should be a major concern for healthcare providers in developing countries. The children of the working class, especially, are basically fed only carbohydrates. Good nutritional proteins happen to be present in the most expensive foods: foods of animal origin, including meats and fish. When advising parents on what types of vegetables to feed infants, it is important to make a clear distinction between vegetables, like carrots and green beans, and legumes, like soybeans and lentils. Table 4 shows the clear nutritional differences between these two groups of foods. Nutritionally speaking, legumes are far superior to vegetables in their protein and iron content and provide more calories. About 90% of the iron present in foods is in the form of iron salts, the

TABLE 3
A Somewhat Eccentric List of Rich Foods to Feed a Toddler with Very Poor Appetite*

Good ice cream
Chicken livers (mothers feel better if they are called liver pate)
Condensed milk (it can be used on top of fruits, desserts, etc.)
Creamy cheeses
Peanut butter and jelly finger sandwiches (also as a spread on fruits, bananas, apples, etc.)
Hamburger meat
Butter or margarine (on breads, cookies, vegetables, etc.)

*In general, the listed foods are very rich in energy (calories) and have high protein and fat content.

TABLE 4
All Values Unless Explained Are in 100 g Protein

VEGETABLES	Energy 100 g Servings	Protein g	Fat g	CHO g	Ca mg	Fe mg
Beets	31	0.9	0.1	7.2	13	0.5
Carrots	31	0.9	0.2	7.1	33	0.6
Peas	71	5.4	0.4	12.1	23	1.8
Potatoes	95	2.6	0.1	21.1	9	0.7
Green beans	15	0.8	0.1	3.4	23	0.6
Spinach	19	2.2	0.4	2.9	122	1.4
Sweet Potatoes	141	2.1	0.5	32.1	40	0.9
LEGUMES						
Kidney beans	118	7.8	0.5	21.4	38	2.4
Chickpeas (Garbanzo)	179	10.2	2.4	30.3	75	3.0
Lentils	106	7.8	—	19.3	25	2.1
Soybeans	130	11	5.7	10.8	73	2.7

All nutrient values taken from *Bowes and Church's Food Values of Portions Commonly Used,* 16th ed., Pennington, J., Ed., Lippincott Co., Philadelphia, 1994.

non-heme iron, which is poorly absorbed. Its absorption is enhanced or blocked by other nutrients in the meal: It is enhanced by vitamin C and meats and inhibited by calcium, bran, phosphate, and polypherols such as tannin, which is found in teas.

Legumes are not a popular food in the Anglo-Saxon world and have never been popular as food staples. They are reputed to be difficult to digest. This is only partially true; in fact, most human beings can digest a meal including legumes without any problems. The excess gas produced by legumes is due to their content of a trisaccharide, raffinose, and a tetrasaccharide, stachyose. These substances are not digested by the small intestinal enzymes and reach the colon, where they are fermented by the local flora.

When possible, cow's milk should not be fed to infants for the first 12 months of life. The reasons why are given in Table 5. When economically possible, a

TABLE 5
Why Cow's Milk Should Not Be Used in the First 12 Months of Life

It contains practically no iron.
The only vitamin added is D.
It has too much sodium.
It has high osmolality.
Large amounts may produce rectal bleeding.

modern, adapted infant formula with iron is the best source of milk for infants this age who are not breast fed, when milk is still a very important part of the infant diet. Infant formula is a complete food. It contains fat, protein, minerals, and vitamins that provide most of the RDAs.

When infant formula is not economically feasible, cow's milk is then the only possible choice. Feeding cow's milk to an infant should be delayed as much as possible, and the daily intake should be restricted to no more than 500 ml daily. Under these circumstances, professional advice on how to provide iron with the diet, especially heme iron, is very useful. Recently in the United States, the feeding of cow's milk to infants has been the subject of considerable controversy among pediatricians and nutrition experts. In a recent review, we stated: "Cow's milk is neither the super food the American Dairy Association led us to believe in the 1950s and 1960s, when 'four or more' glasses of milk a day was the norm for children, nor the 'bad' food some anticholesterol extremists want us to believe today".[16] Recently, Finberg[17] has addressed the pros and cons of cow's milk use in a sensible manner.

V. BRAIN DEVELOPMENT AND THE CRITICAL PERIOD OF BRAIN GROWTH

From early gestation, the brain growth follows three stages of development. The first stage is a period of pure hyperplasia, an increase in the number of cells. In the second stage, both cell numbers and cell size increase. The third stage is a period of pure cell hypertrophy during which cells increase only in size.

In humans, the first period occurs in utero, between the 8th and 25th weeks after conception. This is the period of neuronal growth. By the end of the sixth to seventh month of pregnancy, all of the neurons needed for a lifetime have developed. The supporting structure of the central nervous systems, the neuroglia, develops later and continues growing after the infant's birth. After birth, brain cell division continues, mostly at the neuroglial level, and myelinization of the brain structures continues for years.

By the time an infant is 18 to 24 months of age, the development of the brain is mostly complete. Before this happens, a period of very fast growth occurs from the last trimester of pregnancy to the 18th month of postnatal life. This growth spurt is the critical growth period of the brain. During this period, a proper diet containing all essential macronutrients and iron is critical for the optimal growth of this organ. Dobbing has stated that the human brain has this one chance to grow properly.[18]

The toddler diet should be relatively rich in fats, since the brain is basically a lipid organ. About 60% of the adult brain is fat. Infants receiving human milk receive 50% of their calories from fat. The so-called prudent diet, with 30% or less of the calories coming from fat, is not good for infants and toddlers. Practically all of the authors agree that such a diet may be followed after the first two years of life. For the normal toddler, a diet that provides up to 40% of the calories from fat seems more appropriate.

VI. NON-HEMATOLOGIC COMPLICATIONS OF IRON DEFICIENCY ANEMIA IN INFANTS

Until the last decade, iron deficiency anemia was considered a hematologic disorder, and the classic complications of any anemia were such conditions as pallor and congestive failure. Today it is becoming evident that the most serious complications are the non-hematologic, those that affect behavior and cognition in children. The early pioneer work of Oski and Hoenig[19] opened new horizons for research in this area. Today iron deficiency anemia should be considered a most serious disease for its long-term effects, even after treatment, on cognition and behavior.

Studies to evaluate long-term effects on cognition and behavior are very difficult to design, and the results are difficult to evaluate. First, the human brain, like the rest of the body, has considerable capabilities to recover from moderate injuries and mild nutritional deficiencies. Second, the effect of the environment, per se—"the nurture effect"—on the child's intellectual and psychomotor development is most difficult to judge. Confounding variables such as poverty, family size, parental skills, and nurturing capabilities, chronic infections, and social class play a role in outcomes. The adverse effect of severe poverty in general creates some insurmountable factors that affect the final outcome for the child. Two recent studies conducted in Santiago, Chile, and in San Jose, Costa Rica, were specifically designed to account for all the confounding variables described. Walker and co-workers[20] in Santiago, Chile, studied 196 infants from birth to 15 months of age. When the infants were evaluated at 12 months of age, anemic infants had mean lower mental and psychomotor index than infants in the control group who were not anemic, although some were iron deficient.

These results clearly suggest that iron deficiency anemia, but not iron deficiency, produces adverse effects on mental and motor development. No improvement in these scores was noted after 10 days or 3 months of iron therapy. These authors evaluated the same children when they were 4 1/2 years old with a large battery of developmental and cognitive groups of tests. The deficits found in the formerly anemic infants persisted in these children at school age, despite the effectiveness of iron therapy.

Lozoff[21] and associates conducted a similar study in San Jose, Costa Rica, with children 12 to 23 months of age. The control group for the study consisted of 40 non-anemic infants with hemoglobin values of 12 g/L or higher. Infants in the anemic and intermediate iron deficiency group were treated with iron administered intramuscularly, orally, or by a placebo. After three months of therapy, the children who were iron deficient, even those in whom the deficiency was corrected, had developmental scores near normal but lower than those of the control subjects. Most of the initially anemic infants with severe anemia or chronic iron deficiency continued to score significantly lower on psychomotor and mental development. These authors also retested the children at five years of age, with results similar to those described by Walter and co-workers.

Because of the serious implication of these studies and similar ones conducted in different places in the world, the International Conference on Iron Deficiency and Behavioral Development, organized by the World Health Organization, was held in Geneva on October 12, 1989. More than one dozen papers were presented by different scientists. The proceedings of the symposium were published two years later.[22]

It is interesting to note that after a review of the world literature, those at the symposium found that only seven published studies met the specific criteria of a probability design with appropriate treatment and control groups. All of them were conducted in developing countries, something that must be taken into consideration. The summary and conclusions of the meeting were written by Hass and Wilson Fairchild[23] stating, ''Research in children and experimental animals suggests that iron deficiency anemia is causally associated with less than optimal behavior. It is important that iron deficiency anemia is prevented and treated in all children. Because the specific mechanism and functional significance of this behavior are not completely understood, further studies are essential to clarify the effect of iron deficiency anemia itself, to assess the reversibility of these effects, and to determine the importance of lesser degrees of iron deficiency in children.''

In conclusion, the decision of when to introduce solid foods should be based on sound developmental and nutritional principles. There is no medical reason do to so before infants are four to six months old, and most babies are simply not ready to eat solid foods before that age. The types of foods introduced to infants are strongly influenced by ethical, cultural, and economic forces. Pediatricians should be aware of this and work with parents within this ethical-economic framework.

Today's babies are fed healthier, better foods than ever before, especially in developed countries. The weaning diet must be planned with the nutritional requirements of the brain in its most critical period of growth. After infants are five months old, special attention must be paid to the iron content of foods. Heme iron is the best source of nutritional iron.

Thus, pediatricians and other health professionals must teach parents around the world the importance of the right diet in these critical first two years of life.

With proper nutrition in these years, children everywhere will be able to reach their optimal genetic potential.

REFERENCES

1. Pizarro, F., Yip, R., Dallman, P. R., Olivares, M., Hertrampf, E., and Walter, T., Iron status with different infant feeding regimens: Relevance to screening and prevention of iron deficiency, *J. Pediatr.*, 118, 687–692, 1991.
2. Calvo, E. B., Galindo, A. C., and Aspres, N. B., Iron status in exclusively breast fed infants, *Pediatrics*, 90, 375–379, 1992.
3. Supplement 3 to Codex Alimentarius, Vol. IX, 1988, World Health Organization, Rome, Italy.
4. American Academy of Pediatrics, Committee on Nutrition, Follow-up or weaning formulas, *Pediatrics*, 83, 1067, 1989.
5. ESPGAN, Committee on Nutrition, Comment on the composition of cow's milk based follow-up formulas, *Acta. Paediatr. Scand.*, 79, 250–254, 1990.
6. Commission Directive, On infant formulae and follow-up formulae, *J. Eur. Commun.*, 1991, No. L.
7. American Academy of Pediatrics, Committee of Nutrition, On the feeding of supplemental foods to infants, *Pediatrics*, 65, 1178–1181, 1980.
8. Gryboski, J. and Walter, W. A., *Gastrointestinal Problems in Infancy*, Philadelphia, WB Saunders, 1983.
9. Lebenthal, E., Impact of digestion and absorption in the weaning period in infant feeding practices, *Pediatrics*, 207 (suppl), 1985.
10. National Research Council, *Recommended Dietary Allowances*, Washington, D.C., National Academy Press, 1989.
11. Binberg, I. L., Raw apple in the treatment of diarrheal conditions in children, *Am. J. Dis. Child.*, 45, 18–24, 1933.
12. Vega Franco, L., La manzana: mito, ciencia y realidad en la practica pediatrica, *Bol. Med. Hosp. Infant Mex.*, 44, 726–727, 1987.
13. Kneepkens, C. M. F., Douwes A. C., and Jacobs, C., Apple juice, fructose, and non-specific chronic diarrhea in children, *Eur. J. Pediatr.*, 148, 571–73, 1989.
14. Hyams, J. S., Etienne N. L., Leichtner A. M., et al., Carbohydrate malabsorption following fruit juice ingestion in young children, *Pediatrics*, 82, 64–68, 1988.
15. Hoekstra, J. H., VanKempen A. A. M. W., and Kneepkens, C. M. F., Apple juice malabsorption: fructose or sorbitol? *J. Pediatr. Gastroent. Nutr.*, 16, 39–42, 1993.
16. Hervada, A. R. and Newman, D. R., Weaning: Historical perspectives, practical recommendations, and current controversies, *Curr. Probl. Pediatr.*, 22, 233, 1992.
17. Finberg, L., How good a food for humans is cow's milk? *Am. J. Dis. Child.*, 146, 1432, 1992.
18. Dobbing, J., *The Later Development of the Brain and Its Vulnerability*, 2nd ed., *Scientific Foundations of Pediatrics*, Baltimore, University Park Press, 1981, 744–759.
19. Oski, F. A. and Honig, A. S., The effects of therapy on the developmental scores of iron deficient infants, *J. Pediatr.*, 92, 21–25, 1978.
20. Walker, T., DeAndraca, I., Chadud, P., et al., Iron deficiency anemia: adverse effects on infant psychomotor development, *Pediatrics*, 84, 7, 1989.
21. Lozoff, B., Brittenham, G. M., and Wolf, A. W., Iron deficiency anemia and iron therapy effects on infant development test performance, *Pediatrics*, 79, 981, 1987.
22. International Conference on Iron Deficiency and Behavioral Development, *Am. J. Clin. Nutr.*, 50, 565, 1989.
23. Haas, J. D. and Wilson Fairchild, M., Summary and conclusions of the International Conference on Iron Deficiency and Behavioral Development, *Am. J. Clin. Nutr.*, 50, 703–705, 1989.

Chapter 7

IRON DEFICIENCY

Lloyd J. Filer, Jr.

TABLE OF CONTENTS

I. Background 53

II. Iron Balance 54

III. Current Magnitude of Iron Deficiency in Childhood 55

IV. Iron Deficiency vs. Iron Deficiency Anemia 56

V. Iron Intake 57

VI. Consequences of Iron Deficiency 57

References 59

I. BACKGROUND

Iron deficiency in infancy has been recognized as a public health issue for the past seven decades. Following World War I, Mackay established that 25% of infants from a poor and overcrowded district of the London's East End had hemoglobin concentrations less than 8.25 g/dl at some time during their first two years of life.[1] Providing these infants with a powdered whole milk product that contained iron added as iron ammonium citrate so that each infant received 56 mg of elemental iron per day effectively cured anemia in 80% of infants. Furthermore, Mackay was convinced that general health and resistance to infection improved in those infants given dietary iron.

In spite of this landmark observation of the value of fortifying milk with iron, physicians for the next 40 years relied upon iron-fortified cereal to meet the iron requirements of infants. Unfortunately, the choice of iron compounds added to cereal was such that the iron was nonbioavailable. Bioavailable forms of iron produced off-color and rancidity in the cereal. Egg yolk, considered a good source of dietary iron, was widely used in infant feeding. Subsequently, it was found that the iron in egg yolk was poorly absorbed.[2]

In 1959 Marsh and co-workers reported that the addition of ferrous sulfate to infant formula at a concentration of 12 mg per liter prevented iron deficiency.[3]

0-8493-2764-4/95

Similar results were reported in 1966 by Andelman and Sered who fed the same formula product to a large number of infants followed in the City of Chicago Public Health Clinics.[4] In 1971 the American Academy of Pediatrics (AAP) issued a policy statement that iron-fortified formulas be used for the first 12 months of life.[5] However, many physicians caring for infants were reluctant to translate the results of these studies and recommendations to their practice; thus, some 20 years were to elapse before 85% of physicians were advising mothers to use iron-fortified formula.[6,7]

In 1992 the Committee on Nutrition of the AAP reaffirmed its position on the feeding of iron-fortified formula by stating that "the only acceptable alternative to breast milk is iron-fortified infant formula."[8]

II. IRON BALANCE

Iron deficiency in infancy reflects the balance between iron endowment at birth, dietary intake, and iron requirements for growth and replacement of losses. Low birth weight infants are born with low iron stores and rapidly expand their vascular volume through growth. These infants illustrate two of the components essential for balance.

The feeding of non-iron-fortified formula or whole milk in the first 12 months of life illustrates the other two arms of the equation for balance, i.e., low dietary intake of iron and the potential for excessive gastrointestinal loss of iron, specifically in the young infant fed whole cow milk.[9]

In a 1969 review of public health nutrition problems, Filer[10] cited studies by Schorr and Radel[11] and Gorten and Cross[12] showing that 38% of infants 1.3 to 2.0 kg birth weight developed iron deficiency anemia. In contrast, only 1.7% of infants of comparable birth weight developed iron deficiency anemia when fed an iron-fortified formula.

The relationship between growth and dietary iron absorption from conception to maturity was described by Gorten and co-workers.[13] Iron absorption as measured by ^{59}Fe uptake was increased with change in incremental growth rate.

Many factors in the diet of infants, children, and adults influence the absorption of dietary iron. These range from the amount and chemical form of added iron, the presence in the diet of factors enhancing and/or inhibiting iron absorption, in addition to the iron status of the individual (Table 1). A recent study by Ziegler using a stable isotopic form of iron (^{58}Fe) added to formula or various infant foods serves to illustrate some of these effects (Table 2).[14] Due to the wide variability in amount of isotope incorporated into erythrocytes of individual infants, differences between foods were not statistically significant. Ziegler concluded that rice cereal does not inhibit iron absorption, that grape juice does not enhance iron absorption, and that the small amount of beef in the vegetable beef dinner did not increase iron bioavailability. Iron added as ferrous sulfate or ferrous fumarate was equally bioavailable.

TABLE 1
Determinants of Iron Absorption

Dietary Factors
Enhancers
Ascorbic Acid
Meat
Low pH (Lactic Acid)
Inhibitors
Phytates
Tannins/Polyphenols

TABLE 2
Iron Absorption from Infant Foods
Stable Isotope Study ^{58}Fe[14]

Food System	Iron mg	Ascorbic Acid mg	% Label in Erythrocyte
Rice Cereal/Fruit	2.8	6	5.4
Rice Cereal—Ferrous Sulfate	1.1	5	4.4
Vegetable—Beef	3.2	—	2.5
Grape Juice	2.8	27	4.8
Rice Cereal—Ferrous Fumurate	4.6	4	4.0
Infant Formula	2.0	84	7.9

Woodhead and co-workers have recently described a gender-related difference in iron absorption by preadolescent children.[15] Females approximately 9 years old (Tanner stage 1) incorporated more iron into erythrocytes than males age 10 years (Tanner stage 1). They attributed these differences in gender-related iron absorption to hormonal differences rather than body stores of iron.

III. CURRENT MAGNITUDE OF IRON DEFICIENCY IN CHILDHOOD

Iron deficiency in childhood is usually established by measurement of the prevalence of anemia defined as the number of children in specific age groups with hemoglobin concentrations less than 10.3 g/dl for children age 6 to 23 months or a hemoglobin concentration less than 10.6 g/dl for children 2 to 5 years of age. According to data released by the Pediatric Nutrition Surveillance System (PNSS) of the Centers for Disease Control (CDC), the incidence of childhood anemia among children from low-income families enrolled in the public assistance program WIC (Special Supplemental Food Program for Women, Infants, and Children) declined from 6.8% in 1973 to 3.1% in 1984.[16] If anemia was defined as a hemoglobin concentration less than 11 g/dl, the incidence of anemia would have fallen from 19% in 1973 to 9% in 1984.

Socioeconomic status (SES) has a profound effect on the prevalence of anemia with anemia more prevalent among low SES children. In 1987, Yip estimated an

TABLE 3
Incidence of Iron Deficiency in Children MCV Model

Gender	Age Years	Ethnic Background: White	Black	Mexican American	Cuban	Puerto Rican
Both Sexes	6–11	3.2	4.0	0.5	3.1	2.5
Male	12–15	1.8	2.2	3.5	6.7	2.6
	16–19	0.8	0.9	0.4	0.0	1.2
Female	12–15	2.1	8.0	5.9	—	6.5
	16–19	3.8	13.8	7.9	—	7.9

Data from Life Sciences Research Office, FASEB, *Nutrition Monitoring in the United States*, Government Printing Office, Washington, D.C., DHHS Publication No. (PSH) 89-1255, 1989.

incidence of iron deficiency anemia among children from low-income families of 17% based upon a hemoglobin concentration of 11 g/dl.[17] Study of a stable pediatric practice in a middle-class area of Minneapolis indicates that anemia among 4 to 7 year-olds declined from 5.5% in 1969 to 2.9% in 1986.[16] A prevalence rate of 3% represents the baseline level for distribution of a normal population; thus, one might conclude that there is little or no iron deficiency in this study population.

The incidence of iron deficiency in children 6 to 19 years of age classified on the basis of mean corpuscular volume (MCV) identifies females 12 to 19 years of age, when ethnic background is black, Mexican-American, or Puerto Rican, at greater risk for iron deficiency than white females of the same age (Table 3).[18]

A recent analysis by the Centers for Disease Control of the relationship between iron nutrition and hemoglobin differences between blacks and whites suggests that the observed differences in hemoglobin concentration may reflect genetic and environmental determinants of iron utilization.[19] Until studies designed to explore racial differences in iron absorption, mobilization, and utilization across all age groups are carried out, it will be difficult to interpret the physiological consequences of the data compiled in Table 3.

IV. IRON DEFICIENCY VS. IRON DEFICIENCY ANEMIA

Iron deficiency exists with or without anemia, with the latter reflecting the mild end of the spectrum. Because of the strong association between anemia and iron deficiency, these terms are often used interchangeably. For population groups where the prevalence of anemia is high, most of these cases can be ascribed to iron deficiency. In population groups where the prevalence of iron deficiency is low, most individuals will not manifest anemia.

Yip has advocated a multiple test approach for defining iron deficiency.[17] He advocates use of the ''ferritin model'' or combination of serum ferritin concentration, transferrin saturation, and erythrocyte protoporphyrin concentration to define iron deficiency. Two positive tests from this triad define iron deficiency. However, multiple testing is expensive and not practical for routine screening. As a consequence, surveillance systems to establish the prevalence of iron deficiency in children rely on methods adapted to a field setting, i.e., the determination of hemoglobin concentration or hematocrit. For practical purposes, these relatively inexpensive tests will continue to be the yardstick by which the incidence of iron deficiency is measured.

V. IRON INTAKE

Since 1964 the recommended dietary allowance or RDA for iron intake by infants has ranged from 10 to 15 mg per day. Four national surveys conducted since 1976 indicate that on average, one-year-olds living in the United States receive the RDA for iron when fed an iron-fortified formula in addition to an array of table foods and infant foods, including iron-fortified infant cereal.[20] Much of the improvement in the iron status of infants and children can be attributed to the WIC program. This program serves approximately one-third of annual births with the mandated food package providing iron-fortified formula, infant cereal, and juice.[21] The nutrient profile of this package provides for a daily intake of 26 mg of iron and 75 mg of vitamin C per day. These intakes exceed the 1989 RDA for iron and vitamin C by a factor of 2.6 and 1.9, respectively. Since WIC is targeted to feed infants and children from families whose income does not exceed $22,385 for a family of four as of July 1, 1989, its value in reducing iron deficiency among children becomes evident.

In the past decade, many factors have led to improvement in the iron nutritional status of children living in the U.S. Infant feeding practices have changed, i.e., increased numbers of breast-fed infants, increased use of iron-fortified formula, a reduction in consumption of cow milk, and an increase in the number of families served by public assistance programs (WIC). These changes have resulted in the need to reconsider the value of anemia as an index of iron deficiency. Is routine screening for detection of anemia cost effective? Should such screening be limited to high-risk subpopulations in the U.S., such as low birth weight infants, infants with a history of excessive or early intake of cow milk, infants from low-income families not served by public assistance programs, or infants of specific ethnic backgrounds? As Yip has so clearly stated, the pediatric population has shifted from a position where iron deficiency is a major contributor to childhood anemia to a position where iron deficiency is less well identified by anemia.[17]

VI. CONSEQUENCES OF IRON DEFICIENCY

Iron deficiency has been associated with poor growth, an increase in certain infectious diseases, and reduced physical and intellectual performance. Only recently have these subtle relationships become evident. In the past, pediatricians have associated iron deficiency with such clinical signs as pallor, lethargy, fussiness, infantile obesity, and hemic murmurs.

The relationship between growth in infancy and iron nutritional status was recently reviewed by Owen.[22] He concluded, as Gorten had demonstrated some

25 years earlier, that iron depletion would limit growth. In a study relating total body iron (hemoglobin and tissue iron) to growth of infants cared for in a Catholic Charity orphanage in Baltimore, infants from inner-city Baltimore, and infants raised in rural Pakistan, it was evident that the Pakistani infants whose total body iron was the least grew at the third percentile in contrast to infants reared in Baltimore.[10]

In reviewing the relationship of iron deficiency to infection, Walter and co-workers concluded that feeding infants and children foods fortified with iron did not increase their susceptibility to infection.[23] Their review indicates that iron deficiency has an adverse reaction on human T-cell and phagocyte function, and they cited evidence that iron supplementation had an adverse effect on malaria. Yip has concluded that the adverse effects of iron administration to individuals with malaria is secondary to reticulocyte production following iron therapy.[24] Similar views have been presented by Hershko.[25]

Viteri has reviewed the limited data relating iron nutrition to work capacity and physical performance in children.[26] In contrast to adult populations where hemoglobin concentration has been directly related to maximal work output, children with iron deficiency may not manifest a reduction in work capacity. This may reflect the fact that children have more efficient adaptive mechanisms to compensate for iron deficiency. The inability to relate iron deficiency to physical performance in childhood may also reflect the lack of precise methods for study of work performance in children. According to Sallis, the database relating physical activity or work performance during fitness testing to nutritional indices in childhood is insufficient to draw conclusions related to health outcome.[27]

The effect of iron deficiency and iron deficiency anemia on behavioral development during infancy has been studied in five geographic settings.[28,29,30,31,32] Each of these independently conducted studies has noted that iron deficiency is associated with delays in motor and mental development. These studies from Costa Rica, Guatemala, Chile, the United Kingdom, and the United States were conducted on infant populations with well-defined iron status. Lower mental test scores were observed in four of the groups where the Bayley Scale of Infant Development was used. Motor development was reduced in four of five groups where such testing procedures were employed.

In the studies of Lozoff and co-workers[29] and those of Walter and co-workers,[30] it was observed that reversal of the anemic state had little discernable effect on the indices of mental and psychomotor development when infants were retested three months after correction of their anemia or at age five years. Such adverse short- and long-term sequelae should provide health professionals and those responsible for public health policy incentive to reduce the incidence of iron deficiency during infancy. At a recent international conference on iron deficiency and behavioral development held under the auspices of WHO (World Health Organization), INACG (International Nutritional Anemia Consultative Group), and the UN (United Nations) University, it was concluded that "Iron deficiency anemia (IDA) is causally associated with less than optimal behavior; therefore, it is important that IDA be prevented and treated in all children''.[33]

Prevention of anemia that is estimated to involve 43% of world children, excluding China, under five years of age[26] to assure improved growth, resistance to infection, and improved physical and mental performance are worthy goals for the 21st century.

REFERENCES

1. Mackay, H. M. M., Anemia in infancy: its prevalence and prevention, *Arch. Dis. Child.*, 3, 117–147, 1928.
2. Callender, S. T., Marney, S. R., Jr., and Warner, G. T., Eggs and iron absorption, *Brit. J. Haematology,* 19, 657–665, 1970.
3. Marsh, A., Long, H., and Stierwalt, E., Comparative hematologic response to iron fortification of a milk formula for infants, *Pediatrics,* 24, 404–412, 1959.
4. Andelman, M. B. and Sered, B. R., Utilization of dietary iron by term infants, *Amer. J. Dis. Child,* 111, 45–55, 1966.
5. American Academy of Pediatrics, Committee on Nutrition, Iron-fortified formulas, *Pediatrics,* 45, 55, 1971.
6. Fomon, S. J., Reflections on infant feeding in the 1970s and 1980s, *Am. J. Clin. Nutr.,* 46, 171–182, 1987.
7. Taylor, J. A. and Bergman, A. B., Iron-fortified formulas. Pediatricians' prescribing practices, *Clin. Pediatr.,* 28, 73–75, 1989.
8. American Academy of Pediatrics, Committee on Nutrition, The use of whole cow's milk in infancy, *AAP News,* 8, 18–19, 1992.
9. Ziegler, E. E. and Fomon, S. J., Cow milk, gastrointestinal blood loss, and iron nutritional status of infants, in *Food Intolerance in Infancy: Allergology, Immunology, and Gastroenterology,* Hamburger R. N., ed., New York, Raven Press, 1989, 135–144.
10. Filer, L. J., Jr., The USA today—is it free of public health nutrition problems? *Am. J. Public Health,* 59, 327–338, 1969.
11. Schorr, J. B. and Radel, E., Nutritional anemia in early infancy, IX Congress of the International Society of Hematology Section on Erythrocytes and Erythropathics, Mexico, September 2–5, 1962.
12. Gorten, M. K. and Cross, E. R., Iron metabolism in premature infants. II. Prevention of iron deficiency. *J. Pediatr.,* 64, 509–520, 1964.
13. Gorten, M. K., Hepner, R., and Workman, J. B., Iron metabolism in premature infants. I. Absorption and utilization of iron as measured by isotope studies. *J. Pediatr.,* 63, 1063–1071, 1963.
14. Ziegler, E. E., Bioavailability of iron from infant foods: studies with stable isotopes, in *Dietary Iron: Birth to Two Years,* Filer, L. J., Jr., Ed., New York, Raven Press, 1989, 83–89.
15. Woodhead, J. C., Drulis, J. M., Nelson, S. E., Janghorbani, M., and Fomon, S. J., Gender-related differences in iron absorption by preadolescent children, *Pediatr. Res.,* 29, 435–439, 1991.
16. Yip, R., The changing characteristics of childhood iron nutritional status in the United States, in *Dietary Iron: Birth to Two Years,* Filer, L. J., Jr., Ed., New York, Raven Press, 1989, 37–56.

17. Yip, R., Iron nutritional status defined, in *Dietary Iron: Birth to Two Years,* Filer, L. J., Jr., Ed., New York, Raven Press, 1989, 19–35.
18. Life Sciences Research Office, FASEB, *Nutrition Monitoring in the United States,* Government Printing Office, Washington, D.C., DHHS Publication No. (PSH) 89–1255, 1989.
19. Pery, G. S., Byers, T., Yip, R., and Margen, S., Iron nutrition does not account for the hemoglobin differences between blacks and whites, *J. Nutr.,* 122, 1417–1424, 1992.
20. Ernst, J. A., Brady, M. S., and Rickard, K. A., Food and nutrient intake of 6- to 12-month-old infants fed formula or cow milk: a summary of four national surveys, *J. Pediatr.,* 117, S86–S100, 1990.
21. Batten, S., Hirschman, J., and Thomas, D., Impact of the Special Supplemental Food Program on infants, *J. Pediatr.,* 117, S101–S109, 1990.
22. Owen, G. M., Iron nutrition, growth in infancy, in *Dietary Iron: Birth to Two Years,* Filer, L. J., Jr., Ed., New York, Raven Press, 1989, 103–113.
23. Walter, T., Olivares, M., and Pizarro, F., Iron and infection, in *Dietary Iron: Birth to Two Years,* Filer, L. J., Jr., Ed., New York, Raven Press, 1989, 119–130.
24. Yip, R., Discussion, in *Dietary Iron: Birth to Two Years,* Filer, L. J., Jr., Ed., New York, Raven Press, 1989, 133–139.
25. Hershko, C., Iron and infection, in *Nutritional Anemias,* Nestle Nutrition Workshop Series, Vol. 30, Fomon, S. J., and Zlotkin, S., Eds., New York, Raven Press, 1992, 53–61.
26. Viteri, F. E., Influence of iron nutrition on work capacity and performance, in *Dietary Iron: Birth to Two Years,* Filer, L. J., Jr., Ed., New York, Raven Press, 1989, 141–160.
27. Sallis, J. F., Epidemiology of physical activity and fitness in U.S. children and adolescents, *Critical Reviews in Food Science and Nutrition,* 33, 403–408, 1993.
28. Lozoff, B., Jimenez, E., and Wolf, A. W., Long-term developmental outcome of infants with iron deficiency, *New Engl. J. Med.,* 325, 687–694, 1991.
29. Lozoff, B., Brittenham, G. W., Viteri, F. E., et al., The effects of short-term oral iron therapy on developmental deficits in iron-deficient anemia infants, *J. Pediatr.,* 100, 351–357, 1982.
30. Walter, T., Effect of iron deficiency anemia on infant psychomotor development, in *Dietary Iron: Birth to Two Years,* Filer, L. J., Jr., Ed., New York, Raven Press, 1989, 161–175.
31. Grindulis, H., Scott, P. H., Belton, N. R., and Wharton, B. A., Combined deficiency of iron and vitamin D in Asian toddlers, *Arch. Dis. Childh.,* 61, 843–848, 1986.
32. Oski, F. A., Honig, A. S., Hela, B., and Howamitz, P., Effect of iron therapy on behavior in nonanemic, iron-deficient infants, *Pediatrics,* 71, 877–880, 1983.
33. Haas, J. D. and Fairchild, M. W., Summary and conclusions of the International Conference on Iron Deficiency and Behavioral Development, *Am. J. Clin. Nutr.,* 50, 703–705, 1989.

Chapter 8

FOOD ALLERGY AND ATOPIC DISEASE

Ranjit Kumar Chandra

TABLE OF CONTENTS

I. Introduction 61

II. Immunologic Mechanisms 62

III. Clinical Manifestations 62

IV. Diagnosis 62

V. Management 63

VI. Conclusions 68

References 68

I. INTRODUCTION

The terms *food allergy* and *food hypersensitivity* are used synonymously, and these should be distinguished from the all-inclusive term "food intolerance" which includes all adverse reactions to foods and food additives; the underlying process may be pharmacologic, microbiologic, toxic, idiosyncratic, psychologic, or immunologic. It is only when we can show the existence of any *immunologic* mechanism underlying symptoms and signs related to foods and food additives that the term food *allergy* can be accepted.

The number of dietary antigens that traverse the gastrointestinal tract is enormous. It is not surprising that occasionally this foreign material produces an adverse reaction. Many physicians and most of the lay public believe that the prevalence of food reactions has increased progressively in the last several decades, particularly during the last 20 years. The prevalence of atopic syndromes that may be related to food hypersensitivity, such as eczema and asthma, has gone up dramatically in the last 15 years.[1] The reasons for these epidemiologic findings are not clear.

Children show a higher prevalence of food allergy than do adults.[2] This may be a reflection of the natural history of food allergy since most infants and young children with an allergy become tolerant to the sensitizing foods as they grow

0-8493-2764-4/95

older. The rank order of foods that can cause adverse reactions varies with different population groups with diverse eating patterns. The common occurrence of hypersensitivity to peanuts, eggs, and cow's milk in North America, fish in Scandanavia, and soy in Japan reflects the frequency of their usage in the population.

In this selective review, the immunologic aspects, clinical diagnosis, management, and prevention are discussed. The references should be consulted for more extensive reading and citation of original data.

II. IMMUNOLOGIC MECHANISMS

Food can trigger four types of immunologic reactions. These are the classical reactions described by Gell and Coombs.[3] Type I is immediate hypersensitivity reaction mediated by IgE and is the most serious of reactions, since it can produce severe life-threatening anaphylaxis. Type II involves both antibody and complement which get fixed on the surface of a cell. Type III is immune-complex disease in which antigen-antibody-complex circulates and is trapped in the perivascular site, glomerular basement membrane, or rarely in other parts of the body. Type IV is the only type of immunologic reaction mediated by T lymphocytes and their products such as interleukins. In a given patient, food allergy may be due to any one or more of these reactions.

III. CLINICAL MANIFESTATIONS

Virtually any organ system can be affected. The common manifestations include colic, vomiting, diarrhea, tingling in the mouth, acute dermatitis, eczema, urticaria, angioedema, rhinorrhea, and wheezing. Less frequently, migraine may result from food allergy. Anaphylaxis is the most serious of all clinical manifestations and may be fatal. It can be prevented by careful exclusion of offending foods, taking the precaution of avoiding physical exercise and alcohol in conjunction with ingestion of suspected or proven food allergens, and prompt use of adrenaline. The latter can be carried as a self-administered injection and may be repeated at an interval of 20 minutes on two occasions until more medical help is available.

IV. DIAGNOSIS

Two recent papers have reviewed a practical diagnostic approach for patients suspected to have food allergy.[1,4] Medical history is useful when acute reaction occurs following ingestion of an uncommon food, e.g., angioedema taking place within minutes of eating shellfish or nuts. Occasionally, the amount of food consumed may be an important consideration. A relatively small number of foods are involved in the majority of young children with adverse reactions to foods.

Maintaining a prospective daily diary of foods consumed and examining the temporal relationship to symptoms and signs may yield the correct diagnosis.

Elimination diets may be helpful. It involves the complete exclusion of one or more foods suspected of causing adverse reactions and complete resolution of the clinical problem on such a regimen. For example, in infants who are suspected to have cow's milk allergy, administration of an extensively hydrolyzed formula or a soy or meat-based formula results in relief of symptoms.

In vitro tests for detection of IgE antibodies to foods are useful screening methods in places where ready access to allergists who can do skin tests is limited. The method involves radioallergosorbent or enzyme-linked immunoassays. Using reproducible methods and an appropriate cut-off point, these techniques can achieve a high degree of sensitivity and specificity. Nevertheless, the *in vitro* techniques are less reliable than *in vivo* skin prick tests for the diagnosis of true food allergy.

Skin prick tests using 1:20 dilution of food extracts in glycerine are highly reliable screening procedures. Negative skin tests to a battery of common food allergens virtually exclude significant food hypersensitivity. On the other hand, a positive skin prick test indicates an underlying hypersensitivity but does not prove that the clinical manifestations are causally related to the foods found positive on skin testing. Only exclusion and challenge can prove the existence of food allergy. On average, only 40% of children with positive skin tests are confirmed to have true food allergy.[1] Occasionally, fresh fruits and vegetables may produce a positive skin test, whereas skin tests to stored food solutions may be negative. Although young infants may have false negative reactions, this is rare when a clinically significant food allergy exists.

Other tests occasionally done in research laboratories include basophil and intestinal mast cell histamine release and lymphokine release from lymphocytes stimulated with the suspected food.[1]

The test touted to be the "gold standard" is the double-blind placebo-controlled food challenge. Following the strict elimination of offending food for three weeks, the child is given the suspected food in a hidden form. Unless there is history of previous anaphylaxis, one can start with a dose of 200 mg, and if there is no adverse reaction, the amount can be increased 2–5 times every 30 minutes, until meal-size quantities are achieved. Ideally, the patient should be observed for the next three days to detect delayed reactions. It is prudent to keep resuscitative equipment handy although serious reactions to skin prick tests are rare. In clinical practice, as opposed to academic research setting, a single-blind challenge is generally sufficient to achieve the correct diagnosis.

V. MANAGEMENT

The management of a child with proven food allergy is fairly straightforward. It involves the complete elimination of the foods to which the child is sensitive. If this is a critical food, e.g., cow's milk for a young infant, substitution with other foods must be done under the continued expert guidance of a dietitian. Elimination of several foods from the child's diet would risk the development of malnutrition.

Rarely, oral sodium cromoglycate may provide temporary relief of symptoms. Finally, in severe allergy, a complete substitution with an elemental diet would be needed. In those allergic to cow's milk, an extensively hydrolyzed casein formula should be recommended. Alternatively, a soy formula may be used. However, caution must be exercised when formula changes are made since anaphylactic reaction can occur even when hydrolyzed formulas are given.

There is much recent interest in prevention of allergic disorders in infants who are at high risk because of parental history of atopy. A number of single or multiple strategies have been considered, and there are some data to support the effectiveness of such measures. However, controversy exists, some of it due to apparently conflicting results obtained in different studies. Many of these "differences" are the result of variations in the design and implementation of studies. Secondly, whereas the differentiation between occurrence of atopic symptoms and the proof that these are due to food allergy is of much academic importance, the main finding that patients, their families, and healthcare administration are concerned about is the occurrence, or lack of, clinical manifestations, regardless of what the underlying pathogenetic mechanism might be.

Two main reasons have prompted attempts at prevention. One is the increase in the prevalence of allergic disease, both eczema and asthma. The second is the enormous costs involved in the management of children with different types of allergic disorders. For example, bronchial hyper-responsiveness defined as reduction of 15% in peak expiratory flow after 6 minutes of exercise was seen in 6% of Newfoundland children in 1976 and has now increased to 12% in 1988.[1] Also, there has been an increase in mortality from asthma.

The second main reason for thinking of prevention is the enormous cost of treating children with allergy. These costs can be divided into financial ones, like physician fees, laboratory tests, hospitalization, and medicines. There are also non-economic costs, such as growth failure, school time lost, emotional stress, and decreased participation in sports.

Five major considerations underlie any prevention program. One, sensitive and specific screening tests must be available to identify the high risk group. Two, effective intervention should be available. Three, healthcare organization should be in place for implementation of the program. Four, the program should be accepted by families and by health professionals. Finally, the program should be cost-effective.

Croner[5] has recently reviewed various methods for identification of the infant at high risk of developing allergic disease. He concluded that family history is the best method. If one parent is affected by atopy, there is a 37% chance of the child getting allergic disease. This risk increases to 62% if both parents are affected and to 73% if there is an allergic sibling as well as biparental history. Although its predictive value is not as high as earlier studies suggested, some additional advantage may be gained by cord blood IgE determination, particularly if we use a somewhat higher cut-off point to define elevated levels. Elevated cord blood IgE above 0.9 U/ml carries a 58% risk for allergy in the infant, compared with 18%

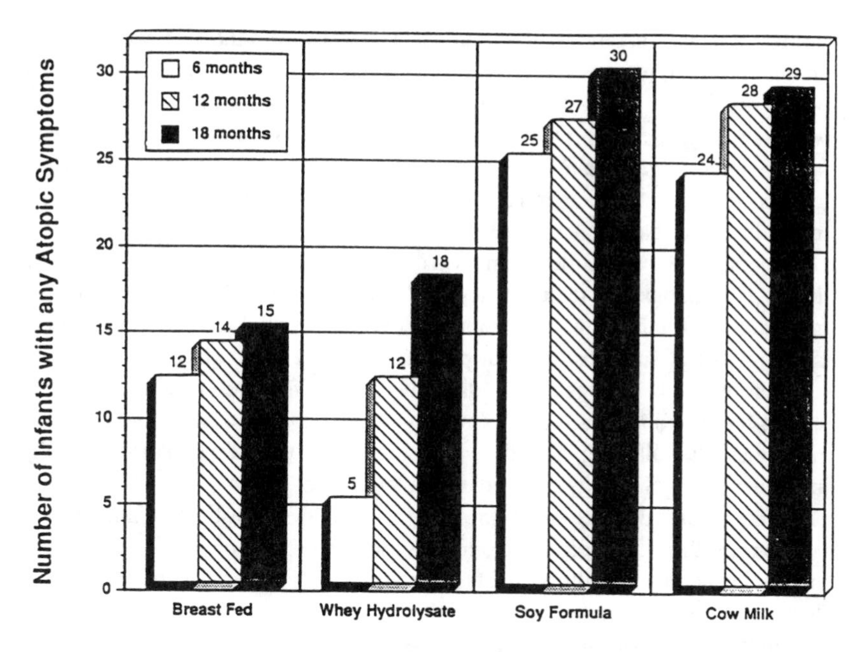

FIGURE 1. Influence of restricted maternal diet (dairy products, egg, peanut, fish, soy) during pregnancy and lactation on incidence of atopic eczema in high-risk infants.

in those whose cord blood IgE is normal. Changes in T cell subsets also have a prognostic significance.[6]

Any discussion of interventions to prevent allergy must be preceded by emphasizing the importance of a proper study design.[7] We should study and analyze separately infants who are at low risk (i.e., no family history, normal cord blood IgE, and normal lymphocyte subsets) and those who are at high risk. Studies should be conducted prospectively rather than retrospectively. If infants are formula-fed, they should be randomized to different groups; there should be adequate sample size, low drop-out rate, and both the parents and observers should be blinded, i.e., they should not know what type of formula is being given to their child. Finally, it is important to have objective criteria of assessment which should include an assessment of severity of symptoms.

Breastfeeding, if it is exclusive and prolonged, is associated with decreased incidence of allergic disease. Breastfeeding reduces exposure to foreign proteins, such as cow's milk, and it promotes maturation of the gut barrier, thereby decreasing absorption of macromolecules. It reduces infection which can act as an adjuvant in producing sensitization.

Mother's diet during pregnancy and lactation can have an influence on the infant's risk of developing food allergy (Figure 1).[8,9] Food antigens are present in amniotic fluid which the fetus swallows all the time. The fetus is capable of forming IgE and mounting other immune responses from the 10th week of gestation; in a small number of cord blood samples, IgE is raised, and specific antibodies are found, and symptoms may occur on first exposure to food antigens within the first 3–5 days of life. In addition, food antigens are present in breast milk. Symptoms can occur on first exposure to foods on weaning; relief of symptoms occurs when the mother's diet is restricted, and these symptoms return when the mother goes back to a normal diet.

The many factors that determine whether or not one sees an effect of maternal dietary precautions during pregnancy and/or lactation on the incidence of allergy in the high-risk infants include the number of foods avoided, duration of avoidance, number of subjects, drop-out rate, and various confounding variables.

There was a highly significant difference in the incidence of allergic disease among breast-fed infants of mothers who avoided common allergenic foods including cow's milk, egg, soy, fish, and peanut, and the infants had a markedly lower incidence of allergy compared with the control group.

If the mother decides not to breast feed, or if the breast-fed infant needs a supplement, then the choice of the formula should depend upon its effectiveness in reducing allergy in high-risk infants, its tolerance, its taste, and very importantly, its cost.

Criteria for demonstrating effectiveness of a formula for prevention of allergy were discussed by the Committee on Nutrition and Allergic Disease, American Academy of Pediatrics, summarized in a recent paper by Kleinman et al.[10] They state: "To support a claim for allergy prevention, infants born from families with history of allergy (biparental or 1 parent or 1 sibling) should be fed the product exclusively from birth for at least 6 months under the conditions of a controlled and randomized study and followed for at least one year thereafter (or until 18 months of age). At that time, they must have a statistically significant lower prevalence of allergy assessed by a clinical scoring system than infants fed a standard milk-based control formula for 6 months and followed until 18 months of age."

In a double-blind study, we randomized 124 infants to receive a conventional cow milk formula, or a soy formula, or an extensively hydrolyzed casein formula.[8] On follow-up until age 18 months, the hydrolysate group had the lowest incidence of allergic disease—9 out of 43 infants, compared with 28 out of 40 in the conventional formula. The severity of eczema as judged by the eczema score was also less in the hydrolysate group. The soy formula group did not show any reduction in the incidence of allergy. Although the extensively hydrolyzed formula was beneficial in the short term, it has a bad taste, is generally poorly tolerated, and is almost four times more expensive than other formulas.

In a second study, we examined the effect of feeding a partial hydrolysate formula, Good Start, in a double-blind randomized study.[11] There were 72 infants in each group. At age 6 months, only two infants showed evidence of IgE antibodies to milk in the whey hydrolysate group, compared with 8 infants in the conventional cow's milk group and 12 in the soy group. This confirmed that the

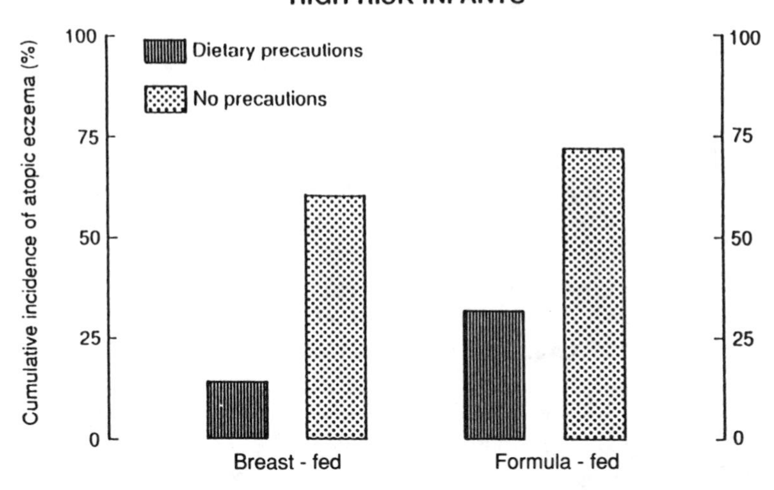

FIGURE 2. Cumulative frequency of atopic disease (eczema, asthma) in high-risk infants either breast-fed (without maternal dietary precautions) or given one of three formulas exclusively for the first 6 months of life. There were 68 infants in the Good Start group, 67 in the Similac group, 68 in the Isomil group, and 60 in the exclusive breast-fed group.

whey hydrolysate was less sensitizing compared with the other two formulas. At the end of 18 months of follow-up, the Good Start-fed infants had the lowest incidence of allergy out of the 3 formula groups, 18 out of 68 babies, compared with 29 out of 67 in the Similac conventional cow's milk formula group, and 30 out of 68 in the Isomil (soy formula) group (Figure 2). These differences are statistically significant. Therefore, using the definition of the American Academy of Pediatrics, we find that the partial whey hydrolysate did indeed prevent allergic disease in high-risk infants and not merely delay its occurrence. At the end of three years, the differences are still significant, the hydrolysate group having the lowest incidence of allergic disease among the three formula groups. The severity of eczema was less in the Good Start group compared with the other formula groups.

These dietary intervention strategies are cost-effective.[12] The exact calculation of cost-benefit ratio will vary in different countries depending upon the cost of health care, e.g., physician's fees, cost of laboratory tests, cost of hospitalization, etc. Also, there are additional costs to the family in terms of costs of medicines, home renovations, child care at home, and lost income. Costs to the child cannot be easily measured, but these include loss of school attendance, reduced physical activity and participation in sports, social isolation, and emotional stress. Compared with a conventional cow's milk formula, those fed on Good Start would be expected to show almost 50% reduction in the cost of treatment of children with allergy. Although an extensively hydrolyzed formula may work in prevention, it is very expensive, so that the overall costs would, in fact, be higher than both the other groups mentioned above.

VI. CONCLUSIONS

Food allergy and atopic disease are common clinical problems in young children. Manifestations can affect any organ system and vary in severity from mild to fatal. The diagnosis is established by history, skin prick tests, and food challenges. For prevention, many intervention strategies can be recommended to prevent allergic disease in high-risk infants. They should preferably be exclusively breast fed, and it would be helpful if the mother restricts common allergic foods from her diet during lactation. If the infant needs a supplement, or the mother decides not to breastfeed, then a hydrolysate formula such as Good Start would be the formula of choice because it is effective in reducing allergic disease and not very expensive. Other helpful factors include delay in introducing egg, fish, and peanut and reducing exposure to dust mite, mold, animals, tobacco smoke, and infection. The topic of food allergy is interesting, complex, and controversial. Reference should be made to recent publications for further information and citations.[1,4,13–17]

REFERENCES

1. Chandra, R. K., Gill, B., and Kumari, S., Food allergy and atopic disease, *Annals Allergy,* 71, 495–506, 1993.
2. Chandra, R. K., Food allergy, *Canad. Med. Assoc. J.,* 146, 367, 1992.
3. Chandra, R. K. and Shah, A., Immunologic mechanisms, In *Food Intolerance,* Chandra, R. K., Ed., New York, Elsevier, 1984, 55–56.
4. Metcalfe, D., Simon, R., and Sampson, H. A., Eds., *Food allergy: adverse reactions to foods and food additives,* London, Blackwell, 1991.
5. Croner, S., Prediction and development of allergy development, *J. Pediatr.,* 121, S58–S63, 1992.
6. Chandra, R. K. and Baker, M., Numerical and functional deficiency of suppressor T cells precede development of atopic eczema, *Lancet,* ii, 1393–1394, 1983.
7. Chandra, R. K., Food allergy 1992 and beyond, *Nutr. Res.,* 12, 93–99, 1992.
8. Chandra, R. K., Puri, S., and Hamed, A., Influence of maternal diet during lactation and use of formula feeds on development of atopic eczema in high risk infants, *Brit. Med. J.,* 299, 228–230, 1989.
9. Hattevig, G., Kjellman, B., Sigurs, N., et al., The effect of maternal avoidance of eggs, cow's milk and fish during lactation upon allergic manifestations in infants, *Clin. Allergy,* 19, 27–32, 1989.
10. Kleinman, R. E., Bahna, S., Powell, G. F., and Sampson, H. A., Use of infant formulas in infants with cow milk allergy, *Pediatr. Allergy Immunol.,* 4, 146–155, 1991.

11. Chandra, R. K. and Hamed, A., Cumulative incidence of atopic disorders in high risk infants fed whey hydrolysate, soy and conventional cow milk formulas, *Ann. Allergy,* 67, 129–132, 1991.
12. Chandra, R. K., Prasad, C., Hamed, A., et al., Five year follow-up of high risk infants with parental history of atopy fed partial whey hydrolysate, soy and conventional cow's milk formulas, *J. Pediatr. Gastro. Nutr.,* (submitted).
13. Esteban, M. M., Ed., Adverse reactions to foods in infancy and childhood, *J. Pediatr.,* 121, S1–S126, 1992.
14. Somogyi, J. C., Muller, H. R., and Ockuizen, T. H., Eds., Food allergy and food intolerance. Nutritional aspects and developments, *Bibl. Nutr. Dieta.,* Basel: Karger, 1991.
15. Hamburger, R. N., Ed., *Food intolerance in infancy, Allergology, Immunology, and Gastroenterology,* New York, Raven Press, 1989.
16. Royal College of Physicians and the British Nutrition Foundation, Food intolerance and food aversion, *J. Roy. Coll. Phys. Lond.,* 18, 83–123, 1984.
17. Bahna, S. L., Breast milk and special formulas in prevention of milk allergy, in *Immunology of Milk and the Neonate,* Mestecky, J., et al., Eds., New York, Plenum, 1991, 445–448.

Chapter 9

NUTRITION THERAPY OF INBORN ERRORS OF METABOLISM

Kimberlee Michals-Matalon and Reuben Matalon

TABLE OF CONTENTS

I. Introduction 71

II. Phenylketonuria 72

III. Maternal PKU 75

IV. Galactosemia 76

V. Maple Syrup Urine Disease 76

VI. Biotinidase 77

VII. Isovaleric Acidemia 77

VIII. Urea Cycle Defects 78

References 78

I. INTRODUCTION

At the turn of the century, Garrod introduced the concept ''inborn error of metabolism'' to describe biochemical defects, primarily in the pathways of intermediary metabolites, causing diseases which occurred in a familial pattern.[1] The increase in understanding of metabolic processes, especially in the pathway of intermediary metabolites, has resulted in the discovery of enzyme defects that can be treated by diet. Understanding the biochemical pathways affected by a specific inborn error of metabolism is therefore essential before attempts to treat a disease can be initiated. Modulation of the intermediary metabolites has become an exciting method of treatment in some of these diseases.

The most practical form of treatment of inborn errors of metabolism is the removal of the component or substrate that is not metabolized. In the event that the compound is essential, then it should be restricted to that needed for optimal growth. In order to prevent the severe manifestations of a treatable inborn error

0-8493-2764-4/95

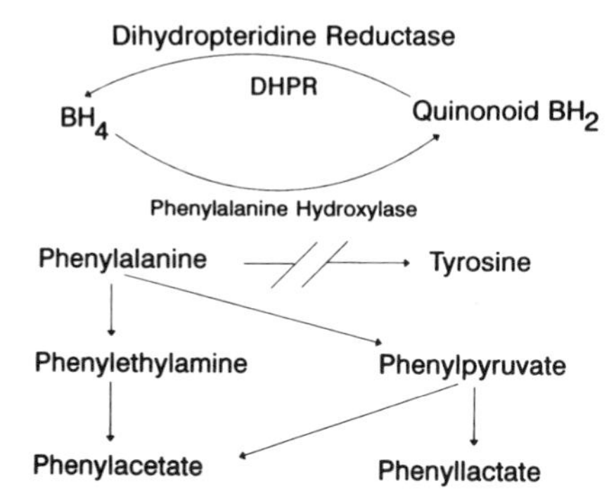

FIGURE 1. Phenylalanine hydroxylase deficiency resulting in elevation of phenylalanine and its metabolites in blood, urine, and CSF. Biopterin (BH_4) is the cofactor in the hydroxylation of the aromatic amino acids phenylalanine, tyrosine, and tryptophan.

of metabolism, dietary treatment must be started before symptoms appear or before irreversible neurological damage has occurred. Early detection of treatable diseases became feasible with the development of the bacterial inhibition assay by Guthrie, which proved to be reliable and cost-effective for newborn screening.[2] As the understanding of the inborn errors of metabolism has expanded, dietary management has become the hallmark for treatment in an effort to ameliorate the severe consequences of many of these genetic defects.

II. PHENYLKETONURIA

Phenylketonuria (PKU) is an autosomal recessive inborn error of phenylalanine metabolism that occurs in approximately one in 10,000 to 15,000 births in the United States.[3] The disease occurs in all ethnic groups but is more common in persons of Northern European background. Folling was the first in 1934 to identify phenylketones in the urine of patients with PKU.[4] Jervis in 1947 identified the defective enzyme as phenylalanine hydroxylase involved in the hydroxylation of phenylalanine to tyrosine as shown in Figure 1.[5] Bickel in 1953 documented that restricting the substrate, phenylalanine, in patients with PKU prevented the severe manifestations of the disease.[6] The successful treatment led to mandatory newborn screening for this disease in the United States.

The clinical features of untreated PKU include developmental delay and mental retardation. Patients also often have seizures, microcephaly, eczema, pigment dilution, and a peculiar odor. At an older age, hyperkinesis becomes a feature of the untreated patient.[3]

The biochemical findings on patients with untreated PKU include elevated levels of phenylalanine in blood, urine, and spinal fluid. A normal concentration of blood phenylalanine is 1–2 mg/dl. Patients with untreated PKU have levels of phenylalanine that often exceed 20 mg/dl. Once an infant with an elevated phenylalanine blood level is identified by newborn screening, confirmatory diagnostic

TABLE 1
Medical Foods Used for Treatment of Phenylketonuria

Product	Company	Recommended Age
Lofenalac	Mead Johnson	Infants
Analog X P	Scientific Hospital Supplies	Infants
Phenex-1	Ross Laboratories	Infants/Toddlers
PKU 1	Mead Johnson	Infants
Maxamaid X P	Scientific Hospital Supplies	1 to 8 years
Maxamum X P	Scientific Hospital Supplies	8 Years/Adults
Phenex-2	Ross Laboratories	Children/Adults
Phenylfree	Mead Johnson	Children
PKU 2	Mead Johnson	Children/Adults
PKU 3	Mead Johnson	Pregnant Women

studies must be completed. Blood should be analyzed for quantitative amino acids to document that the phenylalanine is indeed greater than 6 mg/dl and that the tyrosine blood level is normal. The urine should show elevated phenylalanine and phenylalanine metabolites. Phenylalanine hydroxylase requires the cofactor tetrahydrobiopterin (BH_4) for the conversion of phenylalanine to tyrosine. The cofactor, BH_4, is also utilized in the hydroxylation of tyrosine to dopamine and tryptophan to serotonin. Approximately 2% of infants with hyperphenylalaninemia have BH_4 deficiency. Therefore, this must be ruled out since the treatment for BH_4 deficiency is totally different than that for PKU.[7]

The disease is a clinical spectrum. There can be a wide variation in initial blood phenylalanine levels in part due to clinical variation but also due to the type of formula consumed and the length of time before the diagnosis is confirmed. Patients that have blood phenylalanine levels greater than 2 mg/dl but below 6 mg/dl on normal dietary intake may have benign hyperphenylalaninemia. Benign hyperphenylalaninemia occurs in about 10% of patients with elevated phenylalanine levels. Patients with this condition will not require dietary therapy although occasional monitoring of blood phenylalanine levels is necessary. Blood phenylalanine levels greater than 6 mg/dl require dietary manipulation.

The dietary treatment of phenylketonuria involves the restriction of phenylalanine; however, since phenylalanine is an essential amino acid, adequate amounts should be provided. It is also important to be sure that products beyond the metabolic block are provided in adequate amounts. Phenylalanine cannot be hydroxylated to tyrosine; therefore, tyrosine becomes essential for treatment of patients with PKU. Medical foods are available commercially for treatment of PKU (see Table 1). The medical foods contain no or limited amounts of phenylalanine and provide an adequate amount of tyrosine. The medical foods also provide necessary vitamins and minerals and, in some cases, a significant source of energy.

In infancy an age-appropriate medical food is supplemented with either breast milk or commercial formula that supplies essential phenylalanine. The medical food should provide 80% to 90% of the dietary protein. The infant should be

TABLE 2
Suggested Phenylalanine, Protein, and Energy Intake for Infants, Children, and Adolescents with PKU

	Phenylalanine		Protein	Energy
Age	(mg/day)	(mg/kg)	(g/kg)	(Kcal/kg)
0 to 6 months	200	20 to 60	2.9	95 to 145
6 to 12 months	200	15 to 25	2.2	80 to 135
			(g/day)	(Kcal/day)
1 to 7 years	220	7 to 25	30 to 44	900 to 2300
7 to 11 years	250	7 to 15	32 to 46	1650 to 3300
Females				
11 to 19 years	300	5 to 11	48 to 51	1500 to 3000
Males				
11 to 19 years	350	5 to 13	50 to 64	2000 to 3900

given protein intake that is similar to usual protein consumption of infants taking commercial formulas as shown in Table 2. If the blood phenylalanine level is about 20 mg/dl, then start with 45 mg/kg phenylalanine (see Table 2). There is a great deal of variation in required energy intake (Table 2). Appropriate weight gain will be the best indicator of adequate calorie intake. Poor weight gain may result in tissue catabolism with an associated rise in plasma phenylalanine. The blood phenylalanine needs to be monitored twice weekly and dietary phenylalanine increased or decreased to maintain the blood phenylalanine level in the range of 2–6 mg/dl. As the infant becomes older, part of the breast milk or infant formula is replaced by phenylalanine in baby cereal, fruits, and vegetables. The strained meats are too high in phenylalanine and are not added to the diet. For a toddler, the diet includes measured amounts of fruits, vegetables, and starchy foods along with the age-appropriate medical food.

Children with PKU need an age-appropriate medical food and measured amounts of low phenylalanine foods. Low-protein baked goods and pastas become necessary in order to provide adequate bulk and energy to maintain adequate metabolic control. Blood phenylalanine should be monitored at least once a week to document adequate dietary compliance.

At one time it was theorized that the majority of the brain growth occurs by early childhood and many Centers recommended diet discontinuation at 5 or 6 years of age. The National Collaborative Study of Children with Phenylketonuria was started in 1967 to determine the effectiveness of diet therapy and to evaluate whether nutrition support could be discontinued at 6 years of age.[8] This study eliminated the problems of small sample size and non-standardized tests. The study found a relationship between loss of I.Q. and the age at which dietary control was lost.[9] The recommendation from this study is that the diet should be continued.[10]

Adolescence is a difficult time for management of the phenylalanine restricted diet. Adolescents have unstructured eating habits, often skipping meals, snacking, and eating snack-like meals. The low phenylalanine diet requires a structured

TABLE 3
Initial Phenylalanine Prescription for Pregnant Women with PKU

Age	Trimester 1	Trimester 2	Trimester 3
<19 years	200 mg	200 mg	330 mg
>19 years	180 mg	180 mg	310 mg

eating pattern. Unstructured eating can lead to overeating allowed foods or eating disallowed foods, both of which result in elevated blood phenylalanine levels. Many adolescents discontinued diet therapy because of past clinic policies and later attempted to reinstitute the diet.[11] However, metabolic control is less than optimal in patients that resumed diet as well as in the group that maintained diet.[12] Adolescents would benefit from lower blood phenylalanine levels since response time and visual attention improve with lower blood phenylalanine levels.[13] More recently, abnormal changes in the MRI have been reported in adolescents with PKU that have discontinued diet. The MRI changes often correlate with behavioral or neurological problems.[14]

In order to achieve acceptable long-term dietary compliance, it is important to reinforce that the disease does not go away at a specific age. Improved knowledge and understanding about their disease may lead to improved dietary compliance, especially if coupled with programs that teach self-management.[15] In PKU the diet at one time was discontinued because it was believed that the main deleterious effects of lack of treatment occurred early in life. Therefore, as children with PKU became older, often the expectations for treatment changed, and more liberal goals were allowed. It is now known that brain dysfunction may occur at a later age with poor compliance or discontinued treatment; therefore, long-term improved compliance for PKU is desired. This means stricter control of the blood phenylalanine levels and more frequent monitoring of blood levels.

III. MATERNAL PKU

Elevated blood phenylalanine or its metabolites in pregnant women with PKU act as a teratogen on the fetus. Offspring born to women with untreated PKU have a high incidence of microcephaly, mental retardation, cardiac malformations, and low birth weight.[16] Since most pregnancies are unplanned, the phenylalanine diet must be started or brought under good control as soon as there is suspicion of pregnancy. The dietary prescription should be planned using the lowest phenylalanine requirement for the patient's age and trimester of pregnancy as shown in Table 3.[17] A medical food that meets the requirements for adults should be used as shown in Table 1. The protein intake should start at 75g/day for the first trimester with incremental increases of about 15g/day protein for each trimester. Calories should be adequate to promote weight gain in all women. Recommended weight gain, however, depends on pre-pregnancy weight—for normal weight women, 24 to 32 pounds; underweight women, 28 to 36 pounds; and for overweight women, 20 to 24 pounds.[18] Studies on pregnant women with PKU are currently underway in the United States and Canada.[19,20]

IV. GALACTOSEMIA

Galactosemia is caused by enzyme defects affecting galactose utilization.[21] The disease most commonly associated with galactosemia is deficiency of the enzyme galactose-1-phosphate uridyl transferase. Many states screen for this condition in the newborn period. Patients with galactosemia present with vomiting after ingestion of galactose-containing formula. If galactose ingestion continues, the patient fails to thrive, has deranged liver functions which progress to jaundice, hepatomegaly, and later hemolysis and ascites, with deaths often occurring due to E-coli sepsis. If the patient survives, there can be mental retardation and cataracts.

The relatively simple dietary treatment involves total removal of the unmetabolized substrate galactose, which comes from lactose. Initially, this involves removal of breast milk or commercial formulas based on human milk. Soy formulas or casein hydrolysate are appropriate for treatment of galactosemia. As the infant gets older, all sources of lactose must be eliminated from the diet. The parents must read labels for ingredients such as lactose, milk, milk sugar, milk solids, whey, or whey solids or casein. The substrate galactose-1-phosphate should be monitored in the patient's red blood cells (RBC). Normally, the RBC galactose-1-phosphate is less than 1 mg%. In treated galactosemia, good diet control is indicated when levels remain below 5 mg%. Blood levels of RBC galactose-1-phosphate should be performed frequently in order to document adequate dietary compliance. Once weekly or every other week would be the suggested frequency.

Overall dietary treatment has been successful in galactosemia. The mental retardation and cataracts are not a problem when dietary compliance is good. However, there are reports of increased numbers of learning difficulties and language delay which may be related to the time diet treatment was initiated or poor dietary control.[22] Additionally, large percentages of females with galactosemia have impaired ovarian function.[23] It has been suggested that UDG-galactose may be deficient in patients with galactosemia and may be related to these problems.[24] Studies are now in progress on the effect of uridine on learning disorders and ovarian function in galactosemia.

V. MAPLE SYRUP URINE DISEASE

Maple syrup urine disease (MSUD) is named for the sweet-smelling urine due to ketoacids present in the urine of children with this disease. The enzyme defect is branched-chain ketoacid dehydrogenase complex which leads to elevation of the branched-chain amino acids leucine, isoleucine, and valine.[25] The blood aminogram also has a peak of alloisoleucine present, and the urine has elevated branched-chain ketoacids. The symptoms start within a few days of life and include failure to thrive, vomiting, convulsions, alternating rigidity and hypotonia, stupor, irregular respirations, apnea, and death. Immediately following the diagnosis of MSUD, a diet which is free of leucine, isoleucine, and valine with adequate energy should be started. Daily monitoring of the levels of branched-chain amino acids in the blood is required. Usually isoleucine will normalize within 2–3 days, followed by valine. As these amino acids become normalized, they must be added to the diet as the L-amino acid. Leucine should normalize in 8–10 days, and commercial formula or breast milk can be added to a branched-chain amino acid medical food. The branched-chain amino acid levels should be monitored in the blood weekly. In the acute phase of the disease, dietary management may not be enough. Hemodialysis or peritoneal dialysis may be required.

Some inborn errors of metabolism respond to high doses of vitamins and may be treated by cofactor supplementation. The specific defective enzyme may have altered kinetics, and the high vitamin dose enhances the residual enzyme activity. High doses of thiamine (B_1) may prolong the half-life of the branched-chain ketoacid dehydrogenase complex, the enzyme causing MSUD. Thiamine should be given at 100 mg per day for at least three months.

VI. BIOTINIDASE

Biotinidase deficiency is another example of an inborn error of metabolism that is treated by supplementation cofactor, which in this case is biotin.[26] Biotinidase cleaves biotin from holocarboxylases during metabolic degradation. The liberated biotin can then be reutilized. In patients with biotinidase deficiency, there is not adequate biotin present for the three mitochondrial biotin-dependent carboxylases propionyl-CoA carboxylase, 3-methylcrotonyl-CoA carboxylase, and pyruvate carboxylase. The patients present with hypotonia, seizures, skin rash, and alopecia. The urine contains metabolites, such as 3-hydroxyisovalerate, 3-methylcrotonylglycine, methylcitrate, 3-methylcrotonate, and 3-hydroxypropionate. Treatment involves supplementation of biotin in doses of 10 mg daily.

VII. ISOVALERIC ACIDEMIA

The binding of toxic substances produced by the metabolic block is yet another mode of therapy. Isovaleric acidemia is caused by a deficiency of the enzyme isovaleryl Co-A dehydrogenase which is in the leucine degradative pathway. The abnormal accumulation of isovalerylglycine is diagnostic. Most patients develop vomiting, acidosis, lethargy, and coma in the newborn period, while others have these symptoms only associated with an underlying illness. Isovaleric acidemia responds well to treatment with supplemental glycine (250 mg/kg) which binds with the isovaleric acid and is excreted as isovalerylglycine.[27] Carnitine supplementation (100 mg/kg) may also be added to the treatment regimen as a binding agent.

VIII. UREA CYCLE DEFECTS

Urea cycle defects are inborn errors of metabolism that produce a problem with the synthesis or excretion of nitrogen. Dietary treatment involves using an alternate pathway for the synthesis and excretion of waste nitrogen.[28] Nitrogen may accumulate as glutamate, alanine, or ammonia, but ammonia is the most toxic. The signs and symptoms of the disease usually depend on ammonia intoxification. As ammonia starts to increase, there is lethargy, vomiting, and refusal to eat. As ammonia rises more, the lethargy increases, and there is hyperventilation and grunting respirations. This is followed by coma, respiratory arrest, dilated pupils, and increased intracranial pressure. In severe forms, newborns can be comatose by 2–4 days. In the acute phase, treatment includes peritoneal or hemodialysis. Urea cannot be formed; therefore, precursors are given so the nitrogen can be excreted by an alternate pathway. For example, sodium benzoate is given to bind with glycine which is then excreted as hippuric acid and sodium phenylacetate which binds with glutamate and is then excreted as sodium phenylacetylglutamate. Long-term management includes oral intake of sodium phenylbutyrate to bind and excrete nitrogen. The urea cycle defects also require substrate restriction; therefore, dietary nitrogen is limited. The best outcome is for siblings when they are treated prospectively.[29] All siblings are placed on treatment, and specimens are obtained to determine whether they are affected with the disease. If they are normal, the treatment is discontinued; however, if affected, the treatment continues.

The ability to detect and treat various inborn errors has increased dramatically in recent years. Many of the treatments involve alteration of dietary components. Medical foods and special low-protein diet products are commercially prepared to assist the health professional in treatment of many inborn errors of metabolism. Since treatment for many inborn errors of metabolism is now long term, outcome of the treatment should still be cautiously monitored.

REFERENCES

1. Garrod, A. E., The Croonian lectures on inborn errors of metabolism, *Lancet,* 2, 73, 1908.
2. Guthrie, R. and Susi, A., A simple phenylalanine method for determining phenylketonuria in large populations of newborn infants, *Pediatrics,* 32, 338–343, 1963.
3. Scriver, C. R., Kaufman, S., and Woo, S. L. C., The hyperphenylalaninemias, in *The Metabolic Basis of Inherited Disease,* Scriver, C. R., Beaudet, A. L., Sly, W. S., and Valle, D., Eds., New York, McGraw-Hill, 1989, 495–546.
4. Folling, A., Uber Ausscheidung von Phenylbrenztraubensaure in den Harn als Stoffwechsanomalie in Verbindung mit Imbezillitat, *Z. Physiol. Chem.,* 227, 169–176, 1934.
5. Jervis, G. A., Studies on phenylpyruvic oligophrenia: The position of the metabolic error, *J. Biol. Chem.,* 169, 651–656, 1947.

6. Bickel, H., Influence of phenylalanine intake on phenylketonuria, *Lancet,* ii, 812–813, 1953.
7. Matalon, R., Michals, K., Blau, N., and Rouse, B., Hyperphenylalaninemia due to inherited deficiencies of tetrahydrobiopterin, *Adv. Pediatr.,* 36, 67–82, 1989.
8. Williamson, M., Dobson, J. C., and Koch, R., Collaborative study of children treated for phenylketonuria: study design, *Pediatrics,* 50, 815–821, 1977.
9. Holtzman, N. A., Kronmal, R. A., Van Doornick, W., Azen, C., and Koch, R., Effect of age at loss of dietary control on intellectual performance and behavior of children with phenylketonuria, *N. Engl. J. Med.,* 314, 593–598, 1986.
10. Michals, K., Azen, C., Acosta, P., Koch, R., and Matalon, R., Blood phenylalanine levels and intelligence of ten-year-old children with phenylketonuria in the national collaborative study, *J. Amer. Diet. Assoc.,* 88, 1226–1229, 1988.
11. Schuett, V. E., Brown, E. S., and Michals, K., Reinstitution of diet therapy in PKU patients from twenty-two U.S. clinics, *Am. J. Public Health,* 75, 39–42, 1985.
12. Michals, K., Dominik, M., Schuett, V., Brown, E., and Matalon, R., Return to diet therapy in patients with phenylketonuria, *J. Pediatr.* 106, 933–936, 1985.
13. Krause, W., Halminski, M., McDonald, L., Dembure, P., Salvo, R., Freides, D., and Elsas, L., Phenylketonuria alters the mean power frequency of electroencephalograms and plasma L-DOPA in treated patients with phenylketonuria, *Pediatr. Res.,* 20, 1112–1116, 1986.
14. Thompson, A. J., Smith, I., Kendall, B. E., Youl, B. D., and Brenton, D., Magnetic resonance imaging changes in early treated patients with PKU, *Lancet,* 337, 1124, 1991.
15. Gleason, L. A., Michals, K., Matalon, R., Langenberg, P., and Kamath, S., A treatment program for adolescents with phenylketonuria, *Clin. Pediatr.,* 31, 331–335, 1992.
16. Lenke, R. R. and Levy, H. L., Maternal phenylketonuria and hyperphenylalaninemia, *N. Engl. J. Med.,* 303, 1202–1208, 1980.
17. Acosta, P. B., Austin, V., Castiglioni, L., Michals, K., Rohr, F., and Wenz, E., *Protocol for Nutrition Support of Maternal PKU,* Columbus, Ohio, Ross Laboratories, 1992.
18. Subcommittee on Nutritional Status and Weight Gain During Pregnancy, *Nutrition and Pregnancy,* Washington, D.C., National Academy Press, 1990, 430–431.
19. Koch, R., Hanley, W., Levy, H., Matalon, R., Rouse, B., de la Cruz, F., Azen, C., and Friedman, E. G., A preliminary report of the collaborative study on maternal phenylketonuria in the United States and Canada, *J. Inher. Metab. Dis.,* 13, 641–650, 1990.
20. Matalon, R., Michals, K., Azen, C., Friedman, E. G., Koch, R., Wenz, E., Levy, H., Rohr, F., Rouse, B., Castiglioni, L., Hanley, W., Austin, V., and de la Cruz, F., Maternal PKU Collaborative Study (MPKUCS): The effect of nutrient intake on pregnancy outcome, *J. Inher. Metab. Dis.,* 14, 371–374, 1991.
21. Segal, S., Disorders of galactose metabolism, in *The Metabolic Basis of Inherited Disease,* Scriver, C. R., Beaudet, A. L., Sly, W. S., and Valle, D., Eds., New York, McGraw-Hill, 1989, 453–480.
22. Waggoner, D. D., Buist, N. R. M., and Donnell, G. N., Long-term prognosis in galactosemia: results of a survey of 350 cases, *J. Inher. Metab. Dis.,* 13, 802–818, 1990.
23. Kaufman, F., Kogut, M. D., Donnell, G. N., Koch, R., and Goehelsmann, U., Ovarian failure in galactosemia, *Lancet,* II, 737–738, 1979.
24. Ng, W. G., Wu, Y. K., Kaufman, F. R., and Donnell, G. N., Deficit of uridine diphosphate galactose in galactosemia, *J. Inher. Metab. Dis.,* 12, 257–266, 1989.
25. Danner, D. J. and Elsas, L. J., Disorders of branched-chain amino acid and keto acid metabolism, in *The Metabolic Basis of Inherited Disease,* Scriver, C. R., Beaudet, A. L., Sly, W. S., and Valle, D., Eds., New York, McGraw-Hill, 1989, 671–692.
26. Wolf, B. and Heard, G. S., Disorders of biotin metabolism, in *The Metabolic Basis of Inherited Disease,* Scriver, C. R., Beaudet, A. L., Sly, W. S., and Valle, D., New York, McGraw-Hill, 1989, 2083–2103.
27. Krieger, I. and Tanaka, K., Therapeutic effects of glycine in isovaleric acidemia, *Pediatr. Res.,* 10, 25–29, 1976.

28. Batshaw, B. L., Brusilow, S., Waber, L. J., Blom, W., Brubakk, A. M., Burton, B. K., Cann, H. M., Kerr, D., Mamunes, P., Matalon, R., Myerberg, D., and Schafer, I. A., Treatment of inborn errors of urea synthesis, *N. Engl. J. Med.*, 306, 1387–1392, 1982.
29. Maestri, N. E., Hauser, E. R., Bartholomew, D., and Brusilow, S. W., Prospective treatment of urea cycle disorders, *J. Pediatr.*, 119, 923–928, 1991.

Chapter 10

THE PATHOPHYSIOLOGY OF CHILDHOOD OBESITY

Russell Rising

TABLE OF CONTENTS

I. Introduction 81

II. Possible Causes of Childhood Obesity 82
- A. Metabolic 82
- B. Genetic 83
- C. Environmental 83
- D. Nutritional 84

III. The Environmental Components of Obesity 86
- A. Adults 86
- B. Children 86
- C. Ethnicity 87
- D. Specific Syndromes 88

IV. Criteria for Determining Childhood Obesity 89

V. Problems with Treating Childhood Obesity 91

VI. Nutritional and Metabolic Assessment Before and During Treatment of Childhood Obesity 92

References 95

I. INTRODUCTION

In the early part of the century, people were more physically active at work and during leisure. Today, industrial machinery, the automobile, and home electronic entertainment have contributed to a more sedentary lifestyle. The sedentary lifestyle, combined with a readily available food supply, has led to an increasing prevalence of obesity. In the United States alone, over 34 million adults aged 20 to 74 are considered obese.[43] Obesity is associated with increased risks for non-insulin-dependent diabetes, heart disease, hypertension, and a host of other clinical disorders, and is considered one of the major health problems in the world today.[69] These same societal influences have also contributed to a rapid

0-8493-2764-4/95

increase in childhood obesity with 27% of 6 to 11 and 22% of 12 to 17-year-old children considered obese.[45]

There are many unanswered questions as to the cause of childhood obesity. Some of the physiological changes that lead to obesity in adults may also be present in early childhood and, under certain environmental conditions, may predispose children to body weight gain. These include a lower than average daily energy expenditure[50] and a low ratio of fat to carbohydrate oxidation.[76] Other factors leading to obesity may include differences in basal muscle metabolism,[75] sympathetic nervous system activity,[57] and regulation of body temperature.[54]

II. POSSIBLE CAUSES OF CHILDHOOD OBESITY

A. METABOLIC

The predisposition to obesity may begin at birth due to metabolic variations in energy expenditure.[56] Three-day-old infants were classified as either ''lean'' or ''obese'' based on whether they were born to lean or obese mothers (prepregnancy body weight below or above the 10th or 90th percentile for height and age). At three months of age, all infants had resting metabolic rate (RMR, under non-fasting conditions) and total daily energy expenditure determined by indirect calorimetry and the doubly labeled water method, respectively. They also had body composition determined monthly for 12 months. Although no differences were found in physical characteristics (body weight and length at birth) or RMR between the two groups, total daily energy expenditure was 21% lower in infants that later became overweight when compared to normal infants. A similarity in postprandial metabolic rate in both groups of infants indicated that a lower amount of energy expended for physical activities may be part of the cause of the lower total daily energy expenditure in the infants who later gained weight. This was probably due to infants mimicking the activity patterns of extremely inactive obese parents or siblings.[56]

Children living in a household where one or both parents are obese have an increased risk of becoming obese. The same physiological and environmental influences that led to parental obesity may also lead to childhood obesity. Differences in RMR and physical activity were found in 3- to 5-year-old children who had obese parents. Children with at least one obese parent had a 10% lower RMR in comparison to children of lean parents. Children with lean parents had almost a two-fold greater energy cost for physical activities (371±306 Kcal/d) as opposed to children with at least one obese parent (190±173 Kcal/d, mean±SD), suggesting that children with obese parents are less physically active.[26] In a similar study conducted in the 1930s, the energy cost for physical activities (800 Kcal/d) was 2 to 4 times the values found in the current study.[73] This suggests that energy cost for physical activities has declined over the past several decades. Increasing mechanization has provided children with many non-physical activities, such as watching television or playing table games. Children living a sedentary lifestyle in an environment where food is plentiful are obviously prone to ingesting larger quantities of energy than they expend, thus at risk of becoming obese.

B. GENETIC

In spite of environmental influences, certain individuals with a familial history of obesity may become obese.[8] The obesity phenotype is more likely to be expressed in predisposed individuals living in environments favoring excessive caloric intake and inactivity. This is true even if predisposed individuals are no longer living with their parents. To determine the relationship between parental and adult offspring physical characteristics, Stunkard and colleagues[66] reviewed the Danish adoption register and found a strong relationship ($P<0.01$) between the weight class (thin, medium weight, overweight, and obese) of 540 adult adoptees and body mass index (BMI, weight in kg/height^2 in meters) of their biological parents. No relationship existed between the adoptees and their non-biological parents, suggesting that obesity is partly an inherited trait. A similar situation may exist in very young children. Children who are removed from their biological parents at very young ages and placed in alternative care situations where the environment favors overconsumption and inactivity, and unknowingly have a history of obesity, may become obese before reaching puberty. Knowing the biological family history of body weight maintenance before placing children in alternative care situations may prevent predisposed children from becoming obese. If obesity is partly an inherited trait, this may partly explain why families with obese parents also have obese children.

C. ENVIRONMENTAL

Television is a low-cost entertainment medium that is appealing to children. It is also appealing to parents who find it easier and more convenient to keep their children at home than to send their children to supervised exercise programs within the community. Unfortunately, the hours children spend watching television detracts from exercise-oriented activities. Today children between the ages of 6–11 years old watch 23 hours of television per week, while 12–17-year-old adolescents watch an average of 21 hours of television per week.[1] Fontvieille and colleagues[21] found that television viewing, assessed by a standardized questionnaire, was correlated with percent body fat, suggesting that sedentary activities contribute to childhood obesity. These results concur with Durnin and colleagues[16] who found that physical activity in children has declined over recent decades, suggesting that children are living an increasingly sedentary lifestyle in industrialized countries. In cycles II and III of the National Health Examination Survey, time spent watching television was found to be directly associated with the prevalence of obesity among children and adolescents.[15] The type of program and the amount of time children spend watching television may effect their metabolic rate. Klesges and colleagues[36] measured RMR in obese (weight for height greater than 20% average weight for their age and sex using the National Department of Health,

Education, and Welfare norms[70] and as triceps and subscapular skinfold thicknesses greater than or equal to the 85th percentile)[62] and lean 8- to 12-year-old girls from middle-class families before, during, and after 25 minutes of watching a nonviolent television program in a metabolic laboratory. There was a reduction of RMR in both the obese and lean children during the 25 minutes of television viewing versus both the baseline and post-television viewing periods. Children who spend more of their leisure time watching television request more snack foods from their parents.[68] Since a low level of physical activity and a low RMR have been associated with obesity in adults,[53,50] it is possible that excessive television viewing by children contributes to their obesity. It is also possible that the type of programs children watch may contribute to whether their metabolic rate decreases or increases. Perhaps limiting children to more interactive programs, such as Sesame Street, may prevent the reduction of their metabolic rate and lessen the predisposition to obesity.

Only longitudinal studies can determine if inactivity precedes obesity or if obesity is a consequence of inactivity. The same children that were studied as 3 to 5 year-olds[26] were restudied 12 years later to determine if those who had a low RMR as toddlers became obese teenagers. The teenagers who grew more rapidly had a lower RMR/kg body weight and had at last one obese parent. However, these studies showed that young offspring of obese parents with a low RMR/kg body weight had a precocious pattern of growth but did not become obese.[27] It is possible that overfeeding may first lead to puberty precociously and later on to obesity. Perhaps, if these teenagers were followed up until they were middle-aged adults, they would be overweight, confirming this theory of obese parents having obese offspring.

D. NUTRITIONAL

Energy requirements for children were derived in a time when more physical exertion was necessary for everyday living; thus, energy requirements for children may be overestimated for today's sedentary lifestyle. Utilizing the doubly labeled water method to measure total daily energy expenditure, Prentice and colleagues[48] found that energy requirements recommended by the World Health Organization (WHO)[33] for children aged 0 to 3 years were overestimated by 15%. Furthermore, Goran and colleagues[25] found that energy requirements were overestimated by 25% for 4- to 6-year-old children in Vermont. Resting metabolic rate and the level of physical activity were determined in Caucasian boys and girls ages 5–6 years old.[20] Predicted energy requirements, calculated according to WHO,[33] were 24% above actual total daily energy expenditure as determined by the double labeled water method. These consistent overestimates of childhood energy requirements may be due to children spending more of their leisure time engaged in sedentary activities.

How the different nutrients are utilized can effect future body weight maintenance. The effect of nutrient composition of ingested diets can be determined by consuming meals of known composition and then using indirect calorimetry to measure RMR and the respiratory quotient (ratio of carbon dioxide production to oxygen consumption). The nitrogen content of a 24-hour urine collection is used to calculate non-protein respiratory quotient. This is necessary in order to correct for the effects of protein utilization on the respiratory quotient. A respiratory quotient of 0.71 indicates fat utilization, while a value close to 1.00 indicates carbohydrate utilization. It has been theorized by Flat and colleagues[18] that under sedentary conditions, ingested carbohydrates are quickly metabolized while lipids are stored primarily as body fat. This is due to the small size of the body's carbohydrate stores in relation to fat stores. Nutrient utilization was studied in 7 healthy young non-obese males (23±2 years old). They were fed three different meals of varying nutrient composition followed by measurements of RMR and respiratory quotient for 9 hours. The composition of the meals consisted of a high carbohydrate (62% carbohydrate, 27% protein, and 11% lipid), high fat, with most of the fat provided as long-chain fatty acids (35% carbohydrate, 15% protein, and 50% fat), and high fat, with most of the fat provided as medium-chain fatty acids (35% carbohydrate, 15% protein, and 50% fat) with 9% as long-chain fatty acids and 41% as medium-chain fatty acids. In comparison to the post-prandial respiratory quotient (0.82±0.008), carbohydrate intake promptly increased carbohydrate utilization, as indicated by the rising respiratory quotient (0.82 to 1.00), while fat, ingested as long-chain fatty acids, did not elicit any changes in the respiratory quotient. However, meals containing medium-chain fatty acids elicited only a slight increase in the respiratory quotient, suggesting that medium-chain fatty acids are being metabolized instead of being stored as body fat. Absorption and transport of medium-chain fatty acids occurs by other mechanisms than those involved for long-chain fatty acids.[4] Instead of being incorporated into chylomicrons, medium-chain fatty acids appear in the portal blood as free fatty acids bound to albumin,[64] the form in which all fatty acids are typically made.[18] These results suggests that over the 9-hour period, the body tightly regulates carbohydrate balance and is not effected by changes in the body's fat balance. It is possible that fat balance is regulated over a very long term; for example, it may take several days before the fat balance adjusts to new levels of fat intake as opposed to the carbohydrate balance.

To address the question of long-term regulation of fat balance, 7 young normal weight males (20–26 years old) spent 3 nights in a respiratory chamber for measurements of 24-hour energy expenditure and respiratory quotient and were fed weight maintenance diets containing 50% carbohydrate, 15% protein, and 35% fat for the first 24 hours and then switched to diets containing 37% carbohydrate, 11% protein, and 52% fat for the remainder of the study. The additional dietary fat after the first 24 hours had no effect on the two periods of 24-hour energy expenditure (2783±232 vs. 2820±284 Kcal/d) or respiratory quotient (0.83±0.04 vs. 0.81±0.03), suggesting that the carbohydrate balance is tightly regulated for up to 36 hours while sparing fat utilization.[60] These studies provide evidence that changes in fat balance lag behind changes in fat consumption. Ingesting excess fat calories for long periods of time will lead to a positive fat balance and eventually contribute to body weight gain. This has been further substantiated by the fact that a high respiratory quotient has been found to be a predictor of body weight gain in Pima Indian adults.[74] Additionally, fats containing a large proportion of medium-chain fatty acids may offer an alternative to controlling body weight gain due to excess fat intake. Although the best way to reduce fat deposition is to reduce fat intake, incorporation of medium-chain fatty acids into many of the foods that children tend to consume may offer a way to reduce dietary fat that is channeled to body weight gain. Only future research into nutrient utilization of various diets in children will determine if medium-chain fatty acids are a viable fat substitute for long-chain fatty acids.

III. THE ENVIRONMENTAL COMPONENTS OF OBESITY

A. ADULTS

Socioeconomic status (SES) is an environmental factor that may increase the predisposition to obesity in genetically prone individuals. Socioeconomic status is defined as the amount of income, occupation, and educational level of the primary providers of the family.[65] In westernized societies, obesity is more prevalent among women of lower SES, while women of higher SES, including professional women, have more of an obsession with thinness.[24] However, there is no clear trend between SES and obesity in men and is apparently due to methodological differences in how SES, obesity, or both were determined.[65] Parental lifestyles and beliefs about food consumption may provide a head start for the onset of obesity in children.

In contrast to the trends in developed countries, SES was strongly associated with increasing prevalence of obesity in both women and men in developing countries. This was probably due to individuals of higher SES being able to obtain adequate food supplies. Also, obesity may be a sign of wealth in developing countries.[65] For example, Nigerian pubescent daughters from wealthy families are sent to "fattening huts" before they are married.[11] The prevalence of obesity increases as developing societies enjoy increasing wealth and modernization.[65]

B. CHILDREN

There is variability in the relationship between SES and obesity in children from developing societies. Among the studies of this relationship in girls as revised by Sobal and Stunkard,[65] 35% showed no relationship, 40% found direct relationship similar to that found in adult women, while the remainder found an opposite relationship. In boys, 41% found no relationship, 26% found a direct relationship, while the remainder found a negative relationship. These variable results may be due to the ages at which children were studied, differing measures of SES and obesity, or to differences in data collection procedures.[24]

One of the main problems in elucidating the contribution of genetic and environmental factors in the etiology of childhood obesity is the "inheritance" of both SES and genetic predisposition to obesity from parents. McClendon[41] found that SES and obesity of the parents were the best predictors of SES and obesity in children. Transmission of wealth and other attributes of SES from parents allows children to live the same lifestyle as the parents. Along with monetary wealth, children also receive the parents' beliefs about eating and exercise, as well as attitudes about obesity itself.[65]

Environmental influences on children may start right after birth. Differences in postnatal feeding practices may influence the predisposition to obesity in later life. However, there is still controversy as to whether feeding practices contribute to childhood obesity. For example, Taitz[67] found that bottle-fed infants tended to be heavier at 6 weeks of life than breast-fed infants, while both Holley and Cullen[31] and De Swiet and colleagues[13] found no relationship between feeding practices and body weight gain. Additionally, adolescents who were breast-fed as infants were not obese,[37] while breastfeeding in infancy was related to greater weight gain among children up to 3 years old.[10] The composition of formula fed to infants may also affect their body composition. Co-variate analysis found that feeding practice (bottle- or breast-fed) and type of foods fed (high carbohydrate or high fat) had no influences on body weight, BMI, and the sum of 7 skin folds in 3- to 4-year-old children. Thus, mode of infant feeding and type of foods fed are not changing the predisposition of young children to obesity.[6] It has also been suggested that birth order, birth weight, age of introduction of solid foods, maternal body mass index, marital status of the mother, presence of males in the household, person responsible for child's food intake, and weight status of siblings may also contribute to a greater risk of childhood obesity.[3]

C. ETHNICITY

Ethnic heritage may influence a child's predisposition to obesity. African-American and Caucasian girls have similar rates of obesity from infancy to the beginning of adolescence. By the end of adolescence, the rate of obesity among blacks nearly doubles that of Caucasian adolescent girls.[39,23] The higher prevalence of obesity among African-American families is probably due to the genetic predisposition to obesity being amplified under certain environmental conditions such as low SES.[44] Additionally, Mexican-American children were more obese, according to triceps skin-fold thicknesses being greater than the 85th percentile, than either Puerto Rican or Cuban children.[46] Native-American children have a higher prevalence of obesity than either African-American or Caucasian children.[12] In a study of the prevalence of obesity in Mescalero Apache children, 19.5% were obese according to weight for height being greater than the 95th percentile based on the National Center for Health Statistics (NCHS) reference population,[14,28] while Native-American obese children were 2.5 times more likely to have an obese mother than non-obese children. Additionally, no relationships were found between the onset of obesity and age, sex, number of siblings, birth order, maternal employment, maternal diabetes, or marital status of the mother, suggesting that dietary factors and a low level of physical activity may contribute to the overall high prevalence of obesity in Native-American children.[22] However, there may be other factors that compound ethnicity's role in obesity. This is evident in Pima Indian boys and girls who spent more time watching television than Caucasian boys and girls.[21] These studies suggest that, independently of race, Native-American children are not as physically active as Caucasian children, and this lack of activity may contribute to the high prevalence of childhood obesity within these ethnic groups.

D. SPECIFIC SYNDROMES

Obesity in early childhood may be the result of certain genetic and/or hormonal abnormalities. In specific syndromes, there may be alterations in the proportion of fat-free mass to fat mass and differences in nutrient utilization. Children with Turner syndrome are short and have a characteristic habitus with a webbed neck, shield chests, pigmented nevi, and cubitus valgus[51] and frequently have abnormal carbohydrate metabolism.[63] In a study of fat distribution in obese children (above 120% ideal body weight) with Turner syndrome, triceps, ulnar, subscapular, and paraumbilical skin-fold thicknesses were less than that of normal weight and height matched children, suggesting that increased fat in areas other than the limbs may be mistaken for increased lean body mass. Thus, skin-fold thicknesses are not appropriate for determining body composition in these children, and, perhaps, using radiological or ultrasonic methods to determine body composition would provide a more accurate picture of their total body fat content.[29] Further studies are necessary in order to determine if differences in their metabolic rate contribute to their obesity.

Children with other genetic abnormalities are also prone to obesity. Children with Down's syndrome tend to be mentally retarded, slow growers, and obese. The high prevalence of obesity in these children may be due to excess food intake or a low requirement for energy through reduced physical activity.[47] Children with Prader-Willi syndrome are also characterized by obesity.[9] It was suggested that a low metabolic rate caused the obesity in these children.[74] Schoeller and colleagues[58] utilized the doubly labeled water method to determine body composition and total daily energy expenditure in children with Prader-Willi syndrome. Lower energy requirements of these children was due to less fat-free mass and not to an unusually low metabolic rate. It is important to utilize body composition techniques not dependent on assumptions about fat distribution, such as the doubly labeled water method, in order to prevent misinterpreting the amount of fat-free mass. Overestimating the amount of fat-free mass would predict excessive energy requirements in these children.

Certain abnormalities which occur during adolescence are sometimes associated with excess body weight. Overweight children with precocious pubarche (premature puberty) have increased linear growth along with advanced bone age in comparison to similarly aged normal weight children. However, overweight adolescents with precocious pubarche sometimes exhibit abnormal levels of adrenal hormones which may be indicative of adrenal biosynthetic defects. In a few instances, overweight precocious pubarche patients have been found to exhibit elevated pregnenolone, androstenedione, and dehydroepiandrosterone, while normal levels of these hormones were found in other overweight adolescents with precocious pubarche. The fact that normal levels of adrenal hormones were found in some overweight adolescents with precocious pubarche suggests that excess body weight in these patients is not due to excessive adrenal hormones. Jabbar and colleagues[32] suggest that precocious pubarche is a separate phenomenon, but that children with this condition, along with obesity, may have additional hormonal abnormalities.

IV. CRITERIA FOR DETERMINING CHILDHOOD OBESITY

There are many definitions that are used to define excess body weight in children. These include triceps skin-fold thicknesses greater than or equal to the 85th percentile[42] and weight for height greater than or equal to the 85th percentile as determined from age- and gender-specific growth charts.[28] These criteria correspond approximately to 120% of ideal body weight which is an established definition of obesity in adults.[42] Other criteria used to classify children as obese include body weight for height [(patient's actual body weight/ideal weight for height)*100] greater than 120%,[43] percent excess BMI for age at the 50th percentile [(patient's BMI/mean BMI for age) * 100] greater than 125%,[32] Rohrer's index [(weight in kilograms/height in meters)*10] greater than 160,[35] triceps-to-subscapular skin-fold thickness ratio greater than the 95th percentile for this measurement[46] and accelerated rate of weight gain in proportion to height with a crossing of one major percentile line for weight.[49] All these criteria depend on anthropometric measurements that estimate body composition from previously derived prediction equations and information from age- and gender-specific standardized growth charts. These criteria are not valid if a particular child is not of an age or ethnic group included in the derivation of the prediction equations and standardized growth charts. Also, these standards do not include children with genetic or hormonal abnormalities. Using these criteria to define obesity in children without prior knowledge of the populations used to derive population standards may lead to misclassification of children as obese.

In one example of improper use of these criteria in defining obesity, Hoerr and colleagues[30] estimated body composition in 13-year-old obese African-American and Caucasian girls using underwater weighing, skin-fold thicknesses, and the National Center for Health Statistics (NCHS) growth charts.[43] The Slaughter equation was used to estimate percent body fat utilizing the sum of the triceps and subscapular skin-folds thicknesses.[62] When all obese girls were considered as one group and using underwater weighing as the "gold standard", the Slaughter equation estimated these girls were between 10% to 20% overweight compared with 30% to 50% from the NCHS growth charts. When considered alone, these overestimates were even greater in the African-American girls. These results suggest that NCHS growth charts may underestimate goal weights for adolescent girls

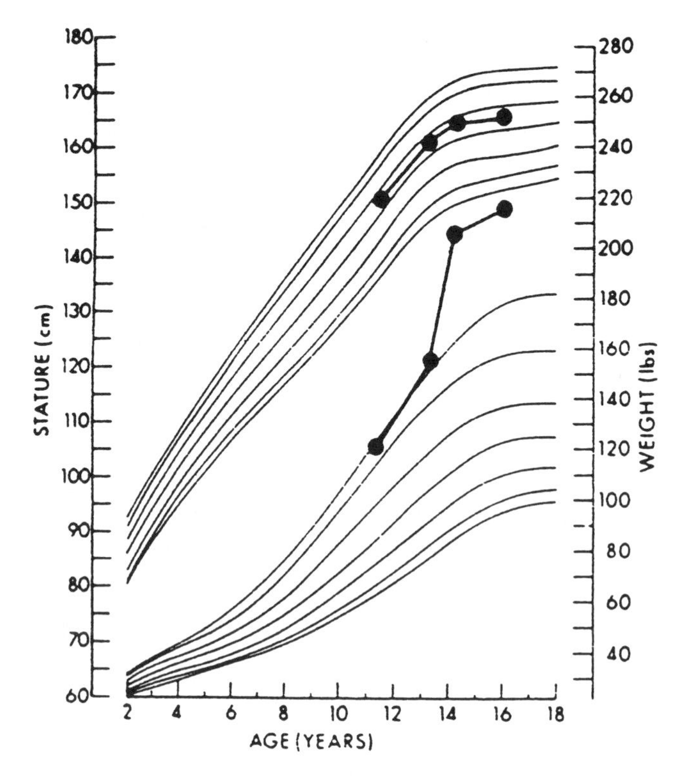

FIGURE 1. Growth progression of an obese girl. (From Pugliese, M.T., Recker, B., and Lifshitz, F., *J. Adol. Hlth. Care*, 9, 181–187, 1988. With permission.)

and that additional measures should be utilized before making inferences about obesity in adolescents, especially in adolescents of specific ethnic groups.

Usually, a child is defined as obese or non-obese from a one-time measurement of body composition during a routine physician's visit. It is important to evaluate growth in children over several years to determine if they are deviating from normal growth patterns before classifying as obese or non-obese. For example, the preponderance of normal and abnormal growth was evaluated in high school students using the NCHS growth charts. Growth data was obtained if each student's medical chart contained at least two visits to a medical professional. Out of more than 1000 students, 24.6% had body weight in excess for height of greater than or equal to 110%. Students in a subset of this population (0.8%) were already above the 95th percentile for weight and crossed at least one percentile during a year. For example, Figure 1 shows the growth progression of an obese girl. Her height was normal but was gaining more than 10 kg/year by age 14.[49] Overall, these studies suggest that growth should be monitored and body composition determined on a yearly basis in order to ensure normal development and maintenance of proper body weight. If accelerated growth or an increase in amount of body fat is detected, the physician or related health professional could initiate treatment for obesity long before it becomes a chronic problem.

V. PROBLEMS WITH TREATING CHILDHOOD OBESITY

Treatment of obesity in children is different than treating obesity in adults. First, children are growing, so they have changing energy requirements during long-term treatment. Second, today's children have more leisure time in comparison to children at the turn of the century, so they have more of an opportunity for spending time involved in non-exercise-oriented activities. Third, parents have to be involved in the treatment of obesity in young children in order to ensure that the children follow all dietary and exercise prescriptions, a problem in families where both parents are employed. Finally, treatment must be long term in order to prevent weight regain. Sometimes financial and time constraints may prohibit many families from participating in any program for reducing their child's body weight. In a one-year treatment program for children, 10- to 15-year-old patients, who were 20% or more overweight according to weight for height tables, were placed on a hypocaloric diet without the encouragement of additional physical activity.[77] After 3 weeks of treatment, RMR, fat-free mass and fat mass were reduced. During the following year, RMR and fat-free mass increased while the percentage of fat mass decreased, suggesting that growth during the treatment period suppresses body fat gain. Unlike adults who have to lose weight to reach a particular goal weight, growing children who are not morbidly obese (greater than 180% ideal body weight) have an additional advantage during treatment because just maintaining body weight during treatment will reduce their relative overweight.[77] Encouraging increased physical activity and weight maintenance in moderately overweight children will help eliminate some of the stress that is part of losing excess body weight. However, growing children who are morbidly obese will still have to lose a large amount of weight in spite of the extra advantage contributed by growth.

In some situations, it is important to consider the ethnic heritage and SES of families that enroll in a treatment program. In some instances, having the mother or father participate along with the child in the treatment process may improve the success rate during treatment and the probability that the child will maintain proper weight. For example, 47 African-American adolescent girls (12 to 16 years old) who were 10 kg overweight for their age, sex, and height participated in an obesity treatment program, along with their mothers, consisting of 16 weekly 1-hour lectures. The group was divided into three different treatments: child participating alone, child and mother meeting in separate rooms concurrently, and child and mother meeting together. All groups were encouraged to become more physically active and to reduce the composition of high caloric foods. Body composition was determined at the start of the program and after 14 weeks of treatment, while body weight was determined weekly. After 14 weeks of treatment, the entire group lost an average of 1.7 kg and significantly reduced their body fat by 2% (37.1% to 35.1%). At the end of treatment, weight loss amounted to 1.6, 3.1, and 3.7 kg, respectively, for the child alone, child and mother separately, and for child and mother together. Thus, the program was most effective when mothers participated along with their children. The combination of modest caloric restriction and increased physical activity produced favorable changes in body weight and body composition. After 6 months of follow-up, most of the children that participated in the study regained their lost weight, suggesting that a continued program is necessary in order to maintain weight loss. Weight regain in this particular study may also be due to a decreased emphasis on thinness and the acceptability of overweight within this ethnic group.[71] Since these children regained their lost body weight, there is a possibility that re-enrolling in short-term obesity treatment programs will perpetuate the ''yo-yo'' syndrome. Only future studies that follow children through repetitive weight loss cycles will determine if children, like adults, are prone to the ''yo-yo'' syndrome.

Monitoring patient compliance to dietary prescriptions in childhood obesity treatment programs has been very difficult. Usually, a 24-hour or a three-day food record is utilized to determine if parents and children are complying with dietary prescriptions. As has been eluded to previously,[52] dietary recalls are subject to error due to the reliance on the patients' self-interpretation of food intake. These errors are further exasperated by the fact that obese adults and obese children under-report their food intake.[40,2] Do obese families under-report their food intake? Currently, there is no evidence to suggest that obese families as a whole under-report their food intake. Only future studies utilizing new technologies involving computerized methods for measuring food intake[52,38] or proper use of present technologies will determine if this phenomenon occurs in obese families.

Similar problems occur when individuals are asked to record their physical activities.[61] Misinterpretation of the amount and duration of physical activities can lead to errors in the calculation of the overall energy cost of physical activities. This problem is further exasperated when parents are asked to record their child's physical activities. A lot of the child's physical activities may be unrecorded, or in older children, the child may overestimate the amount of time spent engaged in vigorous physical activities. Usually, the energy cost of physical activities is calculated by using standardized values for each particular activity and the amount of time engaged in each activity. New techniques, such as the doubly labeled water method, may improve the accuracy in determining physical activity in children.

VI. NUTRITIONAL AND METABOLIC ASSESSMENT BEFORE AND DURING TREATMENT OF CHILDHOOD OBESITY

Improving compliance of children enrolled in obesity treatment programs would greatly increase the success rate and reduce the cost to all patients—a savings that will be reflected in less obesity health-related problems such as heart disease, non-insulin-dependent diabetes, and joint disorders. A thorough approach for treating childhood obesity should include metabolic, nutritional, and behavioral changes that would lead to maintenance of body weight in moderately obese growing children or a gradual reduction of body weight in severely obese young adults. Before initiating dietary treatment, serum vitamins and minerals should be determined, in addition to a complete physical exam, in order to detect any underlying nutritional deficiencies. Food intake should be determined as part of the treatment protocol. The 24-hour dietary recall needs to be conducted by a trained interviewer. This method is the least expensive and most practical for determining food intake in children.[17] It is usually administered to older children or to the parents of young children in order to obtain an idea of current nutrient intake. A more accurate assessment of food intake would require admission to a hospital or metabolic ward. However, this later method of quantitating food intake is usually done by research institutions conducting food intake studies. In order to determine basal energy requirements, some institutions measure RMR using a ventilated hood indirect calorimeter, while others calculate RMR from standard equations. Patients are usually asked to provide a 24-hour urine collection at the time of the RMR measurement. The most common way of assessing body composition in children is by skin-fold thickness measurements. The problem with this method is the need for a trained technician to measure the skin-fold thicknesses and some cooperation by the patient. Single-frequency bioelectrical resistance is another commonly used method for assessing body composition in children in out-patient clinics. The advantages of this method are low cost, safety, and ease of use, and little patient cooperation is required. In the near future, body composition may be assessed by measurement of body water by two new methods: measurement of intra- and extracellular water by multi-frequency bioelectrical impedance and total body water by the dilution space of oxygen-18.[59] Single-frequency bioelectrical impedance and the dilution space of oxygen-18 require knowledge of the hydration of fat-free mass for calculating the amount of fat-free mass and fat mass. The appropriate hydration constant depends on the age and gender of the child. In children from 1 month to 10 years old, hydration of fat-free mass varies from 81% to 75% in boys and from 81% to 77% in girls.[19] In older children, it is assumed that fat-free mass is 73% hydrated. Currently, most treatment programs for childhood obesity utilize a standardized activity questionnaire and standard values for various physical activities for calculating the amount of energy expended for routine physical activities. The advantages of these methods are low cost and ease of use. However, errors arise due to self-interpretation of physical activities.

An initial work-up before beginning treatment for obesity would provide baseline caloric requirements, nutritional status, body fat content, and an indication of the current level of physical activity from which to base initial dietary and exercise prescriptions. Additionally, these procedures would provide a general picture of the overall health of the child prior to entry into an obesity treatment program.

Having an accurate assessment of total daily energy expenditure and the amount of energy expended for routine physical activities would allow health

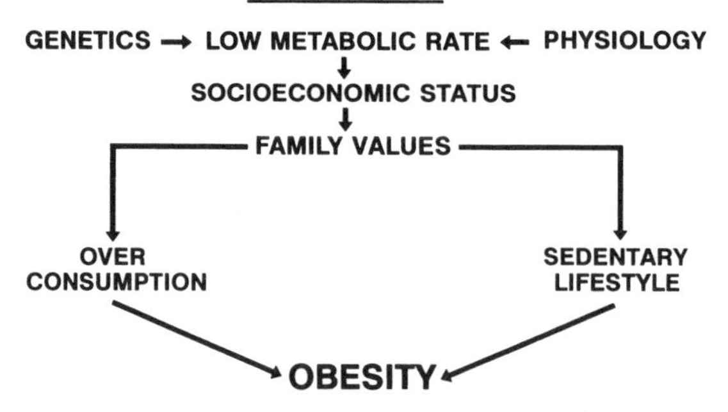

FIGURE 2. Possible causes of obesity in adults.

professionals to detect patients who are not complying with dietary and exercise prescriptions. Recently, the doubly labeled water method was developed to measure total daily energy expenditure accurately under free-living conditions. This method allows accurate assessment of energy requirements and energy expenditure in individuals without the complications associated with dietary recalls or wearing of heavy equipment.[59] The method has been used in infants,[34] young children,[7] and adolescents.[5] This technique involves giving individuals a glass of water containing known amounts of stable, non-radioactive isotopes of oxygen and hydrogen, combined as doubly labeled water. These labels are later determined in the urine over the period of one week, and standard calorimetric formulas[72] are utilized to calculate total daily energy expenditure. The technique is both safe and nonevasive and offers an accurate way to determine total daily energy expenditure in very young children with minimal cooperation. To obtain a measure of the amount of energy expended for routine physical activities, RMR is first determined by indirect calorimetry before consumption of the stable isotopes, and energy expended for routine physical activities is calculated by the difference between total daily energy expenditure and RMR. The problem with calculating the energy expended for routine physical activities is that heavier people expend more energy for the same amount of activity in comparison to individuals of lesser weight. To correct for the effect of body weight, the ratio of energy expended for routine physical activities and body weight or total daily energy expenditure and RMR are utilized.[55]

These new methods may offer a way to recognize non-compliance to all dietary and exercise prescriptions. For example, RMR and total daily energy expenditure would be measured at the initiation and again after two months of treatment. Parents of obese children who are not losing weight would no longer be able to claim that they are following all dietary and exercise prescriptions. Unfortunately, these new methods are still very expensive on a per patient basis; for example, one doubly labeled water study can cost close to a thousand dollars.

In summary, much is known about the pathophysiology of obesity in adults (Figure 2). However, less is known about the pathophysiology of obesity in children. Additional factors have to be considered when conducting research into the causes of childhood obesity. First of all, children are growing, and their nutrient requirements change as they age. Also, children have no control over their environment. They live with their parents and are taught the values and beliefs of their parents, and when in school, they are subjected to whatever nutritional education a particular school offers. The predisposition to obesity may begin right after birth. Future studies of energy metabolism in newborn infants may provide future evidence that childhood obesity begins from the first days of life if not before.

REFERENCES

1. AC Nielsen Company, *Nielsen Report on Television,* New York, Nielsen Media Research, 1990.
2. Alemzadeh, R., Goldberg, T., Fort, P., Recker, B., and Lifshitz, F., Reported dietary intakes of patients with insulin-dependent diabetes mellitus: limitations of dietary recall, *Nutrition,* 8, 87–93, 1992.
3. Alexander, M. A. and Sherman, J. B., Factors associated with obesity in school children, *J. School. Nutr.,* 7, 6–10, 1991.
4. Bach, A. C. and Babayan, V. K., Medium-chain triglycerides: an update, *Am. J. Clin. Nutr.,* 36, 950–962, 1982.
5. Bandini, L. G., Schoeller, D. A., and Dietz, W. H., Energy expenditure in obese and nonobese adolescents, *Pediatr. Res.,* 27, 198–203, 1990.
6. Baranowski, T., Bryan, G. T., Rassin, D. K., Harrison, J. A., and Henske, J. C., Ethnicity, infant-feeding practices, and childhood adiposity, *J. Dev. Behav. Pediatr.,* 11, 234–239, 1990.
7. Blaak, E. E., Westerterp, K. R., Bar-Or, O., Wouters, L. J. M., and Saris, W. H. M., Total energy expenditure and spontaneous activity in relation to training in obese boys, *Am. J. Clin. Nutr.,* 55, 777–782, 1992.
8. Bogardus, C., Lillioja, S., Ravussin, E., Abbott, W., Zawadzki, J. K., Young, A., Knowler, W. C., Jacobowitz, R., and Moll, P. P., Familial dependence of the resting metabolic rate, *N. Engl. J. Med.,* 315, 96–100, 1986.
9. Bray, G. A., Dahms, W. T., Swerdloff, R. S., Fiser, R. H., Atkinson, R. L., and Carrel, R. E., The Prader-Willi syndrome: a study of 40 patients and a review of the literature, *Medicine,* 62, 59–80, 1983.
10. Brodwick, M., The relationship of infant feeding patterns to subsequent obesity in children, Master's thesis, Galveston, TX, University of Texas Medical Branch, 1981.
11. Brown, P. J. and Konner, M., An anthropological perspective on obesity, *Ann. N. Y. Acad. Sci.,* 499, 29–46, 1987.
12. Center for Disease Control, Nutritional surveillance, CDC Surveillance Summaries, 32 (No. 4SS), 23–26, 1983.
13. De Swiet, M., Fayers, P., and Cooper, L., Effect of feeding habit on weight in infancy, *Lancet,* 1, 892–894, 1977.
14. Dibley, M. J., Goldsby, J. B., Staehling, N. W., and Trowbridge, F. L., Development of normalized curves for the international growth reference: historical and technical considerations, *Am. J. Clin. Nutr.,* 46, 736–748, 1987.

15. Dietz, W. H. and Gortmaker, S. L., Do we fatten our children at the television set? Obesity and television viewing in children and adolescents, *Pediatrics,* 75, 807–812, 1985.
16. Durnin, J. V. G. A., Lonergan, M. E., Good, J., and Ewan, A., A cross-sectional nutritional and anthropometric study, with an interval of 7 years, on 611 young adolescent school children, *Brit. J. Nutr.,* 32, 169–179, 1974.
17. Farris, R. P. and Nicklas, T. X., Characterizing children's eating behavior, in *Textbook of pediatric nutrition,* 2nd ed., Suskind, R. M. and Suskind, L. L., Eds., New York, Raven Press, 1993, 505–516.
18. Flatt, J. P., Ravussin, E., Acheson, K. J., and Jequier, E., Effects of dietary fat on postprandial substrate oxidation and on carbohydrate and fat balances, *J. Clin. Invest.,* 76, 1019–1924, 1985.
19. Fomon, S. J., Haschke, F., Ziegler, E. E., and Nelson, S. E., Body composition of reference children from birth to age 10 years, *Am. J. Clin. Nutr.,* 35, 1169–1175, 1982.
20. Fontvieille, A. M., Harper, I. T., Ferraro, R. T., Spraul, M., and Ravussin, E., Daily energy expenditure by 5-year-old children, measured by the doubly-labeled water method, *J. Ped.,* 123, 200–207, 1993.
21. Fontvieille, A. M., Kriska, A., and Pendergrass, C., Leisure activities in Pima Indian and Caucasian children, *Diabetes,* 41 (Suppl. 1): 74A(abstr.) 1992.
22. Gallaher, M. M., Hauck, F. R., Yang-Oshida, M., and Serdula, M. K., Obesity among Mescalero preschool children, *J. Am. Diet. Assoc.,* 145, 1262–1265, 1991.
23. Gartside, P. S., Khoury, P., and Glueck, C. J., Determinants of high-density lipoprotein cholesterol in blacks and whites: the second National Health and Examination Survey, *Am. Heart J.,* 108, 641–653, 1984.
24. Goldblatt, P. B., Moore, M. E., and Stunkard, A. J., Social factors in obesity, *J. Am. Med. Assoc.,* 192, 1039–1042, 1965.
25. Goran, M. I., Carpenter, W. H., and Poehlman, E. T., Total energy expenditure in 4- to 6-year-old children, *Am. J. Physiol.,* 264, E706–E711, 1993.
26. Griffiths, M. and Payne, P. R., Energy expenditure in small children of obese and non-obese parents, *Nature,* 260, 698–700, 1976.
27. Griffiths, M., Payne, P. R., Stunkard, A. J., Rivers, J. P. W., and Cox, M., Metabolic rate and physical development in children at risk of obesity, *Lancet,* 336, 76–78, 1990.
28. Hamill, P. V. V., Drizd, T. A., Johnson, C. L., Reed, R. B., and Roche, A. R., *NCHS Growth Curves for Children Birth to 18 Years, United States, November 1977,* Rockville, Md., National Center for Health Statistics, 1977, Vital and Health Statistics 165 (series II), U.S. Dept. of Health, Education, and Welfare publication (PHS) 78–1650.
29. Hanaki, K., Ohzeki, T., Ishitani, N., Motozumi, H., Ohtahrea, H. M., and Shiraki, K., Fat distribution in overweight patients with Ullrich-Turner syndrome, *Am. J. Med. Gen.,* 42, 428–430, 1992.
30. Hoerr, S. L., Nelson, R. A., and Lohman, T. R., Discrepancies among predictors of desirable weight for black and white obese adolescent girls, *J. Am. Diet. Assoc.,* 92, 450–453, 1992.
31. Holley, D. and Cullen, D., A comparison of weight gain in breast-fed and bottle-fed babies, *Public Health,* 91, 113–116, 1977.
32. Jabbar, M., Pugliese, M., Fort, P., Recker, B., and Lifshitz, F., Excess weight and precocious pubarche in children: alterations of the adrenocortical hormones, *J. Am. Col. Nutr.,* 10, 289–296, 1991.
33. Joint FAO/WHO ad hoc, Expert Committee Report on Energy and Protein Requirements, FAO, *Nutr. Mtgs. Rep. Ser.,* No. 257, 118 pp HMSO, London, 1973.
34. Jones, P. J. H., Winthrup, A. L., Schoeller, D. A., Swyer, P. R., Smith, J., Filler, R. M., and Heim, T., Validation of doubly labeled water for assessing energy expenditure in infants, *Pediatr. Res.,* 21, 242–246, 1987.
35. Keys, A., Fidanza, F., Karvonen, M. J., Kimura, N., and Taylor, H. L. Indices of relative weight and obesity, *J. Chron. Dis.,* 25, 329–343, 1972.
36. Klesges, R. C., Shelton, M. L., and Klesges, L. M., Effects of television on metabolic rate: potential implications for childhood obesity, *Pediatrics,* 91, 281–286, 1993.

37. Kramer, M. S., Do breastfeeding and delayed introduction of solid foods protect against subsequent obesity? *J. Pediatr.*, 98, 883–887, 1981.
38. Kretsch, M. J. and Fong, A. K. H., Validation of a new computerized technique for quantitating individual dietary intake: the Nutrition Evaluation Scale System (NESSy) vs. the weighed food record, *Am. J. Clin. Nutr.*, 51, 477–484, 1990.
39. Kumanyika, S., Obesity in black women, *Epidemiol. Rev.*, 9, 31–50, 1987.
40. Lichtman, S. W., Pisarska, K., Raynes, E., Pestone, M., Dowling, H., Offenbacher, E., Weisel, H., Heshka, S., Matthews, D. E., and Heymsfield, S. B., Discrepancy between self-reported and actual caloric intake and exercise in obese subjects, *N. Engl. J. Med.*, 327, 1893–1898, 1992.
41. McClendon, M. J., The occupational status attainment process of males and females, *Am. Sociolog. Rev.*, 41, 52–64, 1976.
42. National Institutes of Health Consensus Development Conference Statement, Health implications of obesity, *Ann. Int. Med.*, 103, 1073–1077, 1985.
43. National Center for Health Statistics, *Health, United States 1986,* DHHS publication no. (PHS) 87–1232, Hyattsville, National Center for Health Statistics.
44. Ness, R., Laskarzewski, P., and Price, R. A., Inheritance of extreme overweight in black families, *Hum. Biol.*, 63, 39–52, 1991.
45. Obarzanek, E., Methodological issues in estimating the prevalence of obesity in childhood, The New York academy of sciences, conferences on prevention and treatment of childhood obesity, 15(abstr.), 1992.
46. Pawson, I. G., Martorell, R., and Mendoza, F. E., Prevalence of overweight and obesity in US Hispanic populations, *Am. J. Clin. Nutr.*, 53, 1522S–1528S, 1991.
47. Pipes, P. L., Nutrition and individuals with Down's syndrome, *Nutr. News,* 3, 1–4, 1988.
48. Prentice, A. M., Lucas, A., Vasquez-Velasquez, L., Davies, P. S. W., and Whitehead, R. G., Are current dietary guidelines for young children a prescription for overfeeding? *Lancet,* 2, 1066–1069, 1988.
49. Pugliese, M. T., Recker, B., and Lifshitz, F., A survey to determine the prevalence of abnormal growth patterns in adolescents from a suburban school district, *J. Adol. Hlth. Care,* 9, 181–187, 1988.
50. Ravussin, E., Lillioja, S., Knowler, W. C., Christin, L., Freymond, D., Abbott, W. G. H., Boyce, V., Howard, B. V., and Bogardus, C., Reduced rate of energy expenditure as a risk factor for body-weight gain, *N. Engl. J. Med.*, 318, 467–472, 1988.
51. Rimoin, D. L. and Schimke, R. N. *Genetic disorders of the endocrine glands,* St. Louis, C. V. Mosby, 1971, 285–292.
52. Rising, R., Alger, S., Boyce, V., Seagle, H., Ferraro, R., Fontvieille, A. M., and Ravussin, E., Food intake measured by an automated food-selection system: relationship to energy expenditure, *Am. J. Clin. Nutr.*, 55, 343–349, 1992.
53. Rising, R., Harper, I. T., Fontvieille, A. M., Ferraro, R., Spraul, M., and Ravussin, E., Determinants of total daily energy expenditure: variability in physical activity, *Am. J. Clin. Nutr.*, 59, 800–804, 1994.
54. Rising, R., Keys, A., Ravussin, E., and Bogardus, C., Concomitant inter-individual variations in body temperature and metabolic rate, *Am. J. Physiol.*, 263, E730–E734, 1992.
55. Roberts, S. B., Heyman, M. B., Evans, W. J., Fuss, P., Tsay, R., and Young, V. R., Dietary energy requirements of young adult men, determined by using the doubly labeled water method, *Am. J. Clin. Nutr.*, 54, 499–505, 1991.
56. Roberts, S. B., Savage, J., Coward, W. A., Chew, B., and Lucas, A., Energy expenditure and intake in infants born to lean and overweight mothers, *N. Engl. J. Med.*, 318, 461–466, 1988.
57. Saad, M. F., Alger, S. A., Zurlo, F., Young, J. B., Bogardus, C., and Ravussin, E., Ethnic differences in sympathetic nervous system-mediated energy expenditure, *Am. J. Physiol.*, 261, E789–E794, 1991.
58. Schoeller, D. A., Levitsky, L. L., Bandini, L. G., Dietz, W. W., and Walczak, A., Energy expenditure and body composition in Prader-Willi syndrome, *Met.*, 37, 115–120, 1988.

59. Schoeller, D. A., Ravussin, E., Schutz, Y., Acheson, K. J., Baertschi, P., and Jequier, E., Energy expenditure by doubly labeled water: validation in humans and proposed calculation, *Am. J. Physiol.*, 250, R823–R830, 1986.
60. Schutz, Y., Flatt, J. P., and Jequier, E., Failure of dietary fat intake to promote fat oxidation: a factor favoring the development of obesity, 50, 307–314, 1989.
61. Shephard, R. J., Assessment of physical activity and energy needs, *Am. J. Clin. Nutr.*, 50, 1195–1200, 1989.
62. Slaughter, M. H., Lohman, T. G., Boileau, R. A., Horswill, C. A., Stillman, R. J., Van Loan, M. D., and Bemben, D. A., Skinfold equations for estimation of body fatness in children and youth, *Hum. Biol.*, 60, 709–723, 1988.
63. Smith, D. W. and Jones, K. L., XO syndrome, in *Recognizable Patterns of Human Malformation*, Philadelphia, W. B. Saunders, 1982, 72–75.
64. Spector, A. A., Fatty acid binding to plasma albumin, *J. Lipid Res.*, 16, 165–179, 1975.
65. Sobal, J. and Stunkard, A. J., Socioeconomic status and obesity: a review of the literature, *Psychol. Bull.*, 105, 260–275, 1989.
66. Stunkard, A. J., Sorensen, T. I. A., Hanis, C., Teasdale, T. W., Chakraborty, R., Schull, W. J., and Schulsinger, F., An adoption study of human obesity, *N. Engl. J. Med.*, 314, 193–198, 1986.
67. Taitz, L. S., Infantile over nutrition among artificially fed infants in the Sheffield region, *Br. Med. J.*, 1, 315–316, 1971.
68. Taras, H. L., Sallis, J. F., Patterson, T. L., Nader, P. R., and Nelson, J. A., Television's influence on children's diet and physical activity, *Dev. Behav. Pediatr.*, 10, 176–180, 1989.
69. U. S. Department of Health & Human Services, Public Health Service, DHHS, *The Surgeon General's Report on Nutrition and Health*, GPO Stock No. 017–001–0465–1 Washington, D.C., U.S. Government Printing Office, 1988.
70. U. S. Department of Health, Education, and Welfare, *National Center for Health Statistics Growth Curves for Children: Birth to 18 Years*, Washington, D.C., U.S. Government Printing Office, 1977, 32, U.S. Department of Health, Education, and Welfare publication 78–1650.
71. Wadden, T. A., Stunkard, A. J., Rich, L., Rubin, C. J., Sweidel, G., and McKinney, S., Obesity in black adolescent girls: a controlled clinical trial of treatment by diet, behavior modification, and parental support, *Pediatrics*, 85, 345–352, 1990.
72. Weir, J. B., New methods for calculating metabolic rate with special reference to protein metabolism, *J. Physiol.*, 109, 1–9, 1949.
73. Widdowson, E. M., *MRC. Spec. rep. Ser.*, No. 257, HMSO, London, 1947, 96.
74. Widhalm, K., Veitt, V., and Irsigler, K., Evidence for decreased energy expenditure in the Prader-Labhart-Willi syndrome: assessment by means of the Vienna Calorimeter, *Proc. Intl. Cong. Nutr.*, 1981, (abstr.).
75. Zurlo, F., Larson, K., Bogardus, C., and Ravussin, E., Skeletal muscle metabolism is a major determinant of resting energy expenditure, *J. Clin. Invest.*, 86, 1423–1427, 1990.
76. Zurlo, F., Lillioja, S., Puente, A. E. D., Nyomba, B. L., Raz, I., Saad, F. M., Swinburn, B. A., Knowler, W. C., Bogardus, C., and Ravussin, E., Low ratio of fat to carbohydrate oxidation as predictor of weight gain: study of 24-h RQ, *Am. J. Physiol.*, 259, E650–E657, 1990.
77. Zwiauer, K. F. M., Mueller, T., and Widhalm, K. Resting metabolic rate in obese children before, during and after weight loss, *Intl. J. Obes.*, 16, 11–16, 1992.

Chapter 11

NUTRITION FOR SPECIAL NEEDS—IN PEDIATRIC GASTROINTESTINAL DISEASES

Emanuel Lebenthal

TABLE OF CONTENTS

I. Nutrition for the Child with Diarrhea 99
   A. Nutrition for the Child with Acute Diarrhea 99
   B. Nutrition for the Child with Chronic Diarrhea 102
   C. Intractable Diarrhea in Infancy 102

II. Nutrition for the Child with a GI Disease 103

References 104

I. NUTRITION FOR THE CHILD WITH DIARRHEA

Diarrhea is one of the major causes of infant morbidity and mortality worldwide. In the United States, although it is not as frequent as it once was, diarrhea still remains a prevalent cause of morbidity among infants and children.

The main objectives in therapy are (1) identify and correct fluid and electrolyte abnormalities, (2) provide adequate calories and nutrients to the infant to meet both the usual maintenance requirements and the increased need induced by diarrhea, and (3) correct any underlying associated disease entity.

Not long ago, most clinical management of childhood diarrhea stressed fluid and electrolyte therapy and overlooked nutritional therapy. Emphasis has been aimed on the dietary management of diarrhea during the recent few years.

A. NUTRITION FOR THE CHILD WITH ACUTE DIARRHEA

The efficacy of dietary therapy for acute diarrhea should be considered to relate to several dietary factors, including timing to introduce feeding, the total amount of food given, food ingredient, nutrients, etc.

For mothers and doctors alike, the conventional wisdom is ''rest the gut during diarrhea'' believing that ''feeding a child with diarrhea makes the condition worse''. This was apparently not an unreasonable (though entirely erroneous) assumption on their part because an early clinical trial in 1948 comparing a milk formula with fasting during diarrhea (Chung, 1948; Chung and Viscorova, 1948) showed an increase in the volume and frequency of diarrheal stools (which was obvious) yet a beneficial effect for the child because of a net retention of nutrients

0-8493-2764-4/95

(which was not apparent to mothers). However, it is important to emphasize that although stool output may increase by feeding infants with diarrhea, there is a net increase in the absorption of nutrients when infants are fed (Brown and MacLean, 1984; Khin-Maung-U, Nyunt-Nyunt-Wai, Myo-Khin, et al., 1986).

Short-term deprivation of nutrients by fasting during acute diarrhea is serious because a fasting child loses an estimated 1–2% of his/her body weight daily even in the absence of fluid losses due to diarrhea (Rhode, 1978). This becomes more significant when considered in a broader perspective. Children under two years of age in developing countries suffer approximately 6 to 10 episodes of diarrhea per year (Black, Brown, Becker, et al., 1982, 1982a) spending about 16% to 17% of their life having diarrhea. Such recurrent or prolonged diarrhea may lead to Protein Energy Malnutrition through diminished food intake because of anorexia or food withholding (Hoyle, Yunus, & Chen 1980; Martorell, Yarbrough, Yarbrough, et al., 1980; Molla, Molla, Sarker, et al., 1983), impaired absorption and utilization of nutrients (Molla, Molla, Sarker, et al., 1983a), increased catabolic losses through negative nitrogen balance which may amount to 0.9 g/kg/day (Powanda, 1977), and subtle losses of protein in the gastrointestinal tract during diarrhea (Rahaman and Wahed, 1983). Thus, infants and young children are particularly susceptible than older children and adults to the adverse nutritional consequences of diarrhea because of their greater nutritional requirements, lower nutritional reserves, and because in the young diarrhea tends to be frequent and more severe (Black, 1985).

On the other hand, clinical studies have shown that while absorption of all nutrients was impaired during acute diarrhea, substantial absorption still occurred (Molla, Molla, Sarker, et al., 1983, 1983a). Studies carried out in Bangladesh, India, Peru, and Thailand using mixed diets composed of common foods and synthetic formula diets have shown that a nutritionally significant proportion of the diet can be absorbed during enteric infections and diarrhea (WHO, 1985). On the average, 80–95% of dietary carbohydrate, 50–70% of fat, and 50–75% of nitrogen are absorbed (depending on the food source of these nutrients, the amount of intake, and the severity of diarrhea) even during the early phase of illness (Molla, Molla, Sarker, et al., 1983,1983a). In a clinical trial of Peruvian children aged 3 to 36 months with diarrhea, the net absorption of nutrients in the fed child are proportionately related to the amount of dietary energy consumed during diarrhea (Brown, Gastanaduy, Saavedra, et al., 1988). These clinical and community-based research studies therefore demonstrate that diarrhea sufferers should be fed, not fasted—and the sooner the better. Thus, feeding during acute diarrhea should be considered an integral component of Oral Rehydration Therapy.

In the case of an exclusively breast-fed infant, breastfeeding should be continued during the rehydration, maintenance, and convalescent phases of therapy for diarrhea. Infants with acute diarrhea who were breast-fed without interruption during fluid therapy had fewer diarrhea stools, smaller stool volumes, shorter duration of diarrhea, and a tendency towards better weight gain on recovery compared with infants for whom breastfeeding was withheld for an initial 24 hours during fluid therapy (Khin-Maung-U, Nyunt-Nyunt-Wai, Myo-Khin, et al., 1985). Also, breast-fed children with diarrhea have been shown to average a total energy intake 35% greater and a protein intake 250% greater than children who were completely weaned (Chen, Huq, and Huffman, 1980).

As soon as possible, the patient with diarrhea should be provided with an adequate balanced nutrient intake. The milk or formula that the baby has been receiving can be restarted. Soy-based formula can be given full strength; children who were fed soy formula and ORS had significant reductions in stool output compared to those receiving only ORS in the first 24 hours (Santosham, Foster, Reid, et al., 1985). For lactose-based formula, children should be given half-strength formula until diarrhea stops, when full-strength formula can be instituted (Santosham, et al., 1985).

Children who are on solid foods prior to onset of acute diarrhea should be given locally available foods with the highest amount of nutrients and calories relative to bulk, such as fish or meat and cereals. Although digestibility of different staple foods may vary considerably, excellent clinical results have been reported in children fed mixed diets based on rice, banana, milk, and bread, or on wheat noodles and milk (WHO, 1985); thus, there is some basis for expecting common foods to be well tolerated. Also, studies show that small amounts of lactose present in milk-containing mixed diets are generally well tolerated even by known lactose malabsorbers and those during acute diarrhea (Bowie, 1975; Isolauri, Vesikari, et al., 1986). The child should be encouraged to eat frequently as much as he/she wants, about 5 to 7 times a day (about every 3 to 4 hours). In addition, some of these children may also be consuming breast milk or formula in addition to solid foods prior to onset of diarrhea: for these children, continuation of breastfeeding or consumption of formula should be advised in the same manner as those described above.

Feeding during the convalescent period after diarrhea may be of importance in reducing the long-term nutritional impact of diarrheal illnesses. In order to replace nutrients not received during diarrhea and to compensate for ongoing subclinical malabsorption, dietary intake should be greater than normal during convalescence after diarrhea. Older infants (above 4 to 6 months of age) who are exclusively breast-fed prior to illness may require the initiation of complementary feeding to achieve potential growth rates during convalescence and thereafter (WHO, 1985). This high nutritional requirement may be fulfilled by providing meals of high nutrient density (such as fish or meat and cereals) and/or increasing the number of meals offered per day. The adequacy of food intake can be assessed by monitoring the child's weight and length at intervals (e.g., about every 3 months) during convalescence (WHO, 1985).

Foods to be avoided are (1) high fiber bulky foods such as coarse fruits and vegetables, fruit and vegetable peels, and whole grain cereals because they are hard to digest; (2) very dilute soups because, although recommended as fluids, they are not sufficient as foods (and serve only to fill the child without providing sufficient nutrients) and contain very high concentrations of sodium; and (3) foods with a lot of sugar and carbonated beverages since these have high osmolality (because of very high sugar content) which may worsen diarrhea (Wendland, et al., 1979; Snyder, 1982; Dibley, et al., 1984; Mackenzie and Barnes, 1988).

B. NUTRITION FOR THE CHILD WITH CHRONIC DIARRHEA

Chronic diarrhea with prolonged mucosal injury is always associated with malabsorption. Early in the course of diarrhea in an otherwise healthy and nutritionally adequate infant, the standard humanized formula may be tried first. If diarrhea persists for more than 2 weeks in early infancy, one should avoid multiple formula changes by introducing an elemental or semielemental diet. Severe disaccharidase deficiencies and villous atrophy necessitate a trial of an elemental diet. If that fails, parenteral nutrition should be instituted. Worsening of the diarrhea and poor weight gain are reasons to re-evaluate the type of formula being provided.

During mucosal injury, lactase is the first enzyme to be affected and the last to recover. This is the rationale for the use of short-chain polymers of glucose. The studies in our laboratory showed that short-chain glucose polymers of rice comprising two to nine glucose units are hydrolyzed and absorbed in the small intestine faster than isocaloric D-glucose (Azad et al., 1990), and that rice glucose polymers are more rapidly hydrolyzed and absorbed than commercially available corn glucose polymers (Sloven et al., 1990). The benefit of using short-chain glucose polymers is that it has much less osmolarity for the same calorie content of D-glucose.

Although medium-chain triglyceride (MCT—fatty acids vary from 6 to 12 carbons in length) is very advantageous especially in conditions with fat malabsorption, it is recommended for use only in combination with long-chain triglyceride to prevent essential fatty acid deficiency (usually 60:40 ratio). Our studies also showed that infants with protracted diarrhea may be able to tolerate a higher fat intake than is normally provided (Jirapinyo et al., 1990).

During chronic diarrhea, there is an increased intestinal permeability to proteins which can cause dietary protein hypersensitivity to cow's milk protein or soy protein. Therefore, exclusion is recommended during this period. Furthermore, it is preferable to add short polypeptides rather than free amino acids.

The nutritional management of the child with chronic diarrhea includes specific consideration of nucleotides which play an important role in major cellular function, carnitine, vitamin and trace elements.

C. INTRACTABLE DIARRHEA OF INFANCY

This is seen in infants <3 months of age, being characterized by >5 diarrhea stools/24 hours lasting >2 weeks. Here too, correction of dehydration, acidosis, and provision of nutrition are important. If oral therapy is not possible because of emesis, intravenous rehydration and total parenteral nutrition may need to be provided. Feeding should be started as soon as possible, since "resting the gut" perpetuates mucosal atrophy and does not affect intestinal healing. Early continuous nasogastric feeding with human milk or an elemental, isotonic formula may improve weight gain and shorten the duration of diarrhea. Intravenous albumin may need to be given to children with hypoalbuminemia (<2.0g/dl).

II. NUTRITION FOR THE CHILD WITH A GI DISEASE

Congenital (Infantile) Hypertrophic Pyloric Stenosis—In addition to the typical features of projectile vomiting, dehydration, and metabolic abnormalities, the infant may lose weight to a level below that at birth. After surgical relief and correction of metabolic imbalances, oral feedings are begun 4 to 6 hours postoperatively, starting first with 5% glucose in saline 4 ml hourly, then increasing the volume gradually. Normal feedings (breast milk or formula) can be started on the day after the operation when the baby is usually able to be discharged.

Hirschsprung Disease (Congenital Megacolon)—In addition to the typical neonatal obstruction and chronic constipation in the older child, there is often failure to thrive and, sometimes, diarrhea. Here, too, after surgical correction, nutrition rehabilitation is necessary.

Inflammatory Bowel Disease—Children with Crohn's disease have failure to thrive and frequent diarrhea in addition to chronic abdominal pain and other constitutional symptoms. Long-term steroid therapy also introduces nutritional complications such as osteoporosis and growth faltering. Disease activity may need to be arrested by use of total parenteral nutrition or by use of an elemental diet infused by nasogastric tube, particularly in children with delayed growth and those with enteric fistulas.

Food Allergy (Dietary Protein Intolerance)—Prolonged breastfeeding reduces the likelihood later of cow's milk intolerance, but the mother may have to go on a cow's milk elimination diet. Various non-milk-containing formulas such as soy feedings, hydrolyzed milk protein feedings, etc., are available. Sodium cromoglycate may afford suppression of intestinal symptoms in some cases and allow continued ingestion of the food.

Malabsorptive Disorders—These are characterized by defective assimilation of ingested nutrients, presenting as abdominal distension, chronic diarrhea, wasting of muscles, and retarded growth and weight. Common diseases causing malabsorption comprise of the following: cystic fibrosis, protein calorie malnutrition (PCM), chronic pancreatitis, biliary atresia and other cholestatic states, stagnant loop syndrome, massive resection, congenital short gut, chronic giardiasis, immune deficiency, celiac disease, tropical sprue, and idiopathic diffuse mucosal lesions of the small intestines.

Particularly for celiac disease, a gluten-free diet (avoidance of all wheat, rye and barley) is important. Extra fat-soluble vitamins are necessary, and iron- and/or folate-deficient patients need appropriate supplements.

REFERENCES

1. Azad, M. A. K. and Lebenthal, E. L., Role of rat intestinal glucoamylase in glucose polymer hydrolysis and absorption, *Pediatr. Res.* 28:166–179, 1990.
2. Brown, K. B., Gastanaduy, A. S., Saavedra, J. M., Lembcke, J., Rivas, D., Robertson, A. D., Yolken, R., and Sack, R. B., Effect of continued oral feeding on clinical and nutritional outcomes of acute diarrhea in children, *J. Pediatr.*, 112, 191–200, 1988.
3. Bowie, M. D., Effect of lactose-induced diarrhea on absorption of nitrogen and fat, *Arch. Dis. Child.*, 50, 363–366, 1975.
4. Brown, K. B. and MacLean, N. C., Nutritional management of acute diarrhea: an appraisal of the alternatives, *Pediatrics,* 73, 119, 1984.
5. Candy, C. E., Recent advances in the care of children with acute diarrhea: giving responsibility to the nurses and parents, *Journal of Advanced Nursing,* 12, 95–99, 1987.
6. Cash, R. A., A history of the development of oral rehydration therapy, in *Symposium Proceedings—Cereal-based Oral Rehydration Therapy: Theory and Practice,* Northrup, R. S., Ed., *J. Diarr. Dis. Res.*, 5(4), 256–261, 1987.
7. Chung, A. W., The effect of oral feeding at different levels on the absorption of food stuffs in infantile diarrhea, *J. Pediatr.*, 33, 1–13, 1948.
8. Chung, A. W. and Viscorova, B., The effect of early feeding vs. early oral starvation on the course of infantile diarrhea, *J. Pediatr.*, 33, 14–22, 1948.
9. Granger, D. N. and Brinson, R. R., Intestinal absorption of elemental and standard enteral formulas in hypoproteinemic (volume expanded) rats, *J. Parent. and Ent. Nutr.*, 12(3), 278–281, 1988.
10. Greenough, W. B., III, "Super ORT" (editorial), *J. Diarr. Dis. Res.*, 1, 74–75, 1983.
11. Greenough, W. B., III, Status of cereal-based oral rehydration therapy, *J. Diarr. Dis. Res.*, 5(4), 275–278, 1987.
12. Guerrant, R. L. and McAuliffe, J. R., Special problems in developing countries, in *Infectious Diarrhea,* Gorbach, S. L., Ed., Boston, Blackwell Scientific Publications, 1986, 287–301.
13. Hagihira, H., Ogata, M., Takedatsu, N., et al., Intestinal absorption of aminoacids IV. Interference between amino acids during intestinal absorption, *J. Biochem.*, (Tokyo), 47, 139–143, 1960.
14. Hoyle, B., Yunus, M. D., and Chen, L. C., Breast feeding and food intake among children with acute diarrheal disease, *Am. J. Clin. Nutr.*, 33, 2365–2371, 1980.
15. Isolauri, E., Vesikari, T., Saha, P., and Viander, M., Milk vs. no milk in rapid refeeding after acute gastroenteritis, *J. Pediatr. Gastroenterol. Nutr.,* 5, 254, 1986.
16. Jirapinyo, P., Young, C., Srimaruta, N., Rossi, T. M., Cardano, A., and Lebenthal, E., High-fat semielemental diet in the treatment of protracted diarrhea of infancy. *Pediatrics,* 86, 902–908, 1990.
17. Khin-Maung-U, In vitro determination of intestinal aminoacid (14-C-L-Glycine) absorption during cholera, *Am. J. Gastroenterol.,* 81, 536–539, 1986.
18. Khin-Maung-U, Nyunt-Nyunt-Wai, Myo-Khin, Mu-Mu-Khin, Tin-U, and Thane-Toe, Effect on clinical outcome of breast feeding during acute diarrhea, *Brit. Med. J.*, 290, 587–589, 1985.
19. Khin-Maung-U, Nyunt-Nyunt-Wai, Myo-Khin, Mu-Mu-Khin, Tin-U, and Thane-Toe, Effect on clinical outcome of boiled rice feeding in childhood cholera, *Human Nutrition: Clinical Nutrition,* 40(C), 249–254, 1986.
20. Kielmann, A. A. and McCord, C., Home treatment of childhood diarrhea in Punjab villages, *J. Trop. Pediatr. Environ. Child. Hlt.*, 23, 197–201, 1977.
21. Martorell, R., Yarbrough, C., Yarbrough, S., et al., The impact of ordinary illnesses on the dietary intakes of malnourished children, *Am. J. Clin. Nutr.*, 33, 345–350, 1980.
22. Mitchell, J. E., Donald, W.D., and Birdsong, M., A review of 396 cases of acute diarrhea in which early oral feedings were employed in treatment, *J. Pediatr.*, 35, 529–539, 1949.

23. Molla, A. M., Molla, A., Nath, S., and Khatun, M., Food-based oral rehydration salt solutions for acute childhood diarrhea, *Lancet,* (ii), 429–431, 1989.
24. Molla, A. M., Molla, A., Rhode, J., and Greenough, W. B., III, Turning off the diarrhea: the role of food and ORS, *J. Ped. Gastroenterol. Nutr.,* 8, 81–84, 1989a.
25. Molla, A., Molla, A. M., Sarker, S. A., et al., Food intake during and after recovery from diarrhea in children, in *Diarrhea and malnutrition: interactions, mechanisms, interventions,* Chen, L. C. and Scrimshaw, N. S., Eds., New York, Plenum Press, 1983, 113–124.
26. Molla, A., Molla, A. M., Sarker, S. A., et al., Effect of diarrhea on absorption of macronutrients during disease and after recovery, in *Diarrhea and malnutrition: interactions, mechanisms, interventions,* Chen, L. C. and Scrimshaw, N. S., Eds., New York, Plenum Press, 1983a, 143–154.
27. Newey, H. and Smyth, D. H., Transfer system for neutral amino acids in the rat small intestines, *J. Physiol.* (London), 170, 328–342, 1964.
28. Newey, H. and Smyth, D. H., Specificity of carriers in the intestinal transfer of glycine, *J. Physiol.* (London), 171, 74–75, 1965.
29. Powanda, M. C., Changes in body balances of nitrogen and other key nutrients, *Am. J. Clin. Nutr.,* 30, 1254, 1977.
30. Rahaman, M. M. and Wahed, M. A., Direct nutrient loss in diarrhea, in *Diarrhea and malnutrition: interactions, mechanisms and interventions,* Chen, L. C. and Scrimshaw, N. S., Eds., New York, Plenum Press, 1983, 155–160.
31. Rhode, J. E., Preparing for the next round: convalescent care after acute infection, *Am. J. Clin. Nutr.,* 31, 2258–2268, 1978.
32. Rhode, J. E., Selective primary health care: strategies for control of disease in the developing world. XV. Acute diarrhea, *Rev. Inf. Dis.,* 6, 840–854, 1984.
33. Santosham, M., Foster, S., Reid, R., Bertrando, R., Yolken, R., Burns, B., and Sack, R. B., Role of soy-based, lactose-free formula during treatment of acute diarrhea, *Pediatrics,* 76, 292–298, 1985.
34. Scrimshaw, N. S., Taylor, C. E., and Gordon, J. E., *Interactions of nutrition and infection,* Geneva, WHO, WHO Monograph Ser No. 57, 1968.
35. Sloven, D. G., Jirapinyo, P., and Lebenthal, E., Hydrolysis and absorption of glucose polymers from rice compared with corn in chronic diarrhea of infancy, *J. Pediatr.,* 117(3), 502, 1990.
36. Taylor, C. E. and Greenough, W. B., III, Control of diarrheal diseases, *Annu. Rev. Public Health,* 10, 221–244, 1989.
37. World Health Organization, Diarrheal Diseases Control Programme, *WHO Wkly Epidem. Rec.,* 54, 121–123, 1979.
38. World Health Organization, Diarrheal Diseases Control Programme, *Guidelines for the production of oral rehydration salts,* Geneva, 1980.
39. World Health Organization, Recent advances in research on feeding during and after acute diarrhea, Report of the Scientific Working Group on Drug Development and Management of Acute Diarrhea, October 1–3, 1985, CDD/DDM/85.2, WHO CDD Programme, Geneva.

Chapter 12

DIARRHEA AND MALNUTRITION

Andrea Maggioni and Fima Lifshitz

TABLE OF CONTENTS

I. Introduction 107

II. Risk Factors in the Development of Enteric Infections 109
- A. Fecalism 109
- B. Nutritional Practices 110
- C. General Host Factors 110
- D. Immune Function 112

III. Interaction of Diarrhea and Infant Nutrition 113

IV. Effects of Infectious Diarrhea and Malnutrition on the Intestine 114
- A. Infectious Diarrhea 115
- B. Bacterial Overgrowth 116
- C. Malnutrition 117

V. Alteration in Digestion and Absorption in Malnourished Children with Diarrhea 119

VI. Therapeutical Considerations 122

VII. Conclusions 126

References 127

I. INTRODUCTION

The major cause of diarrhea is infection.[1] The morbidity of infectious diarrhea in less-developed countries includes infections with many organisms, the most frequent being rotavirus, campylobacter, enterotoxigenic enteric bacteria, shigella, salmonella, giardia and other parasites (Table 1). These infections account for approximately 750 million episodes of diarrhea per year in children and for approximately 5 million deaths per year among the 3 billion people living in Africa, Asia, and Latin America. Each of the world's 338 million children under the age of 5 years suffers at least two or three episodes of diarrheal illnesses per year. In

0-8493-2764-4/95

TABLE 1
Estimated Morbidity of Infectious Diarrhea in Children 0 to 4 Years Old in Latin America[a,b]

Etiology	Relative frequency (%)	Millions of cases
Rotavirus	10–40	40–160
Enterotoxigenic Escherichia coli	10–30	40–120
Shigella sp.	5–20	30–80
Campylobacter jejuni	8–10	30–40
Giardia lamblia	5	20
Cryptosporidum sp.	5	20
Enteric adenovirus	2–4	8–16
Salmonella sp.	2–4	8–16
Entamoeba histolytica[c]	1–3	4–12
Other	1	1

[a] Modified from Mata, L., Pediatr. Infect. Dis., 5, 117, 1986.
[b] Estimated for 50 million children 0 to 4 years old.
[c] Dientamoeba, Isoospora, Blastocystis, and Balantidium are also included.

the developing world, over 7% of them die.[3–4] In other words, every day diarrhea kills 12,600 children less than 5 years of age. A child living in the poorest areas of the world may have 3 to 10 episodes of diarrhea per year in the first 5 years of life, for a total of 15 to 50 episodes averaging 4 to 6 days each.[5]

Prospective field studies using intensive surveillance techniques have found that 3 to 20% of the diarrhea episodes persist for more than 2 weeks. The incidence of prolonged diarrhea varies greatly in different developing countries, ranging from 11 to 69 episodes per 100 child-years.[6–7] Children who have once had prolonged diarrhea have a three-fold or greater risk of subsequent prolonged diarrheal episodes than the general population.[8] The incidence of prolonged diarrhea generally parallels that of all diarrheal illnesses. Thus, young children, who are especially prone to develop acute diarrhea, are also more likely to develop persistent illness. Similarly, in countries where diarrhea occurs seasonally, the incidence rates of both acute and prolonged diarrhea tend to be elevated at the same time of the year.

In the United States and in other developed countries, diarrheal disease is also an important problem.[9–10] In the U.S. 16.5 million children less than 5 years of age have between 1.3 and 2.3 episodes of diarrhea per year, or 6.5 to 11.5 episodes per child during the first 5 years of life. Although most of these episodes are not severe, about 1.4% of young children require hospitalization annually.[11] A recent ten years' survey demonstrated that about 500 American children, almost all of them from families of low income, died each year of diarrheal disease.[12] In Australia about 10,000 patients are hospitalized, and 20 to 30 deaths occur annually due to gastroenteritis in children.[13] There are no prospective field studies in the U.S. or other developed countries regarding the prevalence of prolonged diarrhea following gastroenteritis, but in a 1985 mail survey among pediatricians, it was

TABLE 2
Incriminating Factors for the Development of Enteric Infections

Fecalism—Contaminated environment
- Increased dose exposure
- Increased frequency of infective agents

Nutritional practices
- Lack of breastfeeding
- Commerciogenic feeding
- Poor hygiene

General host factors
- Nutritional status
- Age
- Antibiotics and other medications

Local host factors
- Gastric acidity
- Bile salts
- Intestinal tonicity and motility
- Antibodies and flora
- Lactoferrin and lysozymes
- Intestinal susceptibility to injury

Immune function
- Immunoglobulins
  - IgG and IgE
  - Secretory IgA, IgM
- White blood cells
  - T lymphocytes
  - Macrophages
  - Mast cells
  - Polymorphonuclear leukocytes

revealed that an average of 13 patients with severe chronic diarrhea in infancy were seen by each pediatrician annually. Therefore, it could be calculated that up to a quarter of a million infants in the United States per year may have severe chronic diarrhea following gastroenteritis.[14]

II. RISK FACTORS IN THE DEVELOPMENT OF ENTERIC INFECTIONS

A. FECALISM

The relationship between the high frequency of enteric infections and poor sanitary living conditions is well known (Table 2). Under these circumstances, there is "fecalism" that leads to an increased exposure to higher doses of infectious agents as well as to an increased frequency of contact with organisms that produce diarrhea. Hand washing before preparing food and feeding children has been identified as an infrequent practice in poor communities. In Nigeria, hand washing with soap was observed in less than 1% of food preparation and feeding episodes among children under five years of age. Hand rinsing with water was a more common practice, observed before 27% of the feeding episodes.[15] In Brazil, the following environmental and behavioral risk factors leading to diarrhea were identified:[16] living in a house with an earthen floor, dirty floor, or with indoor animals; lack of tap water inside the house; lack of refrigerator and food kept at room temperature. Other important relative risk factors included siblings not dressed and the presence of alcohol abuse and mental illness in the households. It is estimated that in certain settings, the promotion of improved personal and domestic hygiene could reduce the incidence of diarrheal episodes by up to 48%.[17]

B. NUTRITIONAL PRACTICES

Of particular importance in the prevalence of diarrhea in infants is the abandonment of the practice of human milk feedings. Breastfeeding not only protects infants from diarrhea, especially when no other foods or liquids are offered in addition to breast milk,[18] but also when the diarrhea occurs, the duration of the illness tends to last less among young infants who are exclusively breast-fed.[19] The beneficial effects of human milk feeding have been related to different mechanisms including decreased contamination. In the unsanitary conditions associated with poverty when human milk is not available, contaminated cow's milk is usually fed to babies. Under these circumstances, cow's milk kept without refrigeration and exposed to flies and poor hygiene becomes a culture medium for bacteria to grow, multiply, and infect the baby.[20] Also human milk feedings are given as continuous feedings, whereas bolus meals are characteristic of bottle feedings. Hospitalized children with chronic diarrhea have reduced purging rates when they receive continuous feedings rather than bolus meals.[21] In addition, human milk has many anti-infective factors such as lactoferrin, lysozyme, and secretory IgA.[22–24] It also includes several known growth factors: epithelial growth factor (EGF), nerve growth factor (NGF), transforming growth factor (TGF),[25] and insulin-like growth factor.[26] These factors may have a role in mucosal repair by their trophic effects on the intestine.[27,28] Therefore, human milk is the single most important nutritional practice which can positively influence diarrheal disease in infancy.

C. GENERAL HOST FACTORS

There are general host factors that may facilitate infection. It is well known that an increased morbidity and mortality from diarrhea is more likely to occur among infants than among older children. It has long been shown that infants less than 3 months of age are particularly susceptible to develop acute gastroenteritis and some of its severe complications such as postinfectious chronic diarrhea and acquired monosaccharide intolerance,[29] multiple food intolerance,[30] and pneumatosis intestinalis.[31]

Hemorrhagic colitis caused by E. Coli 0157:H7 was first recognized in 1982 during outbreaks in Oregon and Michigan.[32] A very young age was a significant risk factor among the reported risk factors for diarrhea caused by this microorganism, other than ingesting contaminated food or water and person-to-person spread. In a recent survey, the risk of infection in day care settings was inversely related to the age of the children who attended.[33] Similarly, the severity of symptoms and the risk of developing serious complication as hemolytic uremic syndrome was higher in children less than five years of age.[34]

In a recent survey among children of a previously healthy population at mountain camps along the Turkey-Iraq border during the 1991 Kurdish refugee crisis, it was evident the rapid progression of morbidity and mortality occurred among young children.[35] Within days in times of disaster, and despite a large-scale emergency relief effort which was promptly started during the initial phase of the crisis, younger children were severely affected by diarrheal disease and nutritional deterioration. They experienced higher mortality and were more likely to develop prolonged diarrhea with subsequent malnutrition.[36,37] There were 199 children aged 5 years or younger who died; the leading cause of death was diarrheal disease (totaling 73%), leading to dehydration and malnutrition. Malnutrition following diarrhea also affected children younger than two years disproportionately, with high mortality rates. This tragic experience reinforces the importance of prevention and underlines the necessity of a prompt intervention in the management of diarrheal disease. Young children need to be cared for first!

Among the general host factors influencing diarrheal disease, the nutritional status of the host is a very important one.[38] Epidemiological evidence from studies in developing countries indicates that preexisting malnutrition, defined by different anthropometric criteria, is associated with an increased incidence of diarrhea, a longer duration of the illness, or both.[39–41] Clinical studies on patients hospitalized for cholera similarly noted that malnourished patients had longer lasting illnesses.[42] Experimental studies in infected animals demonstrated prolonged disease in the malnourished groups vs. the control group.[43]

Gastric acidity is altered in malnutrition, and this alteration may be a very prominent factor that determines the viability of organisms reaching the small intestine and producing disease.[44] In addition to decreased gastric acidity, in malnutrition, gastrin levels are low.[45] Therefore, these infants are not only exposed to increased dosages of infective bacteria, but many more organisms are viable when they reach the small intestine. Viable enterobacteriaceae can be recovered from the gastric juice of malnourished children with low gastric acid secretion; in contrast, these potential enteropathogens are absent from well-nourished children.[46] Furthermore, gastric secretions play an important role in the denaturation and hydrolysis of food antigens.[47,48] Also, the gastric emptying in malnourished infants is delayed with gastric dilatation and a tendency to vomit.[49] The stagnation of gastric fluids may add to the opportunities for bacteria and toxins to proliferate and infect the patient.

Antidiarrheal agents of the type used to inhibit intestinal motility may also facilitate the penetration of infectious organisms by extending the time of exposure of bacteria in the intestine.[50] Additionally, previous treatment with antibiotics might facilitate the development of enteric infection because of overgrowth of bacteria not susceptible to antibiotic treatment or diarrhea due to C. difficile.[51]

D. IMMUNE FUNCTION

Although there are many causes of increased risk of infection in underprivileged, malnourished communities, impaired immunocompetence is now recognized to be an important underlying factor. Because of its widespread occurrence, nutritional deficiency is the most common cause of immunodeficiency worldwide.[52] It should be kept in mind that the widespread and increased prevalence of AIDS may also play a role in immunodeficiency and chronic diarrhea leading to malnutrition in children.[53,54] In primary as well as in secondary malnutrition, the compromised immune function may facilitate the development of gastrointestinal infections. An organism that may be relatively harmless to the well-nourished child may give rise to a severe or even fatal infection in the malnourished child with compromised immunocompetence.[55]

In marasmus and kwashiorkor, lymphoid tissues show a significant atrophy, the size of the thymus is small, there is loss of corticomedullary differentiation, and the Hassal bodies are enlarged, degenerated, and occasionally calcified. In addition, the concentration of thymulin in the thymus of malnourished children has been found to be decreased.[56] Thymic atrophy, which occurs as a result of protein, energy, vitamin, and mineral deficiencies, especially zinc, is associated with a decreased production of thymic hormone, an underlying T-cell deficiency, and an increased susceptibility to disseminated infections.[57]

In the spleen, there is loss of lymphoid cells around small vessels. In the lymph node, the thymus-dependent areas show depletion of lymphoid cells.[58] Delayed cutaneous hypersensitivity responses, both to recall and to respond to new antigens, is markedly depressed; it is common to have complete anergy to a battery of different antigens. There is also a correlation between the degree of immune impairment and the degree of weight loss, with alterations in cellular immunity being normalized as nutritional status improves.[59]

The number of circulating lymphocytes are decreased in malnourished children, particularly the helper phenotype, which is replaced by a proportionate increase in ''null'' lymphoid cells. However, another study reported that not all malnourished patients had decreased circulating T cells.[60] This suggests that specific nutrient deficits associated with malnutrition may be important etiologic factors in the depressed cellular immune response.

The number of B cells are normal or elevated,[61] accounting in part for the normal or elevated immunoglobulin titers. Many authors have speculated that the elevated immunoglobulin levels may be secondary to an increased exposure to various infectious agents to which the malnourished child may not be able to respond. This may not mean that the humoral system is intact; what may be decreased is the antibody affinity,[62] thus accounting for a higher frequency of antigen antibody complexes found in such patients.[63] A significant decrease in secretory IgA and secretory component levels associated with malnutrition has also been reported.[64]

The phagocytosis is also affected in malnutrition. The ingestion of particles by phagocytes is intact, and the subsequent metabolic activation and destruction of

TABLE 3
Effects of Diarrhea on Nutrition

Anorexia
Withdrawal of food
Adverse effects of treatment on nutrition
Impaired absorption
Increased losses
  Stools
  Urine
Increased requirements

bacteria is reduced.[65] Individual complement proteins and hemolytic complement activity in malnourished children are significantly depressed.[66] The increased morbidity-mortality associated with infection and malnutrition is most likely the result of the effect of malnutrition on the immune response. Therefore, in order to combat potentially lethal infections in undernourished children, both antibiotic use and prompt nutritional therapy are essential.

III. INTERACTION OF DIARRHEA AND INFANT NUTRITION

The relationship between infection and malnutrition also encompasses other effects beyond the increased susceptibility to infection. The metabolic consequences of infectious disease imposed by severe gastroenteritis, otitis media, pneumonia, or measles are critical in the development of marasmic-kwashiorkor or kwashiorkor. The catabolic mechanism by which infection leads to a malnourished state include anorexia; replacement of solid foods with a low-energy, low-protein diet; and increased urinary losses of nitrogen, potassium, magnesium, zinc, phosphate, sulfur, and vitamins A, C, and B2[67] (Table 3).

Anorexia is present even with mild subclinical infections,[68] and the loss of appetite makes it very difficult to maintain a constant intake during the acute phase of the illness, although breast milk is rejected to a lesser extent. The anorectic effect of infection may last a few hours or extend for days or weeks. As much as 20 to 70% of the available food may be uneaten during bouts of diarrhea.[69] In addition, when diarrhea strikes, there is usually restriction of solid food intake, especially feedings of animal origin which are replaced with a low-energy, low-protein diet dictated by medical or popular traditions, beliefs, or taboos to "treat" the diarrhea.[20]

Infection is associated with increased urinary nitrogen excretion, resulting mainly from an increased mobilization of amino acids from peripheral muscle for gluconeogenesis in the liver. There is also deamination and excretion of nitrogen in the form of urea; consequently, there is a decrease in whole-blood amino acids after exposure to an infectious agent.[70] In addition to increased gluconeogenesis, there is an increased diversion of amino acids for the synthesis of acute phase proteins as haptoglobin, C-reactive protein, alpha-1-antitrypsin, and alpha-2-macroglobulin, accompanied by a concomitant decrease in visceral protein synthesis.[71] Unless there is an increased dietary intake, the loss of nitrogen will not be compensated for, and a kwashiorkor-like syndrome will result. A major characteristic in the development of kwashiorkor is the decrease in serum albumin, which is closely associated with the appearance of infection. During the infectious process, there is also sequestration or diversion of iron,[72] copper,[73] and zinc[74] from normal metabolic pathways.

Each of the above-mentioned changes associated with infection is felt to be mediated by tumor necrosis factor (TNF) and interleukin-1 (IL-1).[75] Two different types of TNF have been described: TNF-α or cachectin, produced by monocyte macrophages, and TNF-β or lymphotoxin, produced by lymphocytes. Although both TNF-α and TNF-β have similar activities and bind to the same cell receptors, they are distinct molecules that share about 45% homology on the nucleotide level.[76] TNF has been shown to be a potent mediator of systemic metabolic changes which are potentially deleterious to the host.[77] TNF and IL-1 have also been implicated in the acute phase response, which is a reaction produced by the systemic and clinical changes induced during the early stages of an infection or after trauma. An increased TNF production by peripheral blood mononuclear cells has been reported in patients with gram negative septic shock,[78] during acute starvation,[79] and in patients with anorexia nervosa.[80] The recent availability of recombinant TNF (rTNF) and monoclonal antibodies specific to TNF are of great interest, since they may improve the outcome of gram negative sepsis.[81] A better understanding of the role of TNF in patients with acute and chronic infections and its effects on the nutritional status of the host will lead to improved management of infectious diarrhea and its consequences.

In patients with diarrhea, there is additional nutrient malabsorption due to the infectious process and intestinal injury. The presence of severe dehydration and shock induced by diarrhea may in itself induce or exacerbate the intestinal injury of the infection. This may lead to a more prolonged illness, and therefore the patient may be more likely to develop malnutrition. It is now known that early administration of oral hydration solutions to infants with diarrhea results in a better outcome and a more rapid recovery.[82] The presence of primary or associated systemic and/or respiratory infections may lead to parenteral diarrhea and may also contribute to prolongation of the illness,[83] thereby alterating nutritional status of the infant.

IV. EFFECTS OF INFECTIOUS DIARRHEA AND MALNUTRITION ON THE INTESTINE

The majority of the clinical studies of intestinal alterations seen in malnutrition originate from less-developed countries. There the problem is not only the lack of adequate amounts of nutrient intake but also of repeated intestinal infections caused by the heavy microbiological contamination of the environment.[84,85] Enteric infections, chronic environmental enteropathy, and generalized or specific nutrients deficiency are additional factors that contribute to the genesis of the changes present in malnourished children.[86,87] Ideally, one would like to study the effects on the intestinal epithelium of "pure" malnutrition and of factors causing "pure" persistent diarrhea without malnutrition, but in clinical settings, malnutrition and diarrhea are so intimately interconnected that this is not possible.

The pathogenesis of both diarrhea and malnutrition is characterized by the injury to the small intestinal mucosa.[88] It should be pointed out that small bowel lesions may be present before intestinal infection in malnourished infants[89] and may persist after disappearance of the initial infection leading to diarrhea.[90] However, even though there is a strong correlation between the intestinal mucosal injury and the presence and severity of diarrhea, it should be kept in mind that diarrhea can occur without any apparent intestinal mucosal lesion being demonstrated by microscopy or electron microscopy.[91]

A. INFECTIOUS DIARRHEA

In viral diarrhea, there are morphologic and functional changes in the jejunum resulting from penetration of the rotavirus into the intestinal cells. Viral mucosal invasion is associated with loss of the mature absorptive cells, producing a proliferative response in the crypts that results in a repopulation of the intestinal epithelial lining with poorly differentiated crypt cells. Rotavirus infection commonly involves the jejunum, but it can be more diffuse with patchy involvement of the entire small bowel.[92] The extent of the lesions determines the loss of absorptive function, the loss of brush-border membrane disaccharidase activity, and consequently, the severity of the patient's diarrhea. Giardia adhere to the surface of enterocytes whereas Cryptosporidium lodge under the microcalix but outside the cytoplasm.[93] The exact pathogenetic mechanism by which these two enteropathogen parasites cause diarrhea is unknown, but different mechanisms can be involved; they can act as a mechanical barrier to absorption or also directly injure the intestinal mucosa, resulting in accelerated turnover. In addition, they can release parasitic exotoxins or elicit an immunologic response to the parasite by the host.[94] Classic enteropathogenic E. coli organisms tightly adhere to the mucosal surface of both the small and large bowel,[95,96] binding to intestinal epithelial cells, indenting the mucosal surface, and causing dissolution of the glycocalix and flattening of the microvilli. Adherence of E. coli to the epithelium damages the brush-border membrane and decreases the effective absorptive surface area of the intestine. Similar lesions have been seen with cytotoxin-elaborating enterohemorrhagic E. coli.

The dysentery diarrheas are more severe because they are often accompanied by a protein-losing enteropathy and induce toxic manifestations. Diarrhea that results from entero-toxins generally preserve the integrity of the intestinal mucosa, but malabsorption may result from the secretion of water and electrolytes stimulated by the toxins. This leads to dilution of bile salts to a concentration below the critical micellar level, with consequent malabsorption.[97,98] Hypersecretion is associated with alterations in fatty acids, hormones, and neurotransmitters, as well as greater calcium cell permeability induced by mediators.[99] In the case of Shigella, the cytotoxin is a potent inhibitor of protein synthesis as well as a secretagogue and neurotoxins.[100] Another example of cytotoxic mucosal damage is colitis associated with antibiotic use and colonic overgrowth of cytotoxin-producing C. difficile.[51]

In chronic diarrhea of infancy associated with acquired monosaccharide intolerance, the villous atrophy lesions were patchy, the crypt length was increased, and the villi were decreased in size. Many bacteria were observed in contact with the enterocyte surface of the jejunal mucosa. Some of them were overlying the microvilli; others were adherent.[101] It has been postulated that bacteria, not only of ETEC or EPEC type, may be central to the pathogenesis of protracted diarrhea of infants by causing widespread histological and ultrastructural changes which are not confined to the enterocyte surface. The persistence of the bacteria on the surface of the mucosa may be related to the alteration of the immune system, which in turn could be influenced by nutritional status of the host.[102] Several studies have shown that enteroadherent E. coli, especially those with an aggregative pattern of adherence to HEp-2 cells, and cryptosporidium have an unusual capacity to cause persistent diarrhea. They may be associated with approximately half of the cases of persistent diarrhea. This is especially true for children with pre-existing malnutrition who are infected with cryptosporidium.

B. BACTERIAL OVERGROWTH

In malnourished patients, there may also be an alteration in the normal populations of bacterial flora in the intestine that would elicit a strong homeostatic influence on potential pathogenic organisms either by competition for space and nutrients or by elaborating antibacterial metabolites. Small bowel bacterial overgrowth occurs frequently in postinfectious and other diarrheal entities.[103] Bacterial counts in malnourished Third World children can be in the range of 10^7–10^8 bacteria/ml in the jejunum, much greater than the 10^4–10^5 organism/ml found in healthy children. This is likely the result of increased exposure and a higher load of viable organisms as well as a consequence of impairment of IgA secretion, reduced mucosal disaccharidase and mucin production, and mucosal barrier disruption.

It has been suggested that proliferation of enterobacteriaceae and/or anaerobes in the duodenum of some children with acute diarrhea determines whether the illness becomes persistent.[104] Bacterial proliferation may prolong the diarrhea and may correlate with the degree of carbohydrate intolerance.[105] However, a recent review of published studies and the comparison of cultures of duodenal aspirates from Peruvian children with acute and persistent diarrhea and diarrhea-free children did not support this hypothesis. Although many children had enterobacteriaceae and/or anaerobes cultured, there was no correlation with clinical and nutritional outcome. The bacterial counts of these patients were relatively low, and higher numbers of fecal bacteria may produce and be associated with pathology. Age, nutritional status, the environment, and the etiology of the episode were determinants of the duodenal microflora independent of diarrhea.[106] However, under experimental conditions in rats, proliferation of normal fecal and colonic bacteria in the small intestine produced alterations in intestinal function that resemble the abnormalities in intestinal absorption of patients with monosaccharide intolerance.[107]

C. MALNUTRITION

The gastrointestinal tract is particularly susceptible to the effects of nutrient deprivation. The intestine is lined by highly differentiated epithelial cells, which have a high turnover rate. It handles large volumes of fluids and nutrients, and it must act as a barrier against foreign molecules. It also maintains a very complex equilibrium with the resident flora. Primary malnutrition may affect all these variables at the local and systemic levels, leading to the derangement of gastrointestinal functions.[49]

With the advent of the peroral jejunal biopsy technique, the histological features of the small intestine in a malnourished child have been extensively studied. Morphological studies show a striking spectrum of alterations ranging from severe damage to none at all. The picture that emerges from most of the studies is that in marasmus the small intestinal mucosa, although thin, was regarded to be virtually normal.[108] The mitotic index was found to be significantly decreased.[109] In kwashiorkor the mucosal architecture was found to be severely abnormal,[110] although the mitotic index was only moderately reduced.[111] As a result of these findings, it was postulated that cell renewal processes are more sensitive to calorie deficit rather than the overall availability of protein. It should be remembered that food in the intestinal lumen is one of the most important trophic factors which promoted cell renewal, while the decrease in mitotic activity associated with malnutrition may be explained, in part, by the decreased stimulation provided by the limited food intake, especially where anorexia is a feature of the illness.[112]

Ultrastructural studies with the electron microscope in malnourished children have demonstrated the presence of short, sparse microvilli and an increase of multivesicular bodies in the enterocytes.[110,113] It has been suggested that this increase of multivescicular bodies may be indicative of an enhanced absorption of antigenic material by the enterocytes.[114] In patients with kwashiorkor, fat droplets in the supranuclear cytoplasm of the enterocytes have been reported. This is probably due to alterations in the synthesis of apoproteins. This metabolic disturbance has also been proposed as the cause of fatty liver in this form of malnutrition.[115]

In experimental models in rats, it has been shown that the small intestine loses more weight than any other organ during restriction of protein and energy intake.[116] However, in contrast to the marked architectural disruption of villi seen in some clinical studies of malnutrition, the alterations in experimental malnutrition are much more moderate.[117–119] There is a variable degree of mucosal ''thinning'', generally more marked in distal regions of the intestine, although some thinning of even the gastric mucosa was seen in Rhesus monkeys. When villi are shortened, all layers including muscle are affected. Although epithelial cells may appear slightly less elongated than in controls, they are not cuboidal. None of the animal models have shown the severe flat sprue-like villus architecture identified in severe clinical malnutrition. We have shown in a model of protein-calorie malnutrition induced in post-weanling rats that histologic alterations are regional.[120] Thus, in the duodenum, muscle layers, and proximal jejunum, crypts and villi are indistinguishable from well-nourished controls, while in the distal jejunum villi, crypts and muscle are all diminished in thickness. These observations are consistent with the view that a greater availability of nutrient supply spares the most proximal regions of the small intestine. Ultrastructurally, even the distal jejunum that is histologically compromised contains jejunal epithelial cells that are essentially indistinguishable from well-nourished controls. Some epithelial cells do appear to have microvilli that are fused at their base at form twinned structures.[121] This alteration was found in 3–4% observed microvilli in pure protein and protein-energy malnutrition. The frequency of abnormal microvilli is minimal and may only play a small role in causing the reduced absorptive capacity of the intestine in malnutrition. Another study using quantitative stereological techniques in rats with protein-energy malnutrition demonstrated a 60% reduction of total jejunal absorptive surface which occurred despite any obvious change in villous length.[122]

Thus, studies on malnourished experimental animals that are apparently without infection only show modest signs of cellular and histologic damage. These data strongly suggest that infections greatly contribute to the marked structural damage noted in clinical malnutrition in children.

Intestinal permeability is also altered in malnutrition. This affects both the transcellular and paracellular transport mechanism and is dependent on molecular size, suggesting that the damage is probably selective.[123–125] This has to be differentiated from results obtained by means of absorptive tests, which show that the digestion and transport of many nutrients are altered in malnutrition.[126–128] There are data suggesting that bacteria and toxins may have an easier access and an increased ability to penetrate the intestine during malnutrition.[129–131] This means that molecules of different sizes may traverse the epithelium and eventually trigger immune response. Malnutrition affects not only the transport of macromolecules but also that of smaller molecules such as lactulose and mannitol. Intact dietary proteins may also pass into the blood owing to the intestinal damage that results from the infections that produce diarrhea and may sensitize a susceptible host and induce allergy.[132]

The intestine of experimentally malnourished rats shows an increased sensitivity to injury. Under experimental conditions, we have shown that a number of factors present in diarrheal disease might disrupt the intestinal epithelial barrier and allow the passage of macromolecules. Lactose malabsorption and deconjugated bile salts in rats permitted the passage of horseradish peroxidase, a large molecule of 40.000 molecular weight, into the jejunal intestinal epithelial cells.[124,133] There is an enhanced jejunal passage of macromolecules in malnourished rats exposed to pathophysiological concentrations of deoxycholate, and the epithelial cells show signs of ultrastructural damage.[134] These experimental data suggest that malnourished children may be at an increased risk for the intestinal absorption of potential antigens and toxins. Similarly, the jejunal cells of malnourished rats appear more sensitive to hyperosmolality than control rats. After a lactose-induced osmotic load, malnourished rats show an increased jejunal water secretion, mediated by cyclic AMP. There is also an increased sloughing off of epithelial cells in malnourished rats exposed to an osmotic load of lactose.[135]

There are few data regarding the length of time required for regeneration of a damaged intestine. Once offending agents have been eliminated, the time taken to repair the damage to the small intestinal mucosa might be expected to vary depending on the severity of the lesion and the ability of the crypt epithelium to divide and replace the villi. In children recovering from acute gastroenteritis, improvement in mucosal morphology was reported to occur within 8 days,[136] whereas in children with protracted diarrhea of infancy, improvement was seen within 2 to 3 weeks.[137] These observations, concerned with acute intestinal insults, contrast markedly with those of a study in Ugandan children with kwashiorkor in whom only incomplete mucosal repair was documented 12 months after diarrhea, despite clinical and biochemical recovery of malnutrition.[138] In persistent diarrhea accompanied by gross malnutrition, mucosal damage is clearly more refractory to treatment, and alterations in function and structure may continue despite an intensive rehabilitation regimen.

Overall, these data reinforce the concept that the intestinal alterations found in malnourished patients are not merely due to malnutrition alone but are probably the result of an interaction between malnutrition and the enteric infection which triggered diarrhea.

V. ALTERATION IN DIGESTION AND ABSORPTION IN MALNOURISHED CHILDREN WITH DIARRHEA

Malabsorption of nutrients in malnutrition is primarily related to pancreatic insufficiency which develops rapidly when nutrition is compromised.[139] Many studies show that malnutrition has profound structural effects on the exocrine pancreas. There is marked reduction in the number of zymogen granules, cell vacuolization, metaplasia, and cystic dilatation of the ducts. In advanced malnutrition, variable degrees of focal pancreatic fibrosis have been described which can progress to total replacement of the organ with fibrous tissue.[140] The volume of secretions and alkali production did not appear to be altered, but severely malnourished infants had deficient enzyme output. Amylase and chymotrypsin secretions are the most sensitive to malnutrition, decreasing by over 90% of control values. Lipase secretion is also severely altered in patients with kwashiorkor. Trypsin output is the least affected, decreasing by only 60% to 70% of control values.[141] The consequence would be impaired hydrolysis of starch, triglycerides, peptides, and proteins. Decreased digestion of proteins may increase the possibility of absorption of proteins with higher antigenic capacity. Alterations in pancreatic function appear transitory and return to normal as soon as nutritional recovery begins.[142]

Malnourished infants have a decrease in conjugated bile acids, which mainly affects the taurine conjugates.[143] Increased free bile acids in the intestinal lumen exert a negative effect on the formation of the micellar solution during fat absorption and is one of the main causes for the steatorrhea observed in these patients.[144] They also stimulate water secretion in the colon and thus increase the severity of the diarrhea. In turn, the presence of diarrhea and of an abnormal resident flora increases the amount of free bile acids.

The type of fat being consumed may also contribute to steatorrhea, as vegetable fats are better tolerated than animal lipids.[145] The medium-chain triglycerides (MCT), prepared from coconut oil, have a variety of unique features that make them an attractive nutritional substrate in the malnourished patient. Their low molecular weight and relative solubility in water are considered advantageous for digestion and absorption. Lingual and gastric lipase hydrolyze MCT at a rapid rate in comparison to long-chain triglycerides (LCT), and MCT may be directly absorbed through the gastric mucosa. Intraluminal hydrolysis of MCT is nearly complete and the medium-chain fatty acids are absorbed predominantly as free fatty acids, rather than the more slowly absorbed partial glycerides and mixed micelles characteristics of LCT hydrolysis. Of particular advantage is the unique ability of MCT to be directly absorbed by the intestinal mucosa in the setting of pancreatic and/or biliary dysfunction.[146] MCT provide a rapid, easily oxidized fuel source and are only minimally incorporated in body tissues. The rationale of their use in the treatment of diarrheal disease is to provide a high energy source of feeding. However, they do not meet essential fatty acid requirements; therefore, it is crucial that long-chain polyunsaturated fats are appropriately administered for normal growth and development.[147] The fecal energy loss caused by steatorrhea in diarrheal disease is transient and usually of little consequence, but can become critical if caloric intake is limited. Despite the physiological changes in fat digestion and absorption, fat can be offered to and utilized well by patients with malnutrition in practical recovery diets.

Carbohydrate malabsorption has been shown to contribute to the diarrhea in malnourished patients.[148] This complication is more frequent and more severe in malnourished infants, but it may also occur in well-nourished babies with mild diarrhea without dehydration.[149] It is often specific for lactose, but can affect other disaccharides such as sucrose or maltose. We documented that up to one-third of malnourished infants with severe diarrhea may also be unable to tolerate glucose polymers,[150] whereas well-nourished babies may tolerate these carbohydrates more easily.[151] At times carbohydrate intolerance in diarrheal disease may affect all dietary carbohydrates including glucose and fructose, as recognized many years ago in patients with monosaccharide intolerance.[29] Since carbohydrates provide 50% of dietary calories, it must be assumed that carbohydrate malabsorption plays an important role in the development and maintenance of malnutrition.[152] Also, infectious diarrhea and undernutrition during childhood may occur in many populations who are lactose nondigestors per se. In patients with secondary lactose intolerance after gastroenteritis, recovery of malnutrition may take months. Indeed, the presence or absence of lactose intolerance may be more important in determining the final outcome of the illness than the type of diarrhea per se.[153]

In malnutrition, there is an increased frequency and severity of metabolic acidosis in association with diarrhea. This metabolic acidosis is thought to occur in response to a sequence of events in which malabsorbed carbohydrates may play an important role. Carbohydrates are fermented by bacterial flora into short-chain organic acids, some of which can be efficiently absorbed in the colon, conserving a considerable amount of energy.[154] Organic anions are, however, excreted, leaving excess hydrogen ions in the intestine that are neutralized by bicarbonate, eventually resulting in acidosis. Production of fermentative substances, such as deconjugated bile salts and hydroxy fatty acids, in addition to the osmotic pull of malabsorbed carbohydrates, causes decreased sodium absorption, increased fecal potassium loss, and increased colonic fluid, which ultimately leads to diarrhea. For many years, we have known that fecal losses of sodium, potassium, and chloride correlate with the amount of stool produced and may be higher in malnutrition.[155]

Protein malabsorption is a frequent occurrence during malnutrition,[153] and approximately one-third of the nitrogen intake may be lost. There is malabsorption of amino acids, di- and tripeptides,[157] independent of the excessive loss of protein from the gastrointestinal tract induced by the protein-losing enteropathy associated with diarrheal illness.[158] Intestinal peptide hydrolases are depressed, even after nutritional recovery.[159] Results from studies using malnourished animals are again at variance with human studies, showing an increased absorption of certain amino acids.[160,161] Intraluminal amino acids provide fewer disturbances of intraluminal physiology than do residual fats and sugars. Thus, an intake of high-quality dietary protein in excess of individual requirements will allow for the necessary uptake of amino acids for nutritional recovery from protein-energy malnutrition even in the face of impaired absorptive efficiency, and the unabsorbed nitrogenous materials will be relatively innocuous.[162]

Extensive studies on vitamin and trace element absorption are not available, but malabsorption of vitamins A, B_{12}, and folate have been documented. The intestinal parasitoses that often complicate malnutrition can also contribute to a depression in the absorption of vitamin A. Whether the alterations seen in malnutrition change the efficiency of absorption of vitamin K or whether the diarrhea in malnutrition washes out the vitamin-K-synthesizing bacteria to subcritical levels is not fully understood. In severe, malnourished children, circulating levels of vitamin E are diminished. Hypomagnesemia and hypozincemia have been documented in acute enteritis, although there is no definitive documentation of impaired absorption of these minerals.[163]

Despite these disturbances, the intestine of malnourished children retains sufficient absorptive capacity for the repair of the mucosa and for recovery of the patients to begin soon after adequate amounts of nutrients are provided. The intestinal mucosa receives a significant proportion of its nutrition from the intestinal lumen; therefore, it is in an advantageous position to initiate the process of organ repair when feedings are given.

VI. THERAPEUTICAL CONSIDERATIONS

In the child with diarrhea, the maintenance of fluid and electrolytes and the proper dietary treatment remain the essential factors in the successful treatment of this disorder.[164] Oral hydration solutions should be employed to prevent dehydration and therefore must be begun at the first sign of illness and maintained throughout. Dietary elimination of the many ingredients that are not tolerated by the patient with diarrhea is also necessary to achieve a prompt and a smooth improvement of the illness. It is extremely important to prevent the development of prolonged diarrhea, malnutrition, and other complications rather than to remedy them. An aggressive dietary treatment at the onset of the disease is the best prophylaxis and treatment of diarrhea and its complications.

The approach of dietary treatment of acute diarrhea may vary from patient to patient, the age of the child, pre-diarrhea feeding practices, the nutritional status, and the duration of diarrhea prior to medical intervention. The etiology and severity of diarrhea, and the setting in which the patient is treated, also play a role in determining the nutritional intervention. The financial resources of the family and the "culture" will also determine the types of feedings that are given to the child with diarrhea.

The optimal timing of feeding children with acute diarrhea used to be controversial with debates regarding "bowel rest" versus "early feeding".[165–169] However, none of the studies which compared the effect of different times of introduction of the same or similar diets identified adverse clinical effects of early or continuous feeding. Although stool output was increased by feedings during diarrhea, it was also true that early feedings reduced the duration of the illness[170] and improved the nutritional status of the patients.[171,172] These clinical and community-based research studies therefore demonstrate that diarrhea sufferers should be fed, not fasted—and the sooner the better. There is a net increase in the absorption of nutrients when infants with diarrhea are fed. On the average, 80 to 95% of dietary carbohydrate, 50 to 70% of fat, and 50 to 75% of nitrogen are absorbed even during the early phase of illness. Thus, feeding during diarrhea should be considered an integral component of the therapy and be given together with oral rehydration fluids.

There is general agreement that in the case of an exclusively human milk-fed infant, breastfeeding should be continued during the rehydration, maintenance, and convalescent phases of therapy for diarrhea.[173] Although human milk contains a high amount of lactose, it usually is well tolerated.[174] Furthermore, the antiinfective and growth-promoting properties of breast milk might actually shorten the duration of symptomatic illness.[175]

However, the choice of feedings used for the nutritional rehabilitation of non-breast-fed infants who have acute diarrhea remains controversial. The American Academy of Pediatrics has recommended the use of a lactose-free formula[176] as a feeding of choice during an episode of acute diarrhea. Cow's milk and lactose-containing formula may not be well tolerated because their lactose content may exceed the reduced lactase activity of infants with diarrhea,[177–180] causing prolongation of the diarrheal illness. However, the World Health Organization (WHO) recommended the use of diluted cow's milk or cow's milk formula as the first choice for refeeding non-breast-fed infants who have acute diarrhea, and more recently the WHO recommend "to keep on going with the same feeding used before the illness".[181] There are other authors who recommend the use of soy-based formula though there are reports of sensitivity to soy protein in young infants recovering from acute gastroenteritis.[182] Others have advocated protein hydrolysate semi-elemental diets in this period.[183,184] We have shown that there are important beneficial effects in the nutritional management of acute diarrhea with the use of a hypo-osmolar formula containing glucose polymers, medium-chain triglycerides, and casein.[186]

The rational for the recommendation of a specific formula designed for the nutritional therapy of infants with diarrhea is based on the following: carbohydrate intolerance is a frequent complication in young infants after infectious diarrhea,[148] particularly in those infected with Rotavirus[187] and enteroinvasive Escherichia coli.[188] Glucose polymers are better absorbed and better tolerated in diarrheal disease.[150,189] Additionally, there may be fat malabsorption in gastroenteritis,[190] particularly when malnutrition is present;[191] whereas improved fat absorption occurs with medium-chain triglycerides in infants with gastrointestinal disease.[184–192] Protein and nitrogen losses are also high during diarrheal illness, and the absorption of protein may be positively influenced by several dietary factors, including the protein source and its biologic value,[193] as well as the total energy and dilution of the feedings.[194] Casein has a high biologic value, is easily digestible, and has a high performance rate for the treatment of malnourished infants.[195] Finally, a low osmolality of the formula may also exert a beneficial effect; hypertonic feedings enhance intestinal water losses in diarrhea.[196]

Therefore, we believe that, when diarrhea of any type and degree strikes a high-risk patient, namely a young infant less than 3 months of age, or a patient who is malnourished, or was born small for gestational age, he or she should be treated with a specifically designed formula for infants with diarrhea if breast-feedings are not available. This may be the safest choice for the initial refeeding as it compensates for all the possible pathophysiological alterations induced by the illness.[186] An improved absorption of the formula fed during the acute episode of diarrhea may lead to a more rapid recovery with a lesser number of complications than those resulting when the usual feedings and lactose-containing formula are given to an infant with diarrhea. The feeding during the convalescent period may also be of importance since it may reduce the long-term nutritional impact of diarrheal illness. Dietary intake should be greater than normal during convalescence after diarrhea. The high nutritional requirement may be fulfilled by providing meals of high nutrient density and/or increasing the number of meals offered per day of the formula that is designed for infants with diarrhea with the addition of supplements in accordance with the age of the infant and the feedings fed prior to illness.

Recommendations for use of special dietary formulas for infants with diarrhea who are not breast-fed poses a number of interesting considerations. Previous data shows that most infants with acute diarrhea rapidly recover if hydration is preserved, regardless of the dietary treatment given.[82] However, in the United States and in Europe there has been a marked decrease in the number of infants with the severe form of postinfectious chronic diarrhea leading to severe intolerance to several foods.[14] This decline appeared to coincide with an increased usage by pediatricians of specific infant formulas recommended for the treatment of diarrhea. Such formulas have been advocated by many individuals[197] and by the Committee on Nutrition of the American Academy of Pediatrics.[176] A lactose-free, glucose polymers, protein hydrolysate formula has been found to be the most effective treatment of malnourished infants with postinfectious chronic diarrhea.[14,198]

Based on the current available evidence, the practice of cow's milk feeding, diluted or not, during diarrheal illness has to be considered hazardous to the infant's health. Feedings with unprocessed diluted cow's milk appears to pose a special risk to infants with diarrhea. Chronic postinfectious diarrhea, lactose and acquired monosaccharide intolerance, protein sensitivities, multiple food intolerance as well as pneumatosis intestinalis are all well-recognized complications that can occur when cow's milk is fed to a baby with an acute episode of diarrhea.[198] Moreover, even diluted cow's milk feedings have also been associated with frequent dehydration and metabolic acidosis despite administration of hydration solutions.[186] The standard medical practice to initiate feedings of cow's milk to the infant with diarrhea has been shown by most investigators in the field to have detrimental consequences. For the last 25 years, lactose has been found to be a culprit in the evolution of acute diarrhea, in most studies, infants often requiring further formula changes before recovery is elicited. Although most infants with diarrhea improve even while fed cow's milk, the delay in appropriate nutritional rehabilitation of the child with diarrhea can result in a perilious course and in deterioration of the patient and aggravation of the intestinal malabsorption. Even if only a small proportion of infants do not respond to lactose-containing formula, the risk benefit-ratio favors the choice of the most effective treatment. When recommendation is made for oral hydration therapy, the choice of a preparation that reduces stool output is advocated by world authorities . . . why not the same principle for a refeeding regimen?!

Thus, we strongly believe that the recommendations for feeding specially designed diets during diarrheal disease in the high-risk patient should be heeded.

However, we recognize that the use of proprietary formulae designed for the treatment of infants with diarrhea is more difficult to implement where it is more needed.

Of course, special formula preparations are more expensive and generally less available than cow's milk. Nonetheless, the cost of the therapy has to be viewed in relation to the possible benefits derived from appropriate, prompt, nutritional treatment of the infant with diarrhea and malnutrition. The cost of any item can only be ascertained in terms of value. A rough attempt to calculate cost versus value of treatment of such patients with a special infant formula is made below.

In a group of 100 high-risk patients with acute diarrhea, we can estimate that up to 20 of them can develop chronic postinfectious diarrhea[199] and other complications that will require hospitalization. The remaining 80 will improve with rehydration therapy and ordinary feedings, usually containing cow's milk. If we would treat all these 100 high-risk patients with a specific proprietary formula designed for infants with diarrhea for a mean of three days (150 ml/kg/day, for a 5 kg baby, cost for 1 l = $5), the cost of this treatment for all this group of 100 patients would be $1125. In contrast, the treatment cost of the same group of patients with cow's milk (2 l = $1) for the same period of time will be 10 times less, apparently $112.50. However, if we take into consideration that 20 patients of this group may develop serious complications leading to prolongation of diarrhea, malnutrition, and possible hospitalization, the cost would be very different. A mean of 7 days of hospital treatment (average cost $500 per day) would increase the cost of treating this group of patients to $70,000! This amount of money will buy approximately 14,000 l of special infant formula. This amount of formula could be sufficient for 18,600 days of treatment. Since our previous data showed that infants with severe diarrhea fed this kind of formula generally improve within 3 days of treatment, this amount of the "expensive" formula could treat 6222 (2657 patients for 7 days each) patients for the length of time required for diarrhea to improve with a lesser number of complications.

Of course, other dietary treatments specially designed for the treatment of patients with diarrhea are available. Local production of foods that meet the requirements to compensate for the pathophysiological deficits of diarrhea are available, are less expensive, and have been safely used.[173] In most of these studies, local mixtures of foods with cow's milk have been utilized, particularly rice-based. However, these cereal-milk foods are not suitable for infants less than four months of age, a stage in life of high risk for more severe diarrhea and its complications. We can not condone increased infant morbidity of diarrheal disease deaths for economical reasons or governmental policies. The most efficient treatment available must be employed in high-risk infants, as this practice will prove to be cost efficient, too.

However, in older infants with diarrhea and in those not considered at high risk, ordinary treatment with oral hydration and feedings of culturally accepted foods may suffice.

VII. CONCLUSIONS

Infectious diseases, and diarrhea in particular, are the main determinants of wastage and stunting of growth in children in developing countries. Nations that are able to diminish the incidence of diarrhea and other infections clearly exhibit a secular change in improving the nutritional status and growth of children. Children with new or fewer infections have better appetites, and their improved health allows them to be less susceptible to infections. It is known that in certain very poor areas, people who live under primitive conditions and consume minimal food may exhibit remarkably good health.[200] This is because of the maintenance of relatively good sanitary conditions and a high prevalence of breastfeeding, as observed in the State of Kerala in India and among the Amazon Indians in Brazil. However, when these people migrate from rural areas to the periphery of a large urban center, diarrhea and malnutrition usually ensue. These migratory masses have the familiar characteristics of malnourished populations: low income, numerous children, wretched conditions in a contaminated environment. Equally interesting is the observation that in some developing countries, provision of food supplementation made available through food distribution centers has been unsuccessful in combating malnutrition, particularly when there is the continuing presence of poor sanitation, which leads to diarrhea and other infectious diseases. As part of any major policy to prevent malnutrition in less developed countries, attention must be directed toward the control of infectious diarrhea. Only in this way can malnutrition and growth failure be prevented.[198]

It is evident that for certain groups at risk of developing enteric infections, there may be a high cost in nutritional homeostasis and that all efforts for improving economic levels and sanitation may be of paramount importance for the control of malnutrition. Optimal oral therapy as defined by the Centers for Disease Control and Prevention[201] and the World Health Organization[202] includes oral rehydration therapy and early, appropriate feeding but not the use of antidiarrheal drugs. Although several hundred million dollars are spent on these drugs worldwide each year, there is little evidence of their efficacy, and many of the drugs have adverse effects and some possible long-term effects that are not even known yet.[203] However, medical personnel, health authorities, and the patients' families are often less hesitant to incur expenses for ineffective diarrheal medication. Our concern is that emphasizing the use of adjunctive antidiarrheal therapy may divert attention and resources away from the use of oral rehydration therapy and early, appropriate feeding, which are the essential components of effective therapy.

Specially designed infant formulas for the treatment of severe diarrhea may reduce complications and prolonged hospitalization, and prompt dietary treatment may reduce the number of infants with chronic postinfectious diarrhea and malnutrition with concomitant high morbidity and mortality.

REFERENCES

1. Mata, L., Cryptosporidium and other protozoa in diarrheal disease in less developed countries, *Pediatr. Infect. Dis.*, 5, 117, 1986.
2. Snyder, J. P. and Merson M. N., The magnitude of the global problem of acute diarrheal disease. A review of active surveillance data, *Bull. WHO*, 60, 605, 1982.
3. Walsh, J. A. and Warren, R. S., Selective primary health care. An interim strategy for disease control in developing countries, *N. Engl. J. Med.*, 301, 967, 1979.
4. Mason, J. O., Enteric disease and health for all: a public health perspective, *Pediatr. Infect. Dis.*, 5, 57, 1986.
5. Kirkwood, B. R., Diarrhea, in *Disease and mortality in sub-Saharan Africa*, Feachem, R. G. and Jamison, D. T., Eds., Oxford University Press, Oxford, 1990.
6. Black, R. E., Lopez de Romana, G., Brown, K. H., Bravo, N., Grados-Bazalar, O., and Kanashiro, H. C., Incidence and etiology of infantile diarrhea and major routes of transmission in Huascar, Peru, *Am. J. Epidemiol.*, 129, 785, 1989.
7. Mata, L. J., Urrutia, J. J., and Gordon, J. E., Diarrheal disease in a cohort of Guatemalan village children observed from birth to two years of age, *Trop. Geogr. Med.*, 19, 247, 1967.
8. McAuliffe, J. F., Shields, D. S., de Sousa, M. A., Sakell, J., Schorling, J., and Guerrant, R. L., Prolonged and recurring diarrhea in the northeast of Brazil: examination of cases from a community-based study, *J. Pediatr. Gastroenterol. Nutr.*, 5, 902, 1986.
9. Lifshitz, F. and da Costa Ribeiro, H., Diarrheal disease, in *Clinical nutrition*, 2nd ed., Paige, D. M., Ed., St. Louis, C. V. Mosby, 447, 1988.
10. Lifshitz, F., da Costa Ribeiro, H., and Silverberg, M., Childhood infectious diarrhea, in *Textbook of pediatric gastroenterology*, 2nd ed., Silverberg, M. and Daum, F., Eds., New York, Year Book, 284, 1988.
11. Glass, R. I., Lew, J. F., Gangarosa, R. E., LeBaron, C. W., and Ho, M. S., Estimates of morbidity and mortality rates for diarrheal disease in American Children, *J. Pediatr.*, 118, S27, 1991.
12. Ho, M. S., Glass, R. I., Pinsky, P. F., Young-Okoh, N., Sappenfield, W., Buehler, J. W., Gunter, N., and Anderson, L. J., Diarrheal deaths in American children. Are they preventable? *JAMA*, 260, 3281, 1988.
13. Barnes, G., Acute Diarrhea: Diagnosis and management, in *Instructional Seminars in Pediatric Gastroenterology and Nutrition*, 1, 1, 1992.
14. Lifshitz, F., Nutrition for special needs in infancy, in *Nutrition for special needs in infancy: Protein Hydrolysates*, Lifshitz, F., Ed., Marcel Dekker, 1, 1, 1985.
15. The Imo State Evaluation Team, Evaluating water and sanitation projects: Lessons from Imo State, Nigeria, *Health Policy Plan*, 4, 40, 1989.
16. Martinez, J. C., Ashworth, A., and Kirkwood, B., Breastfeeding among the urban poor in southern Brazil: reasons for termination in the first 6 months of life, *Bull. WHO*, 67, 151, 1989.
17. Esrey, S. A., Feachem, R. G., and Hughes, J. M., *Bull. WHO*, 63, 757, 1985.
18. Mata, L., Breastfeeding, diarrheal disease and malnutrition in less developed countries, in *Pediatric nutrition*, Lifshitz, F., Ed., New York, Marcel Dekker, 1982, 355.
19. Brown, K. H., Black, R. E., Lopez de Romana, G., and Kanashiro, H. C., Infant feeding practices and their relationship with diarrheal and other disease in Huascar (Lima), Peru, *Pediatrics*, 83, 31, 1989.
20. Lifshitz, F., Moses Finch, N., and Lifshitz, J., *Children's Nutrition*, Jones and Bartlett Publishers, Boston, 19, 323, 1991.
21. Parker, P., Stroop, S., and Green, H., A controlled comparison of continuous versus intermittent feeding in the treatment of infants with intestinal disease, *J. Pediatr.*, 99, 360, 1981.

22. Macfarlane, P. I. and Miller, V., Human milk in management of protracted diarrhea of infancy, *Arch. Dis. Child.,* 59, 260, 1984.
23. Welsh, J. K. and May, J. T., Anti-infective properties of human milk, *J. Pediatr.,* 941, 1, 1979.
24. Vakil, J. R., Chandan, R. C., Parry, R. M., and Shani, K. M., Susceptibility of several microorganisms to milk lysozyme, *J. Dairy Sci.,* 52, 1192, 1969.
25. Read, L. C., Upton, F. C., Francis, G. L., Wallace, J. C., Dahelnberg, G. W., and Ballard, F. G., Changes in the growth-promoting activity of human milk during lactation, *Pediatr. Res.,* 18, 133, 1984.
26. Koldovsky, O. and Thornburg, W., Hormones in milk, *J. Pediatr. Gastroenterol. Nutr.,* 6, 172, 1987.
27. Carpenter, G., Epidermal growth factor is a major growth-promoting agent in human milk, *Science,* 210, 198, 1980.
28. Oka, Y., Ghishan, F. K., and Green, H. L., Effect of mouse epidermal growth factor/urogastrone on the functional maturation of rat intestine, *Endocrinology,* 112, 940, 1983.
29. Lifshitz, F., Coello Ramirez, P., and Gutierrez-Topete, G., Monosaccharide intolerance and hypoglycemia in infants with diarrhea. I. Clinical course of 23 cases, *J. Pediatr.,* 77, 595, 1970.
30. Lifshitz, F., Food intolerance and sensitivity, in *Advances in pediatric gastroenterology and nutrition,* Lebenthal, E., Ed., Hong Kong, Excerpta Medica, 131, 1984.
31. Coello-Ramirez, P., Gutierrez-Topete, G., and Lifshitz, F., Pneumatosis intestinalis, *Am. J. Dis. Child.,* 120, 3, 1970.
32. Belongia, E. A., Osterholm, M. T., Soler, J. T., Ammend, D. A., Braun, J. E., and Macdonald, K. L., Transmission of Escherichia coli 0157:H7 infection in Minnesota child day-care facilities, *JAMA,* 269, 883, 1993.
33. Spika, J. S., Parson, J. E., and Nordenberg, D., Hemolytic uremic syndrome and diarrhea associated with E. coli 0157:H7 in a day care center, *J. Pediatr.,* 109, 287, 1986.
34. Martin, D. L., MacDonald, K. L., and White, K. E., The epidemiology and clinical aspects of the hemolytic uremic syndrome in Minnesota, *N. Eng. J. Med.,* 387, 1161, 1990.
35. Yip, R. and Sharp, T. W., Acute malnutrition and high childhood mortality related to diarrhea, *JAMA,* 270, 587, 1993.
36. Bern, C., Martines, J., de Zoysa, I., and Glass, R. I., The magnitude of the global problem of diarrheal disease: a ten year update, *Bull. WHO,* 70, 705, 1992.
37. Bhan, M. K., Bhandari, N., and Sazawal, S., Descriptive epidemiology of persistent diarrhea among young children in rural northern India, *Bull. WHO,* 67, 281, 1989.
38. Gordon, J. E., Infectious disease in the malnourished, *Ann. N.Y. Acad. Sci.,* 176, 9, 1971.
39. Tomkins, A., Nutritional status and severity of diarrhea among pre-school children in rural Nigeria, *Lancet,* 1, 860, 1981.
40. Sepulveda, J., Willett, W., and Munoz, A., Malnutrition and diarrhea. A longitudinal study among urban Mexican children, *Am. J. Epidemiol.,* 127, 365, 1988.
41. El Samani, E. F. Z., Willett, W. C., and Ware, J. H., Association of malnutrition and diarrhea in children aged under five years. A prospective follow-up study in a rural Sudanese community, *Am. J. Epidemiol.,* 128, 93, 1988.
42. Palmer, D. L., Koster, F. T., Alam, A. K. M. J., and Islam, M. R., Nutritional status: a determinant of severity of diarrhea in patients with cholera, *J. Infect. Dis.,* 134, 8, 1976.
43. Butzner, D., Butler, D. G., Miniats, P., and Hamilton, J. R., Impact of chronic protein-calorie malnutrition on small intestinal repair after acute viral enteritis. A study in gnotobiotic piglets, *Pediatr. Res.,* 19, 476, 1985.
44. Gainella, R. A., Broitman, S. A., and Zamcheck, N., Salmonella enteritis I. Role of reduced gastric secretion in pathogenesis, *Am. J. Dig. Dis.,* 16, 1000, 1971.
45. Gracey, M., Cullity, G. J., and Suharjono, The stomach in malnutrition, *Arch. Dis. Child.,* 52, 325, 1977.
46. Gilman, R. H., Partanen, R., Brown, K. H., Spira, W. M., Khanam, S., Greenberg, B., Bloom, S. R., and Ali, A., Decreased gastric acid secretion and bacterial colonization of the stomach in severely malnourished Bangladesh children, *Gastroenterology,* 94, 1308, 1988.

47. Kraft, S. C., Rothberg, R. M., and Knauer, C. M., Gastric acid output and circulating antibovine serum albumin antibodies in adults, *Clin. Exp. Immunol.*, 2, 321, 1967.
48. Ruddel, W. S. J., Axon, A. T. R., and Findlay, J. M., Effect of cimetidine on the gastric bacterial flora, *Lancet*, 1, 672, 1980.
49. Brunser, O., Araya, M., and Espinzoa, J., Gastrointestinal tract changes in the malnourished child, in *The Malnourished Child*, Suskind, R. M., Lewinter-Suskind, L., Eds., Raven Press, New York, 19, 261, 1990.
50. Dupont, H. L. and Hornick, R. B., Adverse effects of Lomotil Therapy in Shiegellosis, *JAMA*, 226, 1525, 1973.
51. Lyerly, D. M., Krivan, H. C., and Wilkins, T. D., Clostridium difficile: its disease and toxins, *Clin. Microbiol. Rev.*, 1, 1, 1988.
52. Chandra, R. K., Immunocompetence is a sensitive and functional barometer of nutritional status, *Acta Pediatr. Scand. Suppl.*, 374, 129, 1991.
53. Chelluri, L. and Jastremski, M. S., Incidence of malnutrition in patients with acquired immunodeficiency syndrome, *Nutr. Clin. Pract.*, 4, 16, 1989.
54. Lesbordes, J. L., Chassignol, S., and Ray, E., Malnutrition and HIV infection in children in the Central African Republic, *Lancet*, 2, 337, 1986.
55. Sorensen, R. V., Leiva, L. E., and Kubibidila, S., Malnutrition and the immune response, in *Textbook of Pediatric Nutrition*, 2nd ed., Suskind, R. M. and Suskind, L. Lewinter, Eds., Raven Press, New York, 141, 1993.
56. Jambon, B., Ziegler, O., and Maire, B., Thymulin (factor Thymique serique) and zinc content of the thymus glands of malnourished children, *Am. J. Clin. Nutr.*, 48, 335, 1988.
57. Suskind, L. L., Suskind, D., Murthy, K. K., and Suskind, R. M., The malnourished child, in *Textbook of pediatric nutrition*, 2nd ed., Suskind, R. M. and Suskind, L. L., Eds., Raven Press, New York, 127, 1993.
58. Chandra, R. K., 1990 McCollum Award Lecture, Nutrition and immunity: lessons from the past and new insights into the future, *Am. J. Clin. Nutr.*, 53, 1987, 1991.
59. Law, D. K., Kudrick, S. J., and Abdou, N. I., Immunocompetence of patients with protein-calorie malnutrition, the effects of nutritional repletion, *Ann. Intern. Med.*, 79, 545, 1973.
60. Keush, G. T., Cruz, J. R., and Torun, B., Immature circulating lymphocytes in severely malnourished Guatemalan children, *J. Ped. Gastroent. and Nutr.*, 6, 265, 1987.
61. Suskind, D., Murthy, K. K., and Suskind, R. M., The malnourished child: an overview, in *The Malnourished Child*, Suskind, M. and Suskind, L. L., Eds., Nestle' Nutrition Workshop Series, Vol. 19, Raven Press, New York, 1990.
62. Chandra, R. K., Chandra, S., and Gupta, S., Antibody affinity and immune complexes after immunization with tetanus toxoid in protein energy malnutrition, *Am. J. Clin. Nutr.*, 40, 131, 1984.
63. Chandra, R. K., Nutritional deficiencies and mucosal immunity, in *Textbook of Gastroenterology and Nutrition in Infancy*, 2nd ed., Lebenthal, E., Ed., Raven Press, New York, 1989, 565.
64. Watson, R. R., McMurray, D. N., Martin, P., and Reyes, M. A., Effect of age malnutrition and renutrition on free secretory component and IgA in secretions, *Am. J. Clin. Nutr.*, 42, 281, 1985.
65. Murthy, K. K. and Suskind, R. M., Malnutrition and the Immune Response, in *Textbook of Gastroenterology and Nutrition in Infancy*, 2nd ed., Lebenthal, E., Ed., Raven Press, New York, 1989, 545.
66. Suskind, R., Edelman, R., Kulapongs, P., Sirisinha, S., Pariyanonda, A., and Olsen, R. E., Complement activity in children with protein-calorie malnutrition, *Am. J. Clin. Nutr.*, 29, 1089, 1976.
67. Beisel, W. R., Malnutrition as a consequence of stress, in *Malnutrition and the immune response*, Suskind, R. H., Ed., New York, Raven Press, 1977, 31–36.
68. Beisel, W. R., Metabolic effects on infection, *Prog. Food Nutri. Sci.*, 8, 43–75, 1984.

69. Mata, L., Influence on the growth parameters of children, in *Acute Diarrhea: Its Nutritional Consequences in Children,* Belanti, J. A., Ed., New York, Raven Press, 1983, 85–94.
70. Feigin, R. D., Klainer, A. S., Beisel, W. R., and Hornick, R. B., Whole blood amino acids in experimentally induced typhoid fever in man, *N. Engl. J. Med.,* 278, 293–298, 1968.
71. Beisel, W. R., Cockerell, G. L., and Janssen, W. A., Nutritional effects on the responsiveness of plasma acute-phase reactant glycoproteins, in *Malnutrition and the Immune Response,* Suskind, R. M., Ed., New York, Raven Press, 1977, 395–402.
72. Caballero, B., Solomons, N. W., Batres, R., and Torun, B., Homeostatic mechanisms in the utilization of exogenous iron in children recovering from severe malnutrition, *J. Ped. Gastroenterol. Nutr.,* 5, 740–745, 1986.
73. Castillo-Duran, C., Fisberg, M., Valenzuela, A., Egona, J. Y., and Uahy, R., Controlled trial of copper supplementation during the recovery from marasmus, *Am. J. Clin. Nutr.,* 37, 898–903, 1983.
74. Solomons, N. W., Trace mineral and the underweight child, in *The Underweight Infant, Child and Adolescent,* Cohen, S. A., Ed., Appleton-Century-Croft, Norwalk, CT, 1986, 261–278.
75. Dinarello, C. A. and Wolff, S. M., The role of interleukin-1 in disease, *N. Eng. J. Med.,* 328, 106–113, 1993.
76. Klasing, K. C., Nutritional aspects of leukocytic cytokines, *J. Nutr.,* 118, 1436–1446, 1988.
77. Maury, C. P. J., Tumor necrosis factor—an overview, *Acta Med. Scand.* 220, 387–394, 1986.
78. Parrillo, J. E., Pathogenetic mechanism of septic shock, *N. Engl. J. Med.,* 328, 1471–1477, 1993.
79. Vaisman, N., Schatter, A., and Hahn, T., Tumor necrosis factor production during acute starvation, *Am. J. Med.,* 87, 115, 1989.
80. Shaffner, A., Steinbock, M., Tepper, R., et al., Tumor necrosis factor production and cell mediated immunity in anorexia nervosa, *Clin. Exp. Immunol.* 79, 62–66, 1990.
81. Bone, R. C., The search for a magic bullet to fight sepsis, *JAMA,* 269, 2266–2267, 1993.
82. Hirschoren, N., The treatment of acute diarrhea in children: a historical and physiological perspective, *Am. J. Clin. Nutr.,* 33, 637–663, 1980.
83. Sousa, J. S., Silva, A., and Ribeiro, V. C., Intractable diarrhea of infancy and latent otomastoiditis, *Arch. Dis. Child.,* 55, 937–940, 1980.
84. Mata, L. J., *The Children of Santa Maria Cauque. A Prospective Field Study of Health and Growth,* Massachusetts Institute of Technology Press, Cambridge, 10, 395, 1978.
85. Araya, M., Figueroa, G., Espinoza, J., Zarur, X., and Brunser, O., Acute diarrhea and carrier state in Chilean pre-schoolers of the low and high socio-economic strata, *Acta Pediatr. Scand.,* 75, 645–651, 1986.
86. Baker, S. J., Sub-clinical intestinal malabsorption in developing countries, *Bull. WHO,* 54, 485–494, 1976.
87. Brunser, O., Araya, M., Espinoza, J., Figueroa, G., Pacheco, I., and Lois, I., Chronic environmental enteropathy in a temperate climate, *Hum. Nutr. Clin. Nutr.,* 41C, 251–261, 1987.
88. Lebenthal, E., Prolonged small intestinal mucosal injury as a primary cause of intractable diarrhea of infancy, in. *Chronic diarrhea in children,* Lebenthal, E., Ed., New York, Nestle, Vevey/Raven Press, 1984, 5–29.
89. Rossi, T. M. and Lebenthal, E., Pathogenic mechanism of protracted diarrhea, *Adv. Pediatr.,* 30, 595–633, 1983.
90. Goldgar, C. M. and Vanderhoof, J. A., Lack of correlation of small bowel biopsy and clinical course of patients with intractable diarrhea of infancy, *Gastroenterology,* 90, 527–531, 1986.
91. Lifshitz, F., Nutrition in chronic diarrhea in infancy, in *Feeding the Sick Infant,* Stern, L., Ed., New York, Nestle/Raven Press, 1987, 275–290.
92. Shiner, M., *Ultrastructure of the small intestinal mucosa,* Springer Verlag, Great Britain, 1983, 61–95.
93. Lifshitz, F., Interrelationship of diarrhea and infant nutrition, in *Textbook of Gastroenterology and Nutrition in Infancy,* 2nd ed., Lebenthal, E., Ed., Raven Press, New York, 1989, 657–644.

94. Cohen, M. B., Etiology and mechanisms of acute infectious diarrhea in infants in the United States, *J. Pediatr.*, 118, 534–539, 1991.
95. Sherman, P., Drumm, B., Karmali, M., and Cutz, E., Adherence of bacteria to the intestine in sporadic cases of enteropathogenic Escherichia coli-associated diarrhea in infants and young children: a prospective study, *Gastroenterology*, 96, 86–94, 1989.
96. Rothbaum, R., McAdams, A. J., Giannela, R., and Partin, J. C., A clinicopathologic study of enterocyte-adherent Escherichia coli: a cause of protracted diarrhea in infants, *Gastroenterology*, 83, 441–454, 1982.
97. Bowie, M. D., Effect of lactose-induced diarrhea on absorption of nitrogen and fat, *Arch, Dis. Child.*, 50, 563, 1975.
98. Ringrose, R. E., Thompson, J. B., and Welsh, J. D., Lactose malabsorption and steatorrhea, *Dig. Dis.*, 17, 533–538, 1972.
99. Field, M., Regulations of small intestinal ion transport by cyclic nucleotides and calcium, in *Secretory Diarrhea*, Field, M., Fordtran, J. S., and Shultz, G. S., Eds., Williams and Wilkins, Baltimore, 1980, 21–73.
100. O'Brien, A. D. and Holmes, R. K., Shiga and Shiga-like toxins, *Microbiol. Rev.*, 51, 206–220, 1987.
101. Shiner, M., Putman, M., Nichols, V. N., and Nichols, B. L., Pathogenesis of small intestinal mucosal lesions in chronic diarrhea of infancy. I. Alight Microscopic study, *J. Pediatr. Gastroenterol. Nutr.*, 11, 455–463, 1990.
102. Shiner, M., Nichols, V. N., Barrish, J. P., and Nichols, B. L., Pathogenesis of small intestinal mucosal lesions in chronic diarrhea of infancy. II. An electron microscopy study, *J. Pediatr. Gastroenterol. Nutrition*, 11, 464–480, 1990.
103. Lifshitz, F., The enteric flora in childhood disease-diarrhea, *Am. J. Clin. Nutr.*, 30, 1181–1188, 1977.
104. Bhatnagfar, S., Bhan, M. K., Chechamma, G., Gupta, V., Kumar, R., Bright, D., and Sainis, S., Is small bacterial overgrowth of pathogenic significance in persistent diarrhea? *Acta Pediatr.*, 381 (Suppl.), 108–113, 1992.
105. Coello-Ramirez, P. and Lifshitz, F., Enteric microflora and carbohydrate intolerance in infants with diarrhea, *Pediatrics*, 49, 233, 1972.
106. Penny, M.E., The role of the duodenal microflora as a determinant of persistent diarrhea, *Acta Ped. Scand.*, 381 (Suppl.), 114–120, 1992.
107. Lifshitz, F., Wapnir, R. A., Wehman, H. J., et al., The effects of small intestinal colonization by fecal and colonic bacteria on intestinal function in rats, *J. Nutr.*, 108, 1913, 1978.
108. Brunser, I., Reid, A., Monckeberg, F., Maccioni, A., and Conteras, I., Jejunal mucosa in infantile malnutrition, *Am. J. Clin. Nutr.*, 21, 976–983, 1968.
109. Heyman, M., Boudrd, G., and Sarruti, S., Macromolecular transport in jejunal mucosa of children with severe malnutrition: a quantitative study, *J. Pediatr. Gastroenterol. Nutr.*, 3, 357–363, 1984.
110. Shiner, M., Redmond, A. O. B., and Hansen, J. D. L., The jejunal mucosa in protein-calorie malnutrition. A clinical, histological and ultrastructural study, *Exp. Mol. Pathol.*, 19, 61–78, 1973.
111. Brunser, O., Reid, A. M., Mackberg, F., Maccioni, A., and Conteras, D., Jejunal biopsies in infant malnutrition with special reference to mitotic index, *Pediatrics*, 38, 605, 1966.
112. Sullivan, P. B. and Marsh, M. N., Small intestinal mucosal histology in the syndrome of persistent diarrhea and malnutrition: a review, *Acta Pediat.*, 381 (Suppl.), 72–77, 1992.
113. Brusner, O., Castillo, C., and Araya, M., Fine structure of the small intestinal mucosa in infantile marasmic malnutrition, *Gastroenterology*, 70, 495–507, 1976.
114. Fagundes-Neto, U., Wehba, J., Viaro, T., Machado, N. L., and Patricio, F. R. S., Protracted diarrhea in infancy: clinical aspects and ultrastructural analysis of the small intestine, *J. Pediatr. Gastroenterol. Nutr.*, 4, 714–722, 1985.
115. Theron, J. J., Witman, W., and Prinsloo, J. G., The fine structure of the jejunum in kwashiorkor, *Exp. Mol. Pathol.*, 14, 184–199, 1971.

116. Steiner, M., Effect of starvation on the tissue composition of the small intestine in the rat, *Am. J. Physiol.*, 215, 75–77, 1968.
117. Brunser, O., Effects of malnutrition in intestinal structure and function in children, *Clin. Gastro.*, 6, 341, 1977.
118. Hill, R. B., Jr., Prosper, J., Hirschfield, J. S., and Kern, F. Jr., Protein starvation and the small intestine, in The growth and morphology of the small intestine in weanling rats, *Exp. Molec. Path.*, 8, 66, 1968.
119. Neutra, M. R., Marer, J. H., and Mayoral, L. G., Effects of protein-caloric malnutrition on the jejunal jucosa of tetracycline-treated pigs, *Am. J. Clin. Nutr.*, 27, 287, 1974.
120. Teichberg, S., McGarvey, E., and Lifshitz, F., Quantitative morphology of the rat jejunum during protein-energy malnutrition, *Fed. Proc.*, 39, 767, 1980.
121. Wehman, H. J., Lifshitz, F., and Teichberg, S., Effects of enteric microbial overgrowth on small intestinal ultrastructure in the rat, *Am. J. Gastroenterol.*, 70, 249, 1978.
122. da Costa Ribeiro, H., Teichberg, S., McGarvey, E., and Lifshitz, F., Quantitative alterations in the structural development of jejunal absorptive epithelial cells and their subcellular organelles in protein-energy malnourished rats: a stereological analysis, *Gastroenterology*, 93, 1381–1392, 1987.
123. Behrens, R. H., Lunn, P. G., Northrop, C. A., Hanlon, P. W., and Neale, G., Factors effecting the integrity of the intestinal mucosa of Gambian children, *Am. J. Clin. Nutr.*, 45, 1433–1441, 1987.
124. Teichberg, S., Fagundes-Neto, U., Bayne, M. A., and Lifshitz, F., Jejunal macromolecular absorption and bile salt deconjugation in protein-energy malnourished rats, *Am. J. Clini. Nutr.*, 34, 1281–1921, 1981.
125. Teichberg, S., Jejunal macromolecular absorption in diarrheal disease, in *Carbohydrate Intolerance in Infancy*, Lifshitz, F., Ed., Marcel Dekker, New York, 1982, 173–191.
126. Castillo, C. and Brunser, O., Absorcion en en lactante mariamico, *Rev. Chil. Pediatr.*, 45, 581–586, 1974.
127. James, W. P. T., Sugar absorption and intestinal motility in children when malnourished and after treatment, *Clin. Sci.*, 39, 305–318, 1971.
128. Gurson, G. T. and Sanger, G., d-Xylose test in the marasmic type of protein-calorie malnutrition, *Helv. Paeditar. Acta*, 24, 510–518, 1969.
129. DuPont, H. L., Hornick, S. B., Snyder, M. J., Libonati, J. P., Formal, S. B., and Gangarosa, E., Immunity in shigellosis. II. Protection induced by oral live vaccine or primary infection, *J. Infect. Dis.*, 126, 617–621, 1972.
130. Freter, R., Interaction between mechanisms controlling intestinal microflora, *Am. J. Clin. Nutr.* 27, 1320–1328, 1974.
131. Wolin, M. J., Metabolic interactions among intestinal microorganisms, *Am. J. Clin. Nutr.*, 27, 1320–1328, 1974.
132. Lifshitz, F., Food intolerance and sensitivity, in *Advances in pediatric gastroenterology and nutrition*, Lebenthal, E., Ed., Hong Kong, Excerpta Medica, 1984, 131–140.
133. Teichberg, S., Lifshitz, F., Bayne, M. A., et al., Disaccharide feedings enhance rat jejunal macromolecular absorption, *Pediatr. Res.*, 17, 381–389, 1983.
134. Teichberg, S., Fagundes-Neto, U., Bayne, M. A., and Lifshitz, F., Increased susceptibility of malnourished rats to the effects of deoxycholate on jejunal macromolecular absorption, *Fed. Proc.*, 38, 764, 1979.
135. Lifshitz, F., Teichberg, S., and Wapnir, R., Cyclic AMP-mediated jejunal secretion in lactose-fed malnourished rats, *Am. J. Clin. Nutr.* 41, 1265–1269, 1985.
136. Barnes, G. L. and Townley, R. R. W., Duodenal mucosal damage in 31 infants with gastroenteritis, *Arch. Dis. Child*, 48, 343–349, 1973.
137. Greene, H. L., McCabe, D. R., and Merenstein, G. B., Protracted diarrhea and malnutrition in infancy: changes in intestinal morphology and disaccharidase activities during treatment with total intravenous nutrition or elemental diets, *J. Pediatr.*, 87, 695–704, 1975.

138. Stanfield, J. P., Hutt, M. S. R., and Tuniscliffe, R., Intestinal biopsy in kwashiorkor, *Lancet*, ii, 519–523, 1965.
139. Schneider, R. E. and Viteri, F. E., Luminal events of lipid absorption in protein-energy malnourished children; relationship with nutritional recovery and diarrhea. I. Capacity of the duodenal content to achieve micellar solubilization of lipids, *Am. J. Clin. Nutr.*, 27, 777–787, 1974.
140. Blackburn, W. R. and Vinijchaikul, K., The pancreas in kwashiorkor. An electron microscopic study, *Lab. Invest.*, 20, 305–318, 1969.
141. Barbezat, G. O. and Hansen, J. D. L., The exocrine pancreas and protein-calorie malnutrition, *Pediatrics*, 42, 77–92, 1968.
142. Rossi, T. M., Lee, P. C., and Lebenthal, E., Effect of feeding regimens on the functional recovery of pancreatic enzymes in postnatally malnourished weanling rats, *Pediatr. Res.*, 17, 806–809, 1983.
143. Schneider, R. E. and Viteri, F. E., Studies on luminal events of lipid absorption in protein-calorie malnourished children; relationship with nutritional recovery and diarrhea. II. Alterations in bile acids of the duodenal content, *Am. J. Clin. Nutr.*, 27, 788–796, 1974.
144. Viteri, F. E., Flores, J. M., Alvarado, J., and Behar, M., Intestinal malabsorption in malnourished children before and after recovery. Relation between severity of protein deficiency and the malabsorption process, *Am. J. Dig. Dis.*, 18, 201–211, 1973.
145. Medina, E. and Kaempffer, A. M., An analysis of health progress in Chile, *Bull. Pan Am Health Organ.*, 17, 221–232, 1983.
146. Playoust, M. R. and Isselbacher, K. J., Studies on the intestinal absorption and intramucosa lipolyse of a medium chain triglycerine, *J. Clin. Invest.*, 43, 870–885, 1964.
147. Reif, S. and Lebenthal, E., Nutritional considerations in the treatment of acute and chronic diarrhea, in *Textbook of Pediatric Nutrition*, 2nd ed., Suskind, R. M. and Lewlinter Suskind, L., Eds., Raven Press, Ltd., New York, 1993, 325–339.
148. Lifshitz, F., Coello-Ramirez, P., Guittrez-Topete, G., and Conrdolo-Cornet, M.D., Carbohydrate malabsorption in infants with diarrhea, *J. Pediatr.*, 79, 760–767, 1971.
149. Kumar, V., Chandrasekaran, R., and Bhaskar, R., Carbohydrate intolerance associated with acute gastroenteritis, *Clin. Pediatr.*, 16, 1123–7, 1977.
150. Fagundes-Neto, U., Viari, T., and Lifshitz, F., Glucose Polymer intolerance in infants with diarrhea and disaccharide intolerance, *Am. J. Clin. Nutr.*, 41, 228–234, 1985.
151. Lebenthal, E., Heitlinger, L., Lee, P. C., et al., Corn syrup sugars: in vitro and in vivo digestibility and clinical tolerance in acute diarrhea in infancy, *J. Pediatr.*, 103, 29–34, 1983.
152. Lifshitz, F., Secondary carbohydrate intolerance in infancy, in *Clinical Disorder in Pediatric Gastroenterology and Nutrition*, Lifshitz, F., Ed., New York, Marcel Dekker, 1980, 327–340.
153. Gray, G. M., Walter, W., and Colver, E. H., Persistent deficiency of intestinal lactose in apparently cured tropical sprue, *Gastroenterology*, 54, 552–558, 1968.
154. Perman, J. A., Modler, S., and Olson, A. C., Role of pH in production of hydrogen from carbohydrates by colonic flora: studies in vivo and in vitro, *J. Clin. Immunol.*, 67, 643–650, 1981.
155. Darrow, D. C., The retention of electrolytes during recovery from severe dehydration due to diarrhea, *J. Pediatr.*, 28, 515–520, 1946.
156. Waterlow, J. C., Cravioto, J., and Stephen, J. M. L., Protein malnutrition in man, *Advances in protein chemistry*, 1960, 15:131.
157. Matthews, D. M. and Adibi, S. A., Peptide absorption, *Gastroenterology*, 71, 151, 1976.
158. Ghaddimi, H., Kamar, S., and Abaci, F., Enolofenous amino acid loss and its significance in infantile diarrhea, *Pediatr. Res.*, 7, 161, 1973.
159. Kumar, V., Ghai, V. P., and Chase, H. P., Intestinal dipeptide hydrolose activities in undernourished children, *Archives of Disease in Childhood*, 46, 801, 1971.
160. Wapnir, R. A. and Lifshitz, F., Absorption of amino acids in malnourished rats, *J. Nutr.*, 104, 843, 1974.

161. Kershaw, T. G., Meame, K. D., and Wiseman, G., The effect of semi-starvation on absorption by the rat small intestine in vitro and in vivo, *J. Physiol.,* 152, 182, 1960.
162. Solomons, N. W., Molina, S., and Bulux, J., Effect of Protein-Energy malnutrition on the digestive and absorptive capacities in infants and children, in *Textbook of Gastroenterology and Nutrition in Infancy,* 2nd ed., Lebenthal, E., Ed., Raven Press, New York, 1989, 517–533.
163. Hambidge, K. M., Zinc and diarrhea, *Acta Pediatr.,* 381 (Suppl)., 82–86, 1992.
164. Lifshitz, F., Diarrheal disease, in *Manual of Clinical Nutrition,* Paige, D. M., Ed., Pleasantville, New Jersey, Nutrition Publications, 1983, 291–296.
165. Chung, A. H. and Viscorova, B., The effect of early oral feeding versus early oral starvation on the course of infantile diarrhea, *J. Pediatr.,* 33, 14–22, 1948.
166. Dugdale, A., Lovell, S., Gibbs, V., et al., Refeeding after acute gastroenteritis: a control study, *Arch. Dis. Child.,* 57, 76–78, 1982.
167. Rees, L. and Brooke, C. G. D., Gradual reintroduction of full-strength milk after gastroenteritis in children, *Lancet,* 1, 770–771, 1979.
168. Brown, K. H. and MacLean, W. C., Jr., Nutritional management of acute diarrhea: an appraisal of the alternatives, *Pediatrics,* 73, 799–805, 1984.
169. Plaszek, M. and Walker-Smith, J. A., Comparison of two feeding regimens following acute gastroenteritis in infancy, *J. Pediatr. Gastroenterol. Nutr.,* 3, 245–248, 1984.
170. Santosham, M., Foster, S., Reid, R., et al., Role of soy-based lactose-free formula during treatment of acute diarrhea, *Pediatrics,* 76, 292–298, 1985.
171. Brown, K. H., Gastanaduy, A. S., Saavedra, J. M., et al., Effect of continued oral feeding on clinical and nutritional outcomes of acute diarrhea in children, *J. Pediatr.,* 112, 191–200, 1988.
172. Hjelt, K., Paerregard, A., Petersen, W., Christiansen, L., and Krasilnikoff, P.A., Rapid versus gradual refeeding in acute gastroenteritis in childhood: energy intake and weight gain, *J. Pediatr. Gastroenterol. Nutr.,* 8, 75–80, 1989.
173. Brown, K. H., Dietary management of acute childhood diarrhea: optimal timing of feeding and appropriate use of milks and mixed diets, *J. Pediatr.,* 118, 592–598, 1991.
174. Okuni, M., Okinaga, K., and Baba, K., Studies on reducing sugars in stools of acute infantile diarrhea, with special reference to differences between breast-fed and artificially fed babies, *Tohoku J. Exp. Med.,* 107, 395–402, 1972.
175. Maggioni, G. and Signoretti, A., L'alimentazione del bambino sano e malato, *Il Pensiero Scientifico Ed.,* Roma, 1992.
176. Forbes, G. B. and Woodruff, C. W., Eds., *Pediatric Nutrition Handbook,* 2nd Ed., Chicago, Committee on Nutrition, Academy of Pediatrics, 1985, 209–210, 277–279.
177. Davison, G. P., Goodwin, D., and Robb, T. A., Incidence and duration of lactose malabsorption in children hospitalized with active enteritis: a study in a well-nourished urban population, *J. Pediatr.,* 105, 587–590, 1984.
178. Hyams, J. S., Kraus, P. J., and Gleason, P. A., Lactose malabsorption following rotavirus infection in young children, *J. Pediatr.,* 89, 916–918, 1981.
179. Penny, M. E., Paredes, P., and Brown, K. H., Clinical nutritional consequences of lactose feeding during persistent postenteritis diarrhea, *Pediatrics,* 84, 835–844, 1989.
180. Paige, D. M., Bayless, T. M., Mellitis, E. D., and Davis, L., Lactose malabsorption in preschool black children, *Am. J. Clin. Nutr.,* 30, 1018–1022, 1977.
181. da Costa Ribeiro, H., Member of WHO, Personal communication.
182. Goel, K., Lifshitz, F., Kahn, E., and Teichberg, S., Monosaccharide intolerance and soy protein hypersensitivity in an infant with diarrhea, *J. Pediatr.,* 93, 617–619, 1978.
183. Maclean, W. C., Lopez, C., de Romana, G., Massa, E., and Graham, G., Nutritional management of chronic diarrhea and malnutrition: primary reliance on oral feeding, *J. Pediatr.,* 97, 316–323, 1980.
184. Galeano, N. F., Lepage, G., Lorey, C., Belli, D., Levy, E., and Roy, C. C., Comparison of two special infant formulas designed for the treatment of protracted diarrhea, *J. Pediatr. Gastroenterol. Nutr.,* 7, 76–83, 1988.

185. Lifshitz, F., Fagundes-Neto, U., Castro Ferreira, V., Cordano, A., and da Costa Ribeiro, H., The response to dietary treatment of patients with chronic post-infectious diarrhea and lactose intolerance, *J. Am. Coll. Nutr.*, 9, 231–240, 1990.
186. Lifshitz, F., Fagundes Neto, V., Garcia Olivo, C. A., Cordano, A., and Friedman, S., Refeeding of infants with acute diarrheal disease, *J. Pediatr.*, 118, 99–108, 1991.
187. Sack, D. A., Roads, M., Molla, M., and Wahed, A., Carbohydrate malabsorption in infants with rotavirus diarrhea, *Am. J. Clin. Nutr.*, 36, 1112–1118, 1982.
188. Fagundes Neto, U., Patricio, F. R. da S., Wheba, J., Reis, M. H. L., Gianotic, O. F., and Trabulsi, L. R., An Escherichia coli strain that causes diarrhea by invasion of the small intestinal mucosa induces monosaccharide intolerance, *Arq. Gastroent. Sao Paulo*, 16, 205–208, 1979.
189. Lebenthal, E. and Rong-Bao, L., Glucose polymers in diarrhea: high caloric density nutrients with low osmolality, *J. Pediatr. Gastroenterol. Nutr.*, 11, 1–6, 1990.
190. Jones, A., Avigad, S., Diver-Harver, A. M. S., and Katznelson, D., Disturbed fat absorption following infectious gastroenteritis in children, *J. Pediatr.*, 95, 362–366, 1979.
191. Mann, M. D., Hill, I. D., Peat, G. M., and Bowie, M. D., Protein and fat absorption in prolonged diarrhea in infancy, *Arch. Dis. Child.*, 57, 268–273, 1982.
192. Woolf, G. N., Kurian, R., and Jeejeebhoy, K. N., Diet for patients with a short bowel: high fat or high carbohydrate? *Gastroenterology*, 84, 823–828, 1983.
193. MacLean, W. C., Jr., Klein, G., Lopez de Romana, G., Masse, E., and Graham, G., Protein quality of conventional and high protein rice and digestibility of glutinous and non-glutinous rice by preschool children, *J. Nutr.*, 108, 1740–1747, 1978.
194. Gastanabuy, A., Cordano, A., and Graham, G., Acceptability, intolerance and nutritional value of a rice-based infant formula, *J. Pediatr. Gastroenterol. Nutr.*, 11, 240–246, 1990.
195. Graham, G., Bartl, J. M., Cordano, A., and Morales, E., Lactose-free medium chain triglyceride formulas in severe malnutrition, *Am. J. Dis. Child.*, 126, 330–335, 1973.
196. Klish, W. J., Udall, J. N., Calvin, R. T., and Nichols, B. L., The effect of intestinal solute on water secretion in infants with acquired monosaccharide intolerance, *Pediatr. Res.* 14, 1343–1346, 1980.
197. Lifshitz, F., Perspectives of carbohydrate intolerance in infants with diarrhea, in *Carbohydrate Intolerance In Infancy,* Lifshitz, F., Ed., New York, Marcel Dekker, 1982, 3–20.
198. Fagundes-Neto, U., Dietary management of postinfections chronic diarrhea in malnourished infants, in *Nutrition for Special Needs in Infancy. Protein Hydrolysates,* Lifshitz, F., Ed., New York, Marcel Dekker, 1985, 175–191.
199. Brown, K. H., Epidemiological relationship between malnutrition and chronic diarrhea in infants and children, in *Malnutrition in chronic diet-associated infantile diarrhea,* Lifshitz, C. H. and Nichols, B. L., Eds., Academic Press, 1990, 209–234.
200. Fagundes-Neto, U., Malnutrition and the intestine, in *Clinical Disorders in Pediatric Gastroenterology and Nutrition,* Lifshitz, F., Ed., New York, Marcel Dekker, 1980, 249–266.
201. The management of acute diarrhea in children: oral rehydration, maintenance, and nutritional therapy, *MMWR Morb Mortal Wkly Rep.*, 1992, 41(RR-16).
202. The rational use of drugs in the management of acute diarrhea in children, Geneva, World Health Organization, 1990.
203. Figueroa-Quintanilla, D., Salafar-Lindo, E., Bradley Sack, R., et al., A controlled trial of bismuth subsalicylate in infants with acute watery diarrheal disease, *New Eng. J. Med.*, 328, 1653–1658, 1993.

Chapter 13

NUTRITIONAL COMPLICATIONS OF HIV INFECTION

Donald P. Kotler

TABLE OF CONTENTS

I. Introduction 137

II. Studies of Body Composition 138

III. Pathogenesis of Malnutrition 139

IV. Diagnosis 139

V. Treatment 140

References 141

I. INTRODUCTION

Malnutrition is a common complication of HIV infection and is felt by many to play a significant and independent role in the morbidity and even mortality of the disease.[1–2] Progress in the field of nutrition in HIV infection has been hampered by the large number of other priorities for study and implementation, the wide variety of potential pathogenic mechanisms, a widespread perception of clinical nutrition as market-driven rather than science-driven, and the relative weaknesses of the tools for studying clinical nutrition outside of specialized research centers. However, the possibility of providing clinical benefit to patients using established and available modalities justifies the development and application of the techniques of nutritional monitoring and support.

There are several relevant questions concerning malnutrition in HIV infection. Knowledge of its prevalence, severity, and progression are important to determine in order to evaluate the magnitude of the problem. The respective roles of the pathogenetic mechanisms, involving nutrient intake, nutrient absorption, and intermediary metabolism in producing malnutrition are important to determine to understand the causes of wasting. The effect of malnutrition upon disease course in a patient with AIDS is poorly understood. Nutrient deficiencies might exacerbate the clinical immune deficiency.[3] Severe malnutrition is an important source of morbidity and often results in a markedly diminished quality of life in HIV-infected individuals, in whatever disease it accompanies, or by itself. The impact

0-8493-2764-4/95

of malnutrition extends to society-at-large, since repeated or prolonged hospitalizations or the need for chronic custodial care are costly and compete for available health care resources. The efficacy of nutritional support in HIV-infected individuals, especially in pediatrics, has received little attention.[4] Both the ability to improve nutritional status as well as the secondary effects that might accompany repletion, such as improved quality of life, improved physical and mental performance, improved immune function, and others, require study.

The aims of this presentation are to characterize the wasting processes that occur in HIV-infected individuals; to illustrate potential alterations in nutrient intake, absorption, and metabolism; and to present studies illustrating the potential benefits of nutritional support. The studies of nutritional support to be presented have been performed in patients with overt wasting. It should be remembered that the nutritional management of HIV-infected patients varies with disease stage as well as the presence of specific complications and includes both treatment and prevention. Most published reports have been of studies in adult HIV-infected individuals, though the few reported studies in infants and children suggest that the processes and clinical approaches are substantially the same.

II. STUDIES OF BODY COMPOSITION

Previous studies have defined the characteristics of wasting in AIDS patients. Studies of body composition demonstrated that depletion of body cell mass (protoplasmic mass) was frequent and occurred out of proportion to loss of weight as well as loss of body fat, a situation that differs from starvation and more closely resembles chronic sepsis.[5] Intracellular water volumes also were decreased, while relative extracellular water volumes were high, indicating overhydration. Overhydration may mask progressive weight loss and increase the difficulty in detecting wasting.

The course of wasting illnesses was evaluated by examining the extent of body cell mass depletion as a function of the time before death.[6] The results indicated a progressive depletion of body cell mass. The extent of depletion at death was about 50% of estimated premorbid values, corresponding to a body weight about one-third below ideal. These data are similar to historical studies of starvation in adults and children. The results suggested that the timing of death from wasting in AIDS is related to the degree of body cell mass depletion rather than its specific cause. Other studies also have correlated indices of malnutrition with shortened survival.[7]

Progressive wasting is not an invariable consequence of AIDS. This was shown in a study of clinically stable patients, who were characterized by stable, moderate body cell mass depletion, normal caloric intake, mild-moderate malabsorption of sugars and fats, and hypometabolism.[8] Other stages of the disease and active systemic disease complications are associated with hypermetabolism, however. Thus, wasting must be related to disease complications rather than to the underlying immune deficiency per se.

III. PATHOGENESIS OF MALNUTRITION

The development of malnutrition in AIDS is multifactoral and confounded by interlocking processes.[1–2] Patients may have a variety of symptoms that suggest the presence of a wasting process. Many conditions affecting the oral cavity, pharynx, or esophagus, as well as processes affecting the central nervous system, may impair food intake. Nutrient malabsorption has been noted by several investigators and may be related to intestinal damage often associated with specific enteric infections.[9–10] Small intestinal structure and function are preserved in other patients, implying that the intestines of an AIDS patient should be capable of absorbing sufficient nutrients to maintain nutritional status in the absence of a specific enterocyte infection. Metabolic abnormalities in AIDS patients may be a result of opportunistic infections. They also might occur as a direct consequence of HIV infection. Resting energy expenditure may be elevated in HIV-infected individuals, even in the early stages of the disease.[11] Fasting hypertriglyceridemia is common at several stages of the disease and can be correlated to circulating alpha interferon levels.[2] Thus, metabolic and inflammatory changes might be associated and related to alterations in cytokine activities that occur as a result of infection with HIV or other pathogens.

IV. DIAGNOSIS

Diagnosis of the cause of malnutrition can be determined using an algorithmic approach. Food intake can be estimated by diet history or more carefully by formal calorie counts with the assistance of a dietician. If low, possible causes such as offending medications, local pathology, and focal or diffuse neurologic disease can be evaluated systematically. Since secondary anorexia may be accompanied by malabsorptive or metabolic disorders, their evaluation also may be part of the work-up. The presence of malabsorption, suggested by the complaints of abdominal bloating or other symptoms, needs to be assessed. Noninvasive absorption studies can predict the presence of small intestinal vs. colonic disease as the cause of diarrhea. If small intestinal dysfunction is abnormal and stool examinations are unrevealing, intestinal biopsy may be required to make the appropriate diagnosis. Etiologic diagnosis of malabsorption syndromes should be possible in more than two-thirds of patients and includes such infections as cryptosporidiosis, microsporidiosis, and *Mycobacterium avium intracellulare.*

HIV-infected infants and children have been reported to have a high prevalence of idiopathic enteropathy, though cryptosporidiosis and microsporidiosis have been diagnosed in children. Metabolic derangements that accompany systemic diseases, such as pneumocystis pneumonia, cytomegalovirus infections, MAI, tumors and other complications, typically produce fatigue, tachycardia, or fever. Some metabolic derangements, recently recognized to occur in HIV-infected individuals, may not cause hypermetabolism. Muscle wasting due to loss or endogenous anabolic factors, such as testosterone, may occur in the absence of other disease complications and without detectable changes in nutrient intake or absorption.

V. TREATMENT

The effect of nutritional support in AIDS is an important question, given the numbers of patients seen and the severity of the malnutrition. The goals and general principles of nutritional support in an AIDS patient do not differ from non-HIV-infected individuals. The importance of effective treatment of underlying infections was demonstrated during a longitudinal study of CMV colitis in AIDS patients.[12] The development of CMV colitis was invariably accompanied by progressive tissue depletion before specific treatments were available. In contrast, patients treated with ganciclovir, an agent capable of suppressing the growth of CMV, repleted body mass in the absence of formal nutritional support. Furthermore, other patients, who had coexisting untreated infections, failed to replete body cell mass, even when nutritional support was administered.

Nutritional support frequently is indicated in patients with AIDS who have received a variety of parenteral and enteral approaches. Appetite stimulation has been demonstrated to increase body weight,[13] though specific effects on body cell mass are unknown since the majority of calories deposited appeared to be in adipose tissue. No studies utilizing appetite stimulation in HIV-infected children have been reported. The efficacy of TPN was evaluated and the ability to replete body cell mass was found to be related to the underlying clinical problem.[14] As a group, body cell mass did not change, and TPN was associated with increases in body fat content. However, repletion of body cell mass occurred in patients with eating disorders or malabsorption, while progressive wasting despite TPN was found in patients with systemic infections and their associated metabolic derangements. Very limited studies of TPN in pediatric AIDS patients have been reported. A relatively high incidence of catheter complications has been reported in a pediatric series, with incidence rates of infection and thrombosis about the same as in adult AIDS populations and in cancer patients.[15]

The efficacy of enteral nutrition given via a percutaneously placed gastrostomy tube also was examined in adult AIDS patients with severe eating disorders and relatively preserved absorptive function.[16] Repletion of body cell mass was observed, despite the persistence of systemic infections. Similar data have been obtained, using forced feeding oral regimens, by a few investigators studying pediatric AIDS patients. Of note, body cell mass repletion was associated with increased total lymphocyte counts in peripheral blood in the gastrostomy study. If corroborated by future studies, the results would provide strong suggestive evidence of a potential immune benefit from nutritional support. Studies of the effects of anabolic agents are proceeding. Anecdotal reports suggest that some patients benefit significantly from such therapy, but no studies are being done in pediatric age groups.

Continued progress in the elucidation and treatment of malnutrition in AIDS is needed to make the most effective use of this clinical tool in patients affected by immune deficiencies. The information obtained from treatment of AIDS patients also should be relevant to patients with other chronic, progressive, inflammatory diseases.

REFERENCES

1. Hecker, L. M. and Kotler, D. P., Malnutrition in AIDS, *Nutr. Rev.*, 48, 393–401, 1990.
2. Grunfeld, C. and Kotler, D. P., The wasting syndrome and nutritional support in AIDS, *Semin. Gastrointest. Dis.*, 2, 25–36, 1991.
3. Chandra, R. K., Nutrition, immunity, and infection: present knowledge and future directions, *Lancet,* 1, 688–691, 1983.
4. Nicholas, N., Leung, J., and Fennoy, I., Guidelines for nutritional support of HIV-infected children, *J. Pediatr.* 119, S59–62, 1991.
5. Kotler, D. P., Wang, J., and Pierson, R., Studies of body composition in patients with the Acquired Immunodeficiency Syndrome, *Am. J. Clin. Nutr.*, 42, 1255–65, 1985.
6. Kotler, D. P., Tierney, A. R., Wang, J., and Pierson, R. N., Jr., The magnitude of body cell mass depletion determines the timing of death from wasting in AIDS, *Am. J. Clin. Nutr.*, 50, 444–7, 1989.
7. Chlebowski, R. T., Grosvenor, M. B., Bernhard, N. H., et al., Nutritional status, gastrointestinal dysfunction, and survival in patients with AIDS, *Am. J. Gastroenterol.*, 84, 1288–93, 1989.
8. Kotler, D. P., Tierney, A. R., Brenner, S. K., Couture, S., Wang, J., and Pierson, R. N., Jr., Preservation of short-term energy balance in clinically stable patients with AIDS, *Am. J. Clin. Nutr.*, 57, 7–13, 1990.
9. Yolken, R. H., Hart, W., Oung, I., Shiff, C., Greenson, J., and Perman, J. A., Gastrointestinal dysfunction and disaccharide intolerance in children infected with the human immunodeficiency virus, *J. Pediatr.*, 118, 359–63, 1991.
10. Kotler, D. P., Francisco, A., Clayton, F., Scholes, J. V., and Orenstein, J. M., Small intestinal injury and parasitic disease in the acquired immunodeficiency syndrome (AIDS), *Ann. Intern. Med.*, 113, 444–9, 1990.
11. Grunfeld, C., Pang, M., Shimizu, L., Shigenaga, J. K., Jensen, P., and Feingold, K. R., Resting energy expenditure, caloric intake, and short-term weight change in human immunodeficiency virus infection and AIDS, *Am. J. Clin. Nutr.*, 55, 455–60, 1992.
12. Kotler, D. P., Tierney, A. R., Altilio, D., Wang, J., and Pierson, R. N., Jr., Body mass repletion during ganciclovir therapy of cytomegalovirus infections in patients with the acquired immunodeficiency syndrome, *Arch. Int. Med.*, 149, 901–5, 1989.
13. Von Roenn, J. H., Murphy, R. L., Weber, K. M., Williams, L. M., and Weitzman, S. A., Megesterol acetate for treatment of cachexia associated with human immunodeficiency virus infection, *Ann. Intern. Med.*, 109, 840–841, 1988.
14. Kotler, D. P., Tierney, A. R., Wang, J., and Pierson, R. N., Jr., Effect of home total parenteral nutrition upon body composition in AIDS, *J. Parenter. Enteral. Nutr.*, 14, 454–8, 1990.
15. Gleason-Morgan, D., Church, J. A., Bagnell-Reeb, H., and Atkinson, J., Complications of central venous catheters in pediatric patients with acquired immunodeficiency syndrome, *Pediatr. Infect. Dis. J.*, 10, 11–14, 1991.
16. Kotler, D. P., Tierney, A. R., Ferraro, R., Cuff, P., Wang, J., Pierson, R. N., and Heymsfield, S., Effect of enteral feeding upon body cell mass in AIDS, *Am. J. Clin. Nutr.*, 53, 149–54, 1991.

Chapter 14

NUTRITION AND GROWTH

Fima Lifshitz and Melanie M. Smith

TABLE OF CONTENTS

I. Introduction 143

II. Diagnosis of Nutritional Dwarfing 144

III. Pathophysiology 145

IV. Etiology 150
   A. Dieting in Children 152
   B. The Cholesterol Concern 153

V. Final Considerations 154

References 155

I. INTRODUCTION

Nutrition is essential for normal growth and development, and the adequacy of weight gain and linear growth is an essential measure of children's nutritional status. Although currently the nutritional focus among North American nations is on the reduction of risk factors for chronic disease, the importance of adequate calories and nutrients for growth must not be overlooked.

Worldwide, poverty-related undernutrition is the most important cause of growth retardation.[1] The interdependence of infectious diseases and nutritional status are also of great concern in developing countries. Chronic suboptimal nutrition and/or frequent infections may compromise weight gain and linear growth, resulting in nutritional dwarfing (ND). ND is not necessarily associated with emaciation; short stature and/or poor growth may be the sole manifestations of nutritional inadequacy.[2,3]

Among suburban upper-middle-class adolescents in the United States, ND is most often due to insufficient nutrient intake from a variety of nonorganic causes.[4,5] Organic diseases such as inflammatory bowel disease (IBD), celiac disease, cystic fibrosis, etc. may also be associated with poor growth.[6] However, in these communities, nonorganic ND appears to be more common than classic endocrine disorders of short stature or organic growth failure.[7] It is, therefore,

0-8493-2764-4/95

important to recognize ND even among adolescent populations in which poverty-related malnutrition is rare.

II. DIAGNOSIS OF NUTRITIONAL DWARFING

According to Wellcome Trust Classification (WTC), the diagnostic criteria for ND includes a weight below the mean for age with minimal deficit in weight for height.[8,9] Although this classification covers the majority of ND patients, cross-sectional definitions may fail to distinguish ND from familial short stature and/or constitutional growth delay (FSS/CGD). These normal children also demonstrate weight below the "mean for age" and by definition, their height is also below standards for age; they often do not demonstrate significant weight for height deficits. To avoid misclassification, longitudinal weight and height progression data is essential.[10]

In the differential diagnosis of short stature, the further analysis of body weight progression may be necessary to recognize ND[11] (Figures 1a, 1b, and 1c). ND and FCC/CGD patients clinically appear similar, presenting with short stature, delayed puberty, bone age retardation, without overt malnutrition or biochemical abnormalities of short stature or malnutrition.[10] Calculation of the theoretical weight based on the previous growth pattern may distinguish the poor weight gain of ND (Figure 1a). Theoretical weight is defined as the weight that the patient should have attained at the time of the examination if the patient continued to gain weight at his/her previous rate established during the pre-morbid growth period. A body weight deficit for theoretical weight is characteristic of ND (Figure 1b). On the other hand, adequately nourished short children, i.e., FSS/CGD, usually gain weight along established percentiles, and the theoretical body weight based on the previous growth pattern remains constant (Figure 1c). Thus, weight gain velocity is an important difference between patients with nutritional growth retardation and those with constitutional growth delay.

In addition, ND may also be present in children whose stature is within the normal range and needs to be considered even when there is body weight excess for height[10] if there is a fall-off in weight and height across percentiles. Reduced growth velocity may occur in obese children who are dieting to lose weight.[12] This decrease has been reported to bring children more in line with the heights of the parents,[13] but final height may be reduced compared to nonobese peers.[14] Slight reduction in arm muscle and arm fat area may also be present in ND patients. However, such anthropometric alterations are not diagnostic or found in all patients.[10,11]

Most normal children exhibit minimal deficits or excesses in body weight in proportion to height as they progress along established percentiles.[15] These constitutional variations in body weight are usually within one or two major percentiles for the height; they represent variations in frame size and do not necessarily reflect over- or undernutrition.[16] The body weight and height increments of a high school student with constitutional thinness is depicted to illustrate this point

(Figure 2). Even though his weight was two major percentile lines below his height percentile (more than 20% body weight deficit for height), the adolescent grew normally. A body weight deficit for height which remains constant and permits normal growth to proceed along a set percentile is not malnutrition, as there is a positive balance for continued growth.

The assessment of growth patterns detects ND more accurately than a battery of biochemical or laboratory measurements.[17] ND patients have "adapted" to decreased availability of nutrients, and poor growth is the major expression of the suboptimal nutrition.[18] These patients do not appear to be wasted, and biochemical parameters including serum levels of retinol-binding-protein, prealbumin, albumin, and transferrin levels have not distinguished patients with ND from those with familial or constitutional short stature.[17] In addition, other indices of malnutrition, such as the urinary creatinine height index or the urinary nitrogen/creatinine ratio, do not usually reflect any abnormalities.

Recently, we reported that the erythrocyte Na+-K+ ATPase activity of ND patients was decreased compared to that of familial short stature children.[19] However, until this technique becomes widely available, its clinical application for diagnosis of ND is limited. Insulin-like growth factors (IGF) and their binding proteins (IGFBP) have also been studied in fasting conditions and varying levels of nutritional intake.[20–22] It has been shown that IGF-I is reduced in children with protein-energy malnutrition, prolonged fasting, and/or protein deficient states[23,25] but not in ND.[19] Likewise, serum IGFBP-3 is reduced in protein malnutrition,[25] and the levels improve with nutritional supportive therapy.[23] However, this parameter has not been studied in ND patients.

III. PATHOPHYSIOLOGY

Patients with nutritional growth retardation have reached an equilibrium between their genetic growth potential and their nutritional intake. Growth deceleration is the adaptive response to suboptimal nutrition. Decreased growth brings the nutrient demand into equilibrium with nutritional intake without producing alterations in protein markers of malnutrition or on other biochemical parameters. Of course, there are limits to these adaptive possibilities. Acute malnutrition may be superimposed on the reported short stature if nutritional deprivation becomes more severe. In such patients, anthropometric measurements, such as weight, skinfolds, or biochemical indices, would then also reflect the acute malnutrition.[16]

For many years, it has been known that a diminished energy intake results in a reduced metabolic rate even before there is a loss of body weight. A reduction in energy intake may decrease the rate of protein synthesis since this process is energy expensive and accounts for 10% to 15% of the basal metabolic rate.[25] Protein breakdown is also sensitive to energy deprivation. When dietary energy sources are limited, the nitrogen flux increases, which may result in excessive protein breakdown to provide energy.[26] Nitrogen retention markedly increases during the refeeding of malnourished children.[27] In addition, an increased rate of

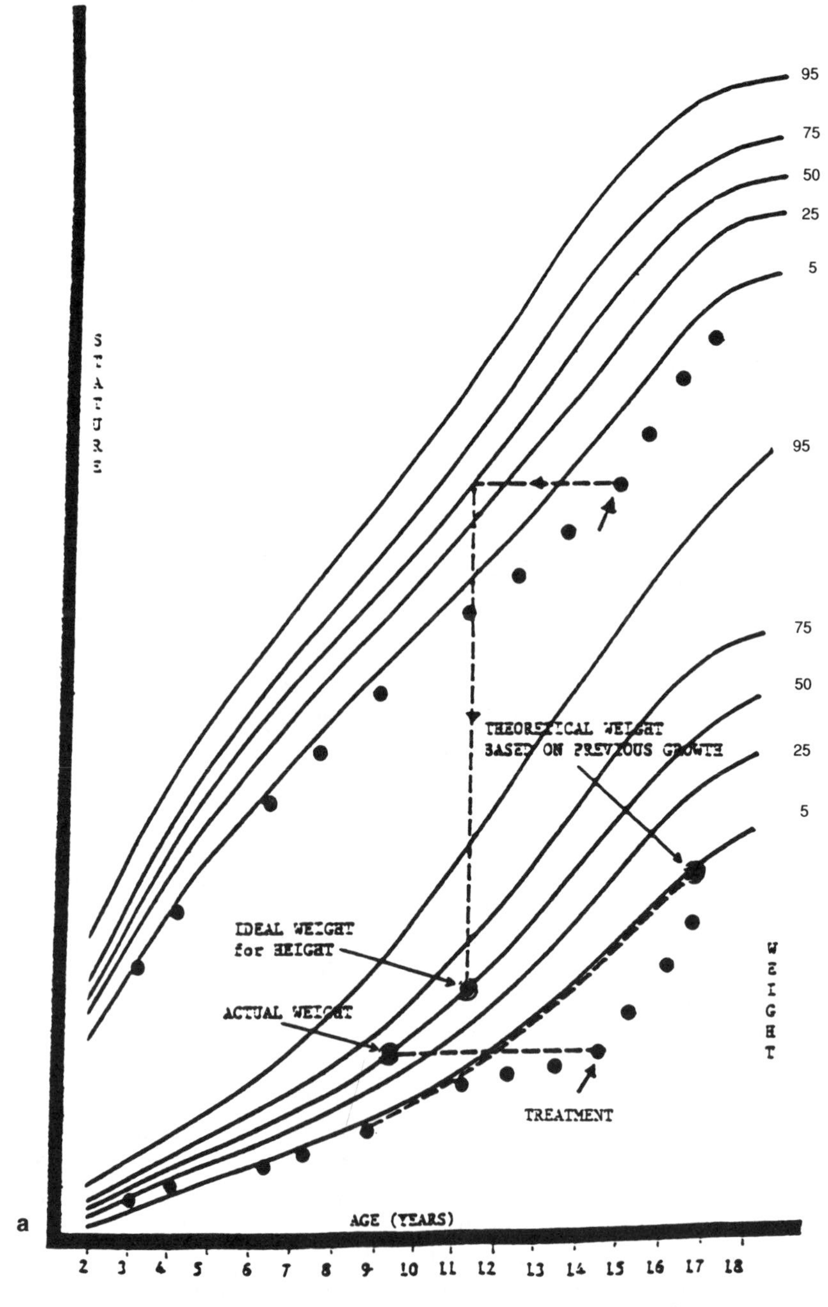

FIGURE 1. Assessment of nutritional dwarfing. To determine nutritional dwarfing, accurate body weight and height measurements are needed. In ND patients, body weight gain decreases in the two patients shown in Figures 1a and 1b around 11 years of age. Calculations of theoretical weight based on preillness growth progression after age 14 years reveals a body weight deficit. In the patient shown in Figure 1a, there is no body weight deficit for height, but when the growth chart is examined and weight progression extrapolated to calculate the theoretical weight, the actual weight deficit is easily

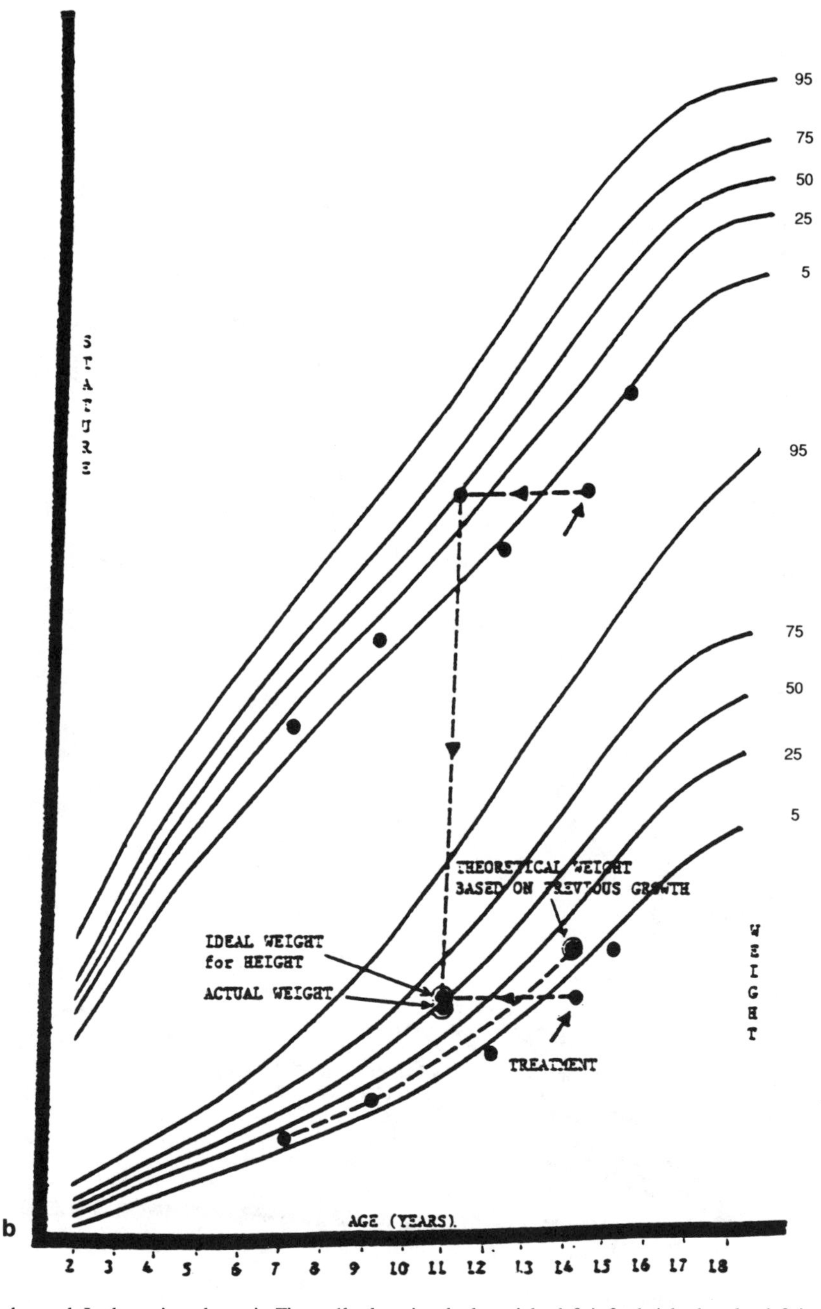

b

observed. In the patient shown in Figure 1b, there is a body weight deficit for height, but the deficit for theoretical weight is more marked. In contrast, the patient shown in Figure 1c reveals lack of nutritional dwarfing. This is a patient with constitutional growth delay who shows that his body weight gain consistently progressed along the lower percentile, with no deviation in growth. Notice that there was no body weight deficit for height or for theoretical weight based on previous growth.

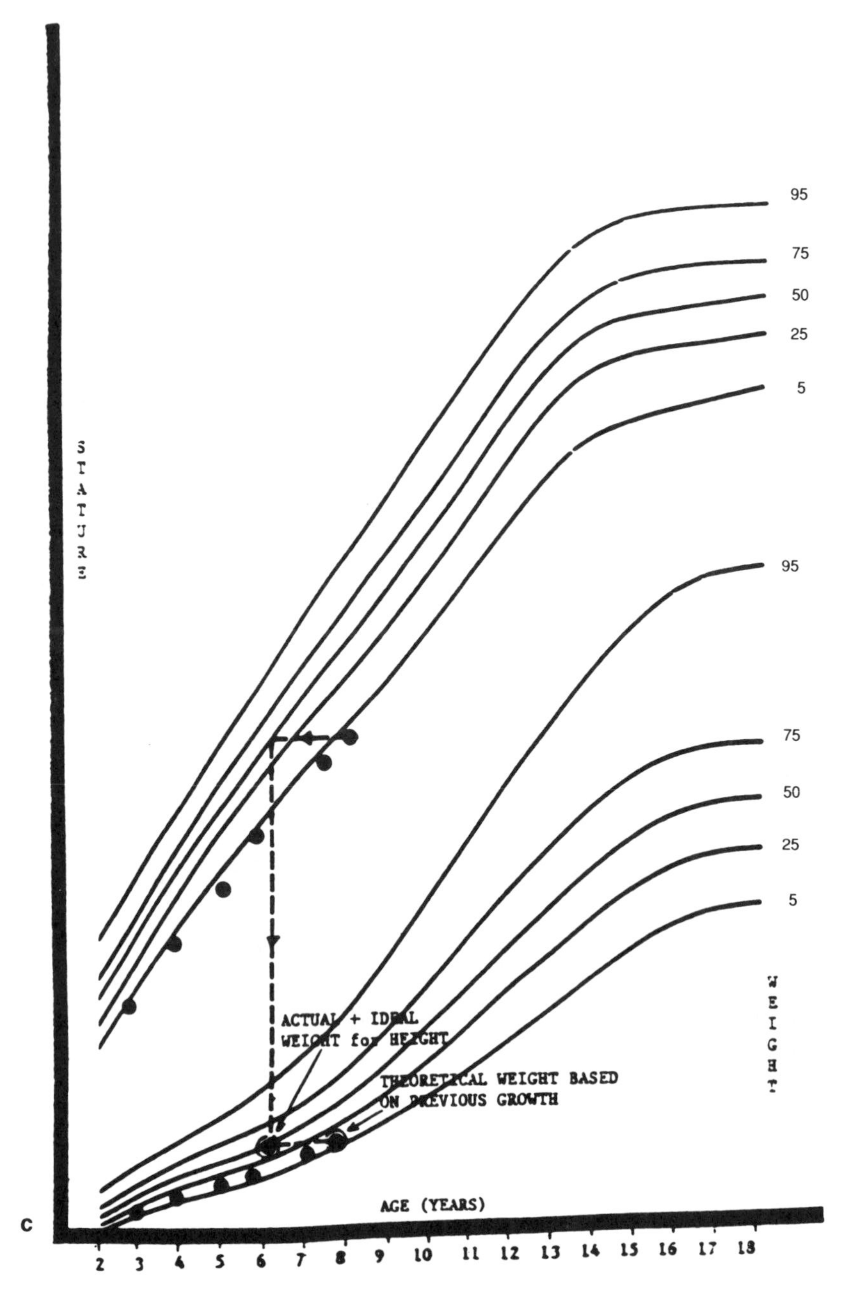

FIGURE 1. (continued)

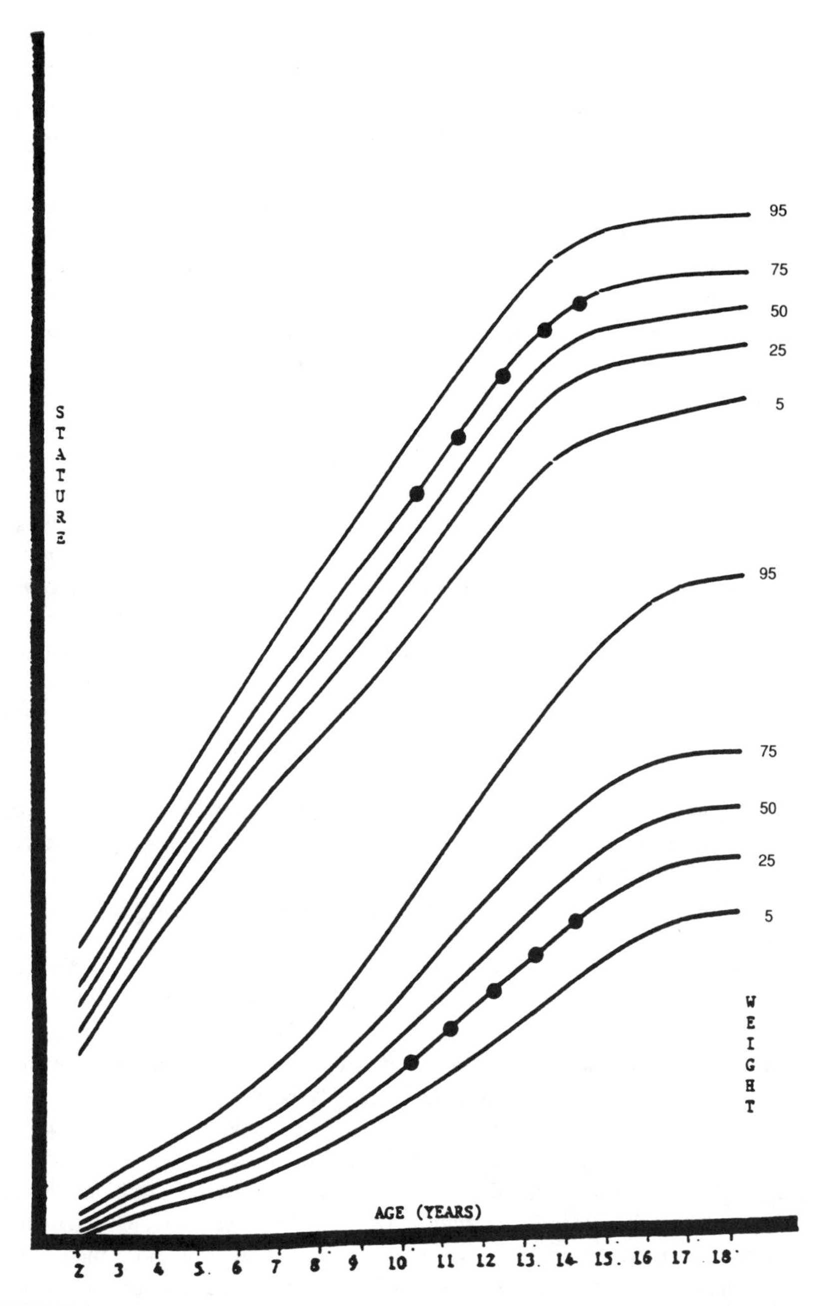

FIGURE 2. Constitutional underweight for height. Both weight and height progress consistently along the same percentiles for at least four years. Even though there is a body weight for height deficit, it is not malnutrition, as positive energy balance has allowed uninterrupted growth.

protein synthesis following improved dietary intake was demonstrated by ^{15}N-glycine measurements in malnourished children,[28] and nutritional recovery normalized the excretion of the radiolabeled amino acid.[26] In ND, the result of the altered rates of protein turnover and nitrogen retention may be the cessation of normal growth as an adaptive response to the decreased intake.

However, there is controversy whether decreased body size is an advantageous adaptation to a limited food supply or whether adverse health and functional impairments result.[28] It has been shown that physical activity is reduced with a 20% decrease in energy consumption,[29] but other measurements of functional compromise are more difficult to elicit. In any event, decreased growth velocity represents functional compromise that should be detected and treated as early as possible.

IV. ETIOLOGY

A variety of pathological entities which lead to decreased nutritional intake and/or malabsorption may lead to ND. Poor nutritional status and impaired growth may result from organic disease. The primary presentation in the so-called occult type of celiac disease is short stature.[30] Linear growth impairment in Crohn's disease is common, and it may precede weight loss, thus being the earliest clinical indication of the disease.[31] The nutritional consequences of these diseases vary in accordance with the severity and duration of the problem before diagnosis and/or intervention. Thus, whenever a child fails to gain weight and/or to grow, investigation to rule out these diseases is warranted.

However, ND is usually the result of nonorganic causes reflective of a voluntary, intentional or unintentional reduction in food intake.[32] Inappropriate eating behaviors, dissatisfaction with body weight and appearance, and unhealthy approaches toward weight control may cause ND among non-impoverished populations.[5]

There may be single or multiple nutritional deficiencies leading to poor growth in these ND patients.[33] Generalized malnutrition with multiple macro- and micronutrient deficits or nutritional alterations of a more specific nature have been observed. For example, ND resulting from decreased energy intake or from specific nutrient deficits including iron and zinc have been reported.[34]

The prevalence of nonorganic ND leading to malnutrition and poor growth in affluent communities is unknown. Only those patients whose height is markedly impaired have been recognized thus far. However, suboptimal nutritional intake may result in a fall-off in height within the normal percentiles which may elude medical attention. We have analyzed the growth patterns of more than 1,017 upper-middle-class students attending a suburban upper-middle-class high school and detected 18 students who had growth alteration associated with poor weight increments.[15] This finding may reflect the estimated prevalence of nutritionally related growth failure in our population (Figure 3).

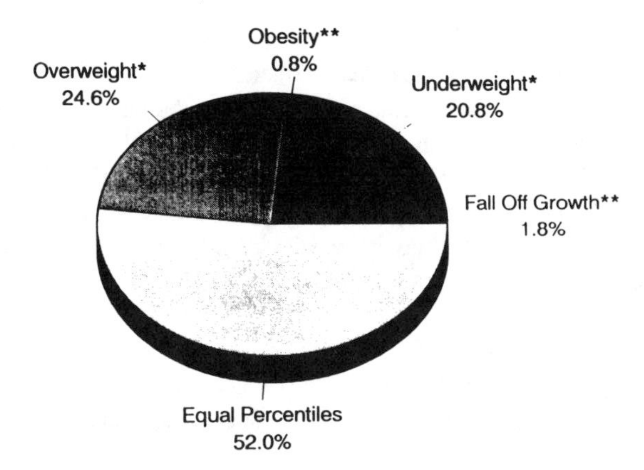

FIGURE 3. Body weight and height progression of 1017 students of a middle-class suburban high school. *Body weight for height excess or deficit of >10%. **Crossing two major percentiles. (Data modified from Lifshitz, F., Moses, N., Pugliese, M., et al., *J. Am. Coll. Nutr.*, 7(Abst.), 417, 1988.)

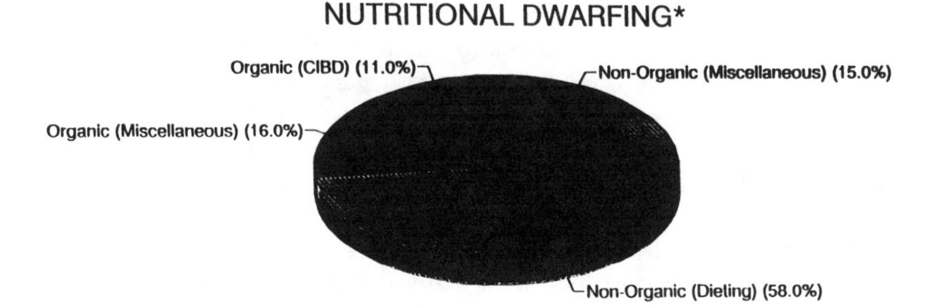

FIGURE 4. The distribution of 306 ND patients diagnosed in a middle-class pediatric endocrine clinic in Manhasset, N.Y. The majority are due to nonorganic causes and specifically due to dieting. CIBD: chronic inflammatory bowel disease.

In a Pediatric Endocrine Clinic located in the same geographic area as the above-mentioned high school, we detected over 300 patients with ND (Figure 4). ND due to nonorganic causes accounted for 73% of the patients with nutritional growth retardation. However, this estimate of ND prevalence in our series is subject to referral and selection bias. Pediatricians and other referral sources know and recognize our interest and expertise in this condition and refer patients who "fit the mold", whereas they may have referred other types of patients to other subspecialties for consultation and treatment.

A. DIETING IN CHILDREN

ND resulting from inappropriate dieting has not been associated with atypical eating disorders.[35] In a prospective study, ND patients were compared to a healthy control group of short-stature patients and did not report more behavioral or cognitive symptoms indicative of an eating disorder. They had typically positive self-perceptions of competence and global self-worth, and neither they nor their parents reported a relative excess of behavior problems.[36] Therefore, most patients with nonorganic ND in a middle-class, suburban population were dieting because of social and cultural factors that dictate eating patterns and diet choices rather than due to specific psychological or behavioral disturbances. The social and cultural ideal of thinness has often been blamed for the dietary behavior.[37] Perhaps this influence has created a fertile environment for the development of inappropriate eating patterns, which in extreme cases, result in anorexia nervosa and bulimia nervosa[38] or in other starvation syndromes.[4,5,39–41] Most ND children simply select inappropriate diet choices which result in insufficient nutrient intake for normal growth and result in ND. These patients, and often their families, expressed preoccupations that revolved around similar issues of body weight and concern with a "healthy" dietary intake. They avoided excess dietary fat and "junk food." Their diets, food preferences, and eating behaviors were described in detail elsewhere.[5] A smaller number of ND patients can be identified with a specific fear or health belief,[4,5,39] such as a fear of the consequences of hypercholesterolemia.[17]

There is a high prevalence of extreme measures taken by high school students to avoid obesity.[37,42,43] Most of this dieting behavior, inappropriate eating habits, and purging does not meet diagnostic criteria for eating disorders such as anorexia nervosa and bulimia. Its frequency indicates the power of the goal to achieve an ideal trim figure in adolescence, yet it occurs during a period of rapid growth and development.[44]

These inappropriate eating behaviors were not related to body weight; even the thinnest students often engage in efforts to lose additional weight.[42] Various types of purging behaviors such as self-induced vomiting were reported in 13% of 10th grade students in other studies.[43]

The medical consequences of inappropriate eating practices can be quite severe. In addition to ND, excessive dieting, binging, and purging may cause other medical complications including electrolyte disturbances, dental enamel erosion, acute gastric dilation, esophagitis, enlargement of the parotid gland, aspiration pneumonitis, and pancreatitis.

B. THE CHOLESTEROL CONCERN

An important determinant in the selection of food is its nutritive value, especially in the prevention of chronic degenerative diseases. Coronary heart disease (CHD), stroke, diabetes, atherosclerosis, and certain types of cancers account for five of the ten most frequent causes of death in the Western world.[45] These nutritionally related diseases may have their roots in childhood; thus, an early start for disease prevention through diet modification has been widely recommended.[46–51] Currently, several expert panels recommend maintenance of ideal body weight and reduction of dietary fat intake to <30% of calories from fat.[50] However, these chronic disorders are complex and poorly understood, and it is not known whether it is necessary to implement the current recommendations in childhood when these diets are potentially harmful to growth and development.[5,17] Interested readers should refer to this book's chapter by Drs. Tarim and Newman regarding cholesterol recommendations in children (Chapter 3).

Additionally, although all experts agree that low-fat, low-cholesterol diets are not indicated for children <2 years, parents, at times with the advice and consent of physicians, have fed infants diets in accordance with adult standards, resulting in failure to thrive.[40] The American Academy of Pediatrics has long recognized the potential nutritional deficits and growth disturbances in children who restrict their intake and advised that all dietary recommendations be followed in moderation and with caution.[49] Nonetheless, high school children are dieting and modifying their nutritional intake without supervision.[42,43] Other investigators demonstrated that as early as the third grade, children are engaging in similar diet behaviors.[52]

Avoiding several of the foods, which constitute the base of our diet without the necessary increase in grains and/or other foods not customarily ingested by our population, can easily result in an inappropriate dietary intake. It should be noted that 3 oz. of lean meat or pasta supply similar amounts of calories; however, much more pasta is needed to provide the same quantities of protein and essential minerals such as iron and zinc for growth.[53] It is difficult to meet the micronutrient needs of children with foods rich in complex carbohydrates that are bulky and filling. Additionally, avoidance of so-called ''junk food'' and substitution of low-calorie snacks such as raw vegetables may not provide sufficient energy. The end result may be insufficient energy sources and/or inadequate mineral intake which will not promote optimal growth and development.[11,17,40]

It must be recognized that a ''prudent diet'' is designed to conform with the desired low-fat and low-cholesterol goals and yet be nutritionally complete providing the necessary nutrient intake for normal growth. However, the recommendations are difficult to interpret into meaningful food choices. The description of <30% of calories from fat requires knowledge of fat and calorie contents of foods which is rarely available. This presentation increases the risk for noncompliance or overzealous elimination of many foods which are familiar and popular to children.[40,53]

V. FINAL CONSIDERATIONS

Children are not "little adults", and their dietary guidelines should differ from those recommended for older persons.[53] The current recommendations by the several national health agencies inappropriately imply that similar dietary needs exist for everyone over 2 years of age.[51] Curtailing dietary intake of fat and cholesterol to <30% of calories may not be beneficial to all children and nutritional growth problems may result in some.[5,39] In Eastern countries and in some specific population groups, adults and children ingest well-balanced, low-fat diets with few or no animal fats because there exists an appropriate "anthropology of eating".[54] These communities have the acquired knowledge, food preferences, and eating habits that support adequate nutrient intake without many animal or commercial products. In North American countries such as the United States, this may be more difficult to accomplish as major changes in lifestyle and dietary patterns and behaviors are required. At present, there is no data that directly suggests that these major adjustments must be implemented immediately for children.

However, the accumulating information linking dietary factors to chronic diseases in adults should not be ignored.[55,56] It is inappropriate to justify a very high intake of saturated fat and cholesterol in children >2 years of age. Recommendations for children should emphasize consumption of a large variety of foods since taste preferences are being developed. Guidelines must highlight moderation and a desirable range of 30 to 35% of calories from fat with special emphasis given to foods low in saturated fats. Preferably, dietary recommendations should focus on reducing saturated fat intake to 8 to 12% of total energy intake. It is the whole diet that is important; single factors such as fat content need to be examined in a larger context. Attention to other nutrients including calories, minerals, and vitamins necessary for normal growth cannot be overlooked.

Although inadequate food supplies and adverse environmental conditions are major problems and the main causes of nutritional growth retardation worldwide, the desire to maintain a "slim and trim figure" and to follow a healthful diet is a frequent problem among American children and adolescents. Appropriate and realistic body weights should be emphasized, and the potential adverse consequences of weight loss attempts and restrictive diets during childhood should be emphasized.

REFERENCES

1. Torun, B. and Viteri, F. E., Protein energy malnutrition, in *Modern Nutrition in Health and Disease,* seventh ed. Shils, M. E. and Young, V. R., Eds., Philadelphia, Lea & Febiger, 1988, 746–773.
2. World Health Organization, Development of indicators for monitoring progress towards health for all by the year 2000, Geneva, "Health for All" Series (4), 1981.
3. FAO/UNICEF/WHO: *Methodology of Nutritional Surveillance,* Report of a joint FAO/UNICEF/WHO Expert Committee, Geneva, Technical Report Series (593), 1976.
4. Lifshitz, F., Nutrition and Growth, in *Clinical Nutrition. Nutrition and Growth Supplement,* (4) Paige, D. M., Ed., St. Louis, C. V. Mosby, 1985, 40–47.
5. Lifshitz, F., Moses, N., Cervantes, C., et al., Nutritional dwarfing in adolescence, *Semin. Adolesc. Med.,* 3, 255–256, 1987.
6. Kelts, D. G., Grand, R. J., Shen, G., et al., Nutritional basis of growth failure in children and adolescents with Crohn's disease, 76, 720–727, 1979.
7. Lifshitz, F., Nutritional dwarfing in adolescence, *Growth, Genetics, & Hormones,* 3(4), 1–5, 1987.
8. Keller, W. and Fillmore, C. M., Prevalence of protein-energy malnutrition, *World Health Stat. Q.,* 36, 129–167, 1983.
9. Abdenur, J. E., Pugliese, M. T., Cervantes, C., et al., "Alterations in spontaneous growth hormone secretion and the response to growth hormone releasing hormone in children with nonorganic nutritional dwarfing, *J. Clin. End. Metab.,* 75, 930–934, 1992.
10. Towbridge, F. L., Marks, J. S., DeRomana, G. L., et al., Body composition of Peruvian children with short stature and high weight for height: implication for the interpretation for weight for height: Implication for the interpretation for weight-for-height as an indicator of nutritional status, *Am. J. Clin. Nutr.,* 46, 411–418, 1987.
11. Lifshitz, F. and Moses, N., Nutritional growth retardation, in *Pediatric Endocrinology,* Lifshitz, F., Ed., New York and Basel, Marcel Dekker, Inc., 1991, 111–133.
12. Dietz, W. H and Hartung, R., Changes in height velocity of obese preadolescents during weight reduction, *Am. J. Dis. Child,* 139, 704–708, 1985.
13. Epstein, L. H. and Rena, R. W., Long-term effects of family-based treatment of childhood obesity. *J. Consult. Clin. Psychiatry,* 1, 91–95, 1987.
14. Merrit, R. J., Obesity, *Curr. Probl. Pediatr.,* 12, 1–58, 1982.
15. Pugliese, M., Recker, B., and Lifshitz, F., A survey to determine the prevalence of abnormal growth patterns in adolescence, *J. Adoles. Health Care,* 9, 181–187, 1988.
16. Lifshitz, F., Moses, N., Pugliese, M., et al., Criteria for diagnosis of malnutrition in children with nutritional dwarfing, *J. Am. Coll. Nutr.,* 7, 417 (Abstract), 1988.
17. Lifshitz, F. and Moses, N., Growth failure: A complication of dietary treatment of hypercholesterolemia, *Am. J. Dis. Child.,* 143, 537–542, 1989.
18. Viteri, F. E. and Torun, B., Protein-energy malnutrition, in *Clinical Nutrition,* 2nd ed., Paige, D. M., Ed., C. V. Mosby, St. Louis, 1988, 531–546.
19. Lifshitz, F., Friedman, S., Smith, M. M., et al., Nutritional dwarfing: a growth abnormality associated with reduced erythrocyte Na+ K+ ATPase activity, *Am. J. Clin. Nutr.,* 54, 1–7, 1991.
20. Clemmons, D. R. and Underwood, L. E., Nutritional regulation of IGF-I and IGF binding proteins, *Ann. Rev. Nutr.,* 11, 393–412, 1991.
21. Thissen, J. P., Underwood, I. E., Maiter, D., et al., Failure of IGF-I infusion to promote growth in protein-restricted rats despite normalization of serum IGF-I concentrations, *Endocrinology,* 128, 885–890, 1991.
22. Manson, J. M. and Wilmore, D. W., Positive nitrogen balance with growth hormone and hypocaloric intravenous feeding, *Surgery,* 100, 188–197, 1986.

23. Baxter, R. C. and Martin, J. L., Binding proteins for insulin-like growth factors: structure, regulation and function, *Prog. Growth Factir. Res.,* 1, 49–68, 1989.
24. Phillips, L. S. and Unterman, T. G., Somatomedin activity in disorders of nutrition and metabolism, *Clin. Endocrinol. Metab.,* 13, 145–189, 1984.
25. Clemmons, D. R., Klibanski, A., Underwood, L. E., et al., Reduction of plasma immunoreactive somatomedin-C during fasting in humans, *J. Clin. Endocrinol Metab.,* 53, 1247–1250, 1981.
26. Poehlman, F., Melby, C. L., and Badylak, S. F., Resting metabolic rate and post-prandial thermogenesis in highly trained and untrained males, *Am. J. Clin. Nutr.,* 47, 793–798, 1988.
27. Read, W. W., McLaren, D. S., Tchalian, M., et al., Studies with ^{15}N-labelled ammonia and urea in the malnourished child, *J. Clin. Invest.,* 48, 1143–1149, 1969.
28. Waterlow, J. C., Golden, M. H., and Garlick, P. J., Protein turnover in man measured with ^{15}N=Comparison of end products and dose regimens, *Am. J. Physiol.,* 235 (2); E 165-E 174, 1978.
29. Beaton, G. H., The significance of adaptation in the definition of nutrient requirements and for nutrition policy. In Baxter with, Waterlow, K., Eds., *Nutritional Adaptation in Man,* John Libley, London, 1985, 219–232.
30. Viteri, F. E. and Torun, B., Nutrition, physical activity and growth, in *The Biology of Normal Human Growth,* Ritzer, M., Apsia, A., and Hall, K., Eds., Raven Press, New York, 1981, 269–273.
31. Groll, A., Candy, D. C. A., Preece, M. A., et al., Short stature as the primary manifestation of coeliac disease, *Lancet,* 21, 1097–1099, 1980.
32. Kanof, M. E., Lake, A. M., and Bayless, T. M., Decreased height velocity in children and adolescents before the diagnosis of Chron's disease, *Gastroenterology,* 95, 1523–1527, 1988.
33. Lifshitz, F. and Moses N., Nutritional dwarfing: growth, dieting and fear of obesity, *J. Am. Coll. Nutr.,* 7, 368–376, 1988.
34. Lifshitz, F. and Nishi, Y., Mineral deficiencies during growth, in *Pediatric Diseases Related to Calcium,* Anast, C. and DeLuca, H., Eds., Elsevier, North-Holland, New York, 1980, 305–322.
35. Prasad, A. S., Zinc in growth and development and spectrum of human zinc deficiency, *J. Am. Coll. Nutr.,* 7, 377–384, 1988.
36. Button, E. J. and Whitestone, A., Subclinical anorexia nervosa, *Psychol. Med.,* 11, 509–516, 1981.
37. Sandberg, D. E., Smith, M. M., Fornari, V., et al., Nutritional dwarfing: is it a consequence of disturbed psychosocial functioning? *Pediatrics,* 88, 926–33, 1991.
38. Storz, N. S. and Greene, W. J., Body weight, body image, and perception of fad diets in adolescent girls, *J. Nutr. Ed.,* 15, 15–18, 1983.
39. Herzog, D. B. and Copeland, P. M., Eating disorders, *N. Engl. J. Med.,* 313, 295–303, 1985.
40. Pugliese, M. T., Lifshitz, F., Grad, G., et al., Fear of obesity: a cause of short stature and delayed puberty, *N. Engl. J. Med.,* 309, 513–518, 1983.
41. Pugliese, M. T., Weyman-Daum, M., Moses, N., et al., Parental health beliefs as a cause of non-organic failure to thrive, *Pediatrics,* 8, 179–182, 1987.
42. Smith, N. J., Excessive weight loss and food aversion in athletes simulating anorexia nervosa, *Pediatrics,* 66, 139–142, 1980.
43. Moses, N., Banilivy, M., and Lifshitz, F., Fear of obesity among adolescent females, *Pediatrics,* 83, 393–398, 1989.
44. Killen, J. D., Taylor, C. B., Telch, M. J., et al., Self-induced vomiting and laxative and diuretic use among teenagers: precursors of the binge-purge syndrome, *JAMA,* 255, 1447–1449, 1986.
45. Lifshitz, F., Tarim, O., and Smith, M. M., Nutrition in adolescence, in *Endocrinology and Metabolism Clinics of North America,* Rosenberg, R., Ed., 22, 673–683, 1993.
46. Byrne, G., Surgeon General takes aim at saturated fats, *Science,* 241, 651, 1988.
47. Consensus Conference: Lowering blood cholesterol to prevent heart disease, *JAMA,* 253, 2080–2086, 1985.

48. American Cancer Society, Nutrition and Your Health: Dietary Guideline for Americans, *Home and Garden Bulletin,* No. 232, Washington, D.C., Government Printing Office, 1985.
49. American Heart Association, Nutrition committee: dietary guidelines for healthy American adults, *Circulation,* 74, 1465A–1468A, 1986.
50. American Academy of Pediatrics, Committee on Nutrition: Prudent life-style for children: dietary fat and cholesterol, *Pediatrics,* 78, 521–525, 1986.
51. Expert Panel: Report of the national cholesterol Education program expert panel on detection, evaluation, and treatment of high blood cholesterol in adults, *Arch. Intern. Med.,* 148, 36–69, 1988.
52. Expert Panel: Report of the national cholesterol Education program expert panel on children and adolescents, *Pediatrics,* 89, 525–584, 1992.
53. Maloney, M. J., McGuire, J., Daniels, S. R., et al., Dieting behavior and eating attitudes in children, *Pediatrics,* 84, 482–487, 1989.
54. Lifshitz, F., Moses, N., and Lifshitz, J., Children's Nutrition, Boston, Jones and Bartlett Publishers, 1991.
55. O'Connell, J. M., Dibley, M. J., Sierra, J., et al., Growth of vegetarian children. The farm study, *Pediatrics,* 84, 475–481, 1989.
56. Muldoon, M. F., Manuck, S. B., and Mathews, K. A., Lowering cholesterol concentrations and mortality: a quantitative review of primary prevention trials, *Br. Med. J.,* 301, 309–313, 1990.
57. National Academy of Sciences, Diet and Health. Implications for Reducing Chronic Disease Risk, Washington, D.C., National Academy Press, 1989.

Chapter 15

TRACE ELEMENTS IN PARENTERAL NUTRITION*

Adib A. Moukarzel and Marvin E. Ament

TABLE OF CONTENTS

I. Introduction 159

II. Iron 161

III. Zinc 162

IV. Copper 164

V. Chromium 166

VI. Manganese 169

VII. Molybdenum 170

VIII. Selenium 171

IX. Aluminum 173

X. Iodine 173

XI. Fluoride 174

XII. Other Trace Elements 174

References 175

I. INTRODUCTION

In humans, more than 60 minerals compose approximately 4% of body weight.[1] Seven of these are macrominerals (calcium, chlorine, magnesium, phosphorus, potassium, sodium, and sulfur) and have body stores not greater than

*Partially funded by Maimonides Research Foundation.

0-8493-2764-4/95

15 g. The other essential microminerals are present in minute or trace amounts, sometimes called ultratrace, quantities of less than 100 mg.[2–4] The fifteen minerals recognized as trace elements are iron, zinc, copper, manganese, nickel, cobalt, molybdenum, selenium, chromium, iodine, fluorine, tin, silicon, vanadium, and arsenic.[2] Trace elements are considered essential if their deficiency consistently results in impairment of function.[5] Trace element deficiency may produce both specific as well as nonspecific effects. Conversely, an excess of a trace metal, which arises either through a failure of intrinsic control mechanisms or through excessive exposure, may produce toxicity, which is either immediately apparent or latent and, ultimately of potential significance.[6] Trace metals in either human tissues, serum, or urine undergo pronounced alterations in response to infection, stress, malignancy, hormones, and drugs.[7–20] Usually, use of trace elements in parenteral nutrition is limited to chromium, cobalt, copper, iodine, iron, manganese, molybdenum, selenium, and zinc. In other parts of the world, silicon has also found its way into the parenteral regimens.[11] One of the early investigators to appreciate the need to add trace elements to parenteral regimens was Ellis.[21] He proposed the addition of zinc to dextrose large-volume parenteral solutions, but this practice was not supported by the AMA Department of Food and Nutrition.[3] Shils[22] stated that trace element deficiency may not be clinically evident until patients receive approximately two months of unsupplemented intravenous solutions, but he suggested that trace elements be used in all parenteral nutrition regimens. Early reports from Hinkins[23] and Solomons[24] alerted the clinician that patients receiving TPN were not receiving adequate concentrations of trace elements and suggested that cobalt, copper, iodine, manganese, and zinc be incorporated into intravenous parenteral nutrition solutions. In 1976, Hoffman and Ashby[25] indicated that trace element concentrations in certain intravenous parenteral nutrition solutions might be significant, and later, Gettin[26] (by both atomic absorption and neutron activation analysis methods) reported that intravenous solutions themselves contained trace element contamination. These studies corroborated earlier work done by Davis and Turco.[27–29] Shear and Bozian[30] showed, however, that both parenteral amino acids and protein hydrolysates (which have not been commercially available since the late 1970s) contain insufficient trace elements for recommended supplementation. Hauer and Kaminski[31] further supported the inadequate trace element content of parenteral amino acid and protein hydrolysate, as well as dextrose and lipid systems. With the advent of the use of parenteral nutrition solutions in critically ill patients (who may be particularly prone to trace element losses), Odne,[32] followed by Fliss and Lamy,[33] provided major directions that led to the eventual addition of trace elements to parenteral nutrition solutions. The AMA Department of Foods and Nutrition[3] later submitted guidelines for parenteral trace element supplementation (Table 1). In spite of the recommendations of the AMA for single-additive parenteral solutions, combination trace element solutions are available from several manufacturers. The AMA panel also recommended a large, single-additive, multidose vial with a volume sufficient to provide 10 to 30 daily adult doses for use in hospital pharmacies with

TABLE 1
Suggested Daily Intravenous Intake of Essential Trace Elements

	Pediatric[a] (μg/kg)	**Adult**		
		Stable	**Acute Catabolic[b]**	**GI Losses**
Zinc	300[c] 100[d]	2.5 to 4 mg	Additional 2mg	Add 12.2 mg/L small-bowel fluid lost; 17.1 mg/kg of stool or ileostomy output[e]
Copper	20	0.5 to 1.5 mg	—	—
Chromium	0.14 to 0.2	10 to 15 μg	—	20 μg
Manganese	2	0.15 to 8 mg	—	—

[a] Limited data are available for infants weighing less than 1500 g. Their requirements may be more than the recommendations because of their low body reserves and increased requirements for growth.

[b] Frequent monitoring of blood concentrations in these patients is essential to provide proper dosage.

[c] Premature infants (weight less than 1500 g) up to 3 kg of body weight. Thereafter, the recommendation for full-term infants applies.

[d] Values derived by mathematical fitting of balance data from a 71-patient-week study in 24 patients.

[e] Mean from balance study.

Data from American Medical Association, Department of Foods and Nutrition, *JAMA*, 241, 2051, 1979.

a large demand for parenteral feeding and smaller vials with no more than three maximum adult doses for limited use. Last, the panel stipulated that daily intravenous intake of essential trace elements represents (for the stable adult) more than 20 mL of additional fluid intake[3] to respect volume restrictions. Early literature[34–37] principally highlighted copper, zinc, and chromium (three of the four AMA-recommended trace elements);[34–37] still other trace elements have received attention.[38–40]

II. IRON

Iron[18] is known as an important component of porphyrin-based compounds which are bound in protein such as hemoglobin and myoglobin. In smaller amounts within the body tissue, iron is associated with enzymes and mitochondria which have significant metabolic functions. In addition, it is found in storage or transport forms bound to protein as ferritin and transferrin. Hemosiderin is a protein form of storage iron. When the body's iron concentrations are lower, more iron is bound to ferritin, and when body concentrations are high, hemosiderin is the predominant form of iron.[19] Very little iron is excreted from the body. 0.2 to 0.5 mg/day is typically excreted in the feces,[20] 0.2 mg in the urine, and a variable amount in sweat.[14,41] In healthy men, losses average 0.6 to 1.0 mg/day. In women, menstrual losses increase iron losses by an additional 0.5 to 0.8 mg/day or more. In normal humans, iron stores are governed by the control of absorption from the gastrointestinal tract. Normal losses of iron occur from the gastrointestinal tract when it is inflamed. It may also be lost from bile drainage. Normal males require 1 mg of absorbed iron per day and menstruating females require 2 mg per day. Patients who are receiving TPN may require even more iron, depending on whether or not they have abnormal losses or frequent blood drawn.

Intravenous iron needs to be provided in patients who are unable to absorb it through the enteral route. Since iron is absorbed proximally in the duodenum and jejunum, a trial of an oral iron preparation should always be given first. Dilute forms of iron dextran have been added to TPN solutions without any serious consequences. If one limits the amount of iron in a single day to what the normal requirements are, there is unlikely to be any problem with its administration. We have seen no consequences with providing 0.5 to 1 mg of an iron dextran preparation daily.[17] In order to document that sufficient iron is being provided, serum iron, total iron binding capacity, and percent saturation should be followed. This should prevent secondary hemochromatosis. We are aware of cases in which patients have been given an overdose of iron dextran when doses have been incorrectly administered. This resulted in a development of a secondary hemochromatosis with resultant cirrhosis of the liver.

III. ZINC

Zinc is a necessary element needed for the function of more than 70 enzymes from different species. Examples of such enzymes are carbonic anhydrase, carboxypeptidase, alkaline phosphatase, oxydoreductases, tranferases, ligases, hydrolases, lysases, and isomerases. In the presence of a zinc deficiency, there is a marked effect on nucleic acid metabolism, therefore influencing protein and amino acid metabolism. Zinc is a major constituent of DNA polymerase, reverse transcriptase, RNA polymerase, RNA synthetase, and the protein chain elongating factor. In the presence of zinc deficiency, growth arrest occurs. Many functions which are dependent on protein synthesis are suppressed by zinc deficiency. These obviously include growth cellular immunity and fertility. In addition, hair growth, wound healing, and plasma protein levels may be affected. Zinc is found through all the soft tissues of the body as well as in blood cells and teeth. Zinc is bound to protein firmly, and during the deficiency state the concentration of zinc in tissues does not change in a major way. Endogenous stores of zinc are mobilized in the fasting state, but in the presence of anabolism, do not meet the body's needs. This is because the net movement of zinc is into the tissues during anabolism and not out of it. Zinc which directly enters the circulation via the parenteral route is bound by albumin or an alpha-2-macroglobulin and carried to the liver. Zinc is excreted mainly in the feces, with a smaller amount in the urine. Urinary excretion is not influenced by intake. Significant losses of zinc are said to occur in the tropics; however, losses of zinc diminish with deficiency.[41]

Diarrhea and stomal and fistula losses are major sites of increased or abnormal losses of zinc from the body. Patients who are hypercatabolic may also have increased losses of zinc in the urine.[42] Zinc is taken up by the liver and other tissues. An infection results in increased uptake of zinc by the liver. The assessment of zinc status and requirements may be difficult. The most typical way to determine zinc status is to measure the circulating plasma levels, because it is a method of analysis which can be readily obtained in most communities. Although hair zinc levels are low when there is a low-grade chronic deficiency, this type of determination is not readily available in most communities. Leukocyte enzyme levels are a reliable indicator of zinc deficiency, which is not a measurement easily obtained.

Investigators[34,37] believe that patients receiving parenteral nutrition should receive a minimum of 2.5 mg/day if no diarrhea is present. In the presence of diarrhea and major gastrointestinal losses, the requirements[39] of zinc may increase substantially. Determination of zinc in small intestinal fluid has been as high as 12 mg/l and in some as high as 17 mg/l. Patients who receive parenteral nutrition and are having large losses from the intestinal tract have to be given more than the maintenance levels.

Infants have a unique property when compared to adults. They are growing. They may require maintenance levels plus additional zinc for growth. The preterm infant has the greatest requirement of zinc, because it is especially during the last trimester that the infant's zinc is transferred from the mother to the infant. It has been said that anywhere from 0.5 to .75 mg of zinc per day is taken up during the last 3 weeks of gestation and the first three weeks of post-natal life. Some investigators have reported a requirement of as much as 300 mcg/kg per day to maintain balance. In older children, 50 mcg/kg per day has been said to maintain serum levels and promote growth. Our experience using 100 mcg/kg/day shows that level is excessive.

Recently, some investigators have shown that serum zinc levels and urinary zinc outputs before and after major operative trauma show dramatic alterations in serum zinc concentration.[43] These studies have shown that serum zinc concentration drops markedly 6 hours after surgery in both supplemented and unsupplemented groups. It gradually returns to near normal without exogenous supplies of zinc. Provision of therapeutic dose of zinc was unable to prevent the abrupt dip of serum zinc level. In a study of adult patients receiving total parenteral nutrition, Tagaki[43] showed that administration of 3.9 mg/day of zinc was adequate to prevent plasma zinc levels from falling and maintained it in the normal range. None of the patients who received this dose of zinc developed signs or symptoms of zinc deficiency or showed abnormally elevated plasma zinc levels. In none of the subjects studied did the erythrocyte concentration of zinc move out of the normal range. Therefore, the erythrocyte zinc levels are not a sensitive indicator of body zinc status.

Infants may require an even larger amount of zinc on a milligram/kilogram basis because of their growth. In premature infants at the time of birth, there is a greater need for zinc in those who require parenteral nutrition.[44] James and MacMahon found that infants require as much as 300 mcg/kg/day to maintain balance.[45] At least 100 mcg/kg/day should be given to infants beyond the first six months of life. After adolescence, 50 mcg/kg/day should be sufficient, provided the patient does not have any extraneous losses.[44–47]

IV. COPPER

The human body contains approximately 75 mg of copper. Two percent of the total copper turns over daily. This meant that approximately 1.5 mg of copper is needed each day. In normal human beings, 25–40% of copper that is taken in food and drink is absorbed. The rate of absorption is regulated by the intestinal mucosa. Copper is excreted via the bile into the intestinal tract.[48] The absorption of copper throughout the intestinal tract is regulated in part by sulphur-rich proteins in the mucosal cells.[49] These proteins are either identical with or closely related to the threonine moiety of metallothionein. Following administration of copper into the bloodstream, it forms a complex with albumin and free amino acids and is transported to the liver. This fraction of plasma copper (so-called direct-reacting copper) normally makes up 10% or less of the total copper in plasma. The rest is firmly bound to ceruloplasmin. Although many of the key functions of this protein are unknown, we know that it catalyzes oxidative reactions and transports copper to tissues. Protein fraction or apoceruloplasmin is synthesized in the liver, where copper is added to form the holoprotein. Ceruloplasmin reaches a maximum in plasma 24 hours after the dose of copper is given. Intravenous copper is not accumulated by extrahepatic tissue until after its appearance in ceruloplasmin; tissues tend to take up ceruloplasmin copper in preference to cupric iron.[50] Ceruloplasmin is more efficient in restoring cytochromic oxidase activity than is cupric chloride.

Copper is a part of a variety of enzymes including cytochrome-C-oxidase, superoxide dismutase, dopamine-B-hydroxylase, monamine oxidase, and lysyloxidase.[49] Major manifestations of copper deficiency are shown through the consequences of ceruloplasmin and lysyloxidase deficiencies, although some have described abnormalities in catecholamine metabolism.[51] Ceruloplasmin is intimately related to iron metabolism. Ceruloplasmin is an iron oxidase. When red cells are broken down, the iron released is ingested by macrophages. It is ultimately released by the macrophages and comes down as transferrin for transport to iron-storing and iron-requiring tissues. Iron storage in the liver also is in equilibrium with transferrin. Ceruloplasmin oxidizes ferrous iron and supports the transfer of iron from stores to transferrin.[52] Vitamin B2; riboflavin, reduces the iron to the ferrous form in order to cross the cell membrane. After this has occurred, it is reoxidized to the ferric form to bind to transferrin.[53] As a result of copper deficiency, a secondary iron deficiency develops.

Two of the body's supportive proteins, collagen and elastin, are dependent on the copper-containing enzyme lysyloxidase for their formation. In the deficiency state of copper, poorly formed collagen and elastin are synthesized or are reduced in production. This can result in poor wound healing and weakening of tissues.

Because of this, patients with Menke's disease are susceptible to aneurysm formation. The most commonly recognized deficiency of copper is secondary leukopenia.[54]

The greatest concentrations of copper are in the liver and brain, with smaller amounts in the heart, kidney, spleen, and skeleton.[55] Two-thirds of the body's copper is found in the liver and the brain. Most of the copper circulates in the blood bound to ceruloplasmin; a small amount is found in albumin.[56] The albumin-bound copper is the true transport copper which exchanges with tissue. In normal individuals, the body copper is controlled by absorption and biliary secretion.[57,58] Formally, copper absorption is enhanced by deficiency and is depressed by ascorbic acid, cadmium, and phytates in the diet. Interestingly, zinc inhibits copper absorption by promoting intestinal metallothionine synthesis. It is this increased metallothionine production which binds copper and prevents its transfer to the circulation when the body has normal or excessive amounts of copper present.[59] Balance studies show that the dietary requirement is in the range of 2–5 mg/day. Of this amount, approximately one-third is absorbed (0.6–1.6 mg/day).[60] Other studies have shown that 1.2 mg/day is more than adequate.[61] Between 0.5 to 1.3 mg of copper is excreted each day in the bile. There is no enterohepatic circulation of copper.[60] Only 10–60 mcg/d are excreted in the urine. Infusion of copper intravenously does not increase its losses.[61] Small amounts of copper may be lost in the perspiration each day; the most commonly cited amount is 0.34–0.58 mg/day.[62] Diarrhea increases copper losses, but it is not proportional to the volume of diarrhea. Liver disease and reduced liver function reduces the amount of copper that is lost from the body. Cholestasis markedly reduces the amount of copper lost. Corticosteroids in excess increase urinary copper excretion, whereas deficiency results in the converse.[63] Trauma results in major increases in urinary copper excretion, just as it does for other trace metals. Studies have shown that the mean urinary copper excretion rate rises to 256 mcg/day.[64]

Copper in circulation is bound to albumin. Other amino acids that are bound to copper are histidine, threonine, and glutamine. Copper is taken up by the liver and bone marrow. Copper in the liver is incorporated in the ceruloplasmin and is released into the circulation or excreted into the bile. Within the bone marrow, it is incorporated into erythrocuprine and released in the red cells. Ceruloplasmin has been recognized as donating copper for incorporation into enzymes, such as superoxide dismutase, lysyl oxidase, and cytochrome oxidase.[65]

Plasma copper is obviously reduced in copper deficiency and is affected by a variety of factors that alter the serum concentration of ceruloplasmin. Protein calorie malnutrition results in a decrease in serum ceruloplasmin level; nephrosis also results in reduced levels of copper.[66] Infections, inflammatory conditions, and Hodgkin's disease all increase the level of serum copper. Oral contraceptives also increase plasma copper levels. Studies have shown the range of plasma copper in patients taking oral contraceptives to increase to 300 μg/dL. This represents an increase from normal range of 118 μg/dL in women who are not taking oral contraceptives. Measurement of copper in hair is not a reliable index of this element's deficiency, since hair is not a major place of deposition of copper.[67] A typical Western diet supplies 2 to 4 mg of copper per day, and with this intake copper deficiency has never been seen in an adult. During pregnancy, the major portion of the fetus' copper is obtained during the last 10 weeks of gestation.[68] During the first month of life, a premature or term infant typically will retain 100 to 130 μg/kg/day. Malnourished infants have been shown to require anywhere from 40 to 135 μg/kg/day to correct their deficiencies.

Studies[38] in adult patients receiving parenteral nutrition have shown that a minimum of 0.3 mg/day was necessary to meet the needs of patients, as long as they did not have abnormal gastrointestinal losses. Patients with diarrhea required at least two-thirds more. Patients who have large ostomy losses should receive additional copper besides that deemed necessary for normal maintenance. Patients with abnormal liver function may require as little as 0.10 mg/day.[61] In critically ill patients, it is recommended that at least 0.50 mg/day be provided.[69] In infants, the intravenous requirement for copper ranges between 10–50 μg/kg/day.[70] Some investigators have shown that a dosage of 50 μg/kg/day resulted in a reversal of leukopenia and neutropenia within a period of 5 days and a normalization of the copper level in a comparable period of time.[71] Anemia need not be present at the same time as leukopenia and neutropenia. A greater deficiency of copper may be required for this to take place. Copper deficiency may also be responsible for scurvy-like changes in long bones. The typical bony changes include osteoporosis, metaphyseal, and soft tissue calcification.[72] These changes are essentially those of scurvy. Patients who receive parenteral nutrition should no longer become copper deficient if physicians pay attention to the total needs of their patients. Children who receive 100 μg/kg/day have serum copper levels in the normal range.

V. CHROMIUM

Chromium is similar to the other essential trace elements; it is a transition element in the periodic table. The fact that this element is able to form coordination compounds and chelates is a characteristic which makes essential metals available in living systems.[73] Chromium is recognized in our diet in two forms: an inorganic chromium ($Cr3+$) in a biologically active molecule which seems to be a dinicotinachromium $3+$ complex, which is coordinated with amino acids to stabilize the molecule.[74]

Studies done throughout the 1950s and 1960s showed that there was an insulin potentiating factor found in brewer's yeast, which when fed to rats produced glucose tolerance. Chromium was recognized as the active ingredient in the glucose tolerance factor and was reported to be essential for optimal function of insulin in mammalian tissue.[75] It has been shown that the glucose tolerance factor which has been extracted from brewer's yeast binds to and potentiates insulin.[76] Chromium so far has not been recognized to function in enzyme systems, in metallo protein complexes, and/or in related structures. The exact role of chromium and how it functions has not been elucidated. It is uncertain how well chromium is absorbed in humans.[77] It is thought that organically complexed chromium is more immediately available for metabolic needs, but inorganic chromium requires incorporation into biologically active molecules.

Chromium binds competitively to transferrin; this suggests that it may be transported and stored with ferric iron.[78] Whereas chromium affects glucose disposal, there is no evidence that iron does. Chromium is released into the blood along with insulin following a glucose challenge.[79] Surprisingly, this release has not been as evident in subjects with abnormal glucose tolerance.[77] Ninety-five percent of chromium excretion takes place in the kidney and averages 1 to 10 μg/day.[80,81,85] Excretion of chromium is into the urine, and urinary losses are enhanced by glucose loading in diabetic subjects.[84]

Chromium seems to be a very safe trace metal, even in pharmacologic doses. A dose of 250 mcg given to malnourished, chromium-deficient children apparently showed no side effects.[82]

Human chromium deficiency has been difficult to document. Most studies involve individuals who have abnormal glucose intolerance and/or elevated lipid levels. Because chromium levels have often been difficult to measure, it has been very difficult in many instances to interpret results of studies. Chromium has been distributed throughout the human body, and its concentration declines with age.[83] This therefore correlates that glucose tolerance decreases with age. It would seem logical that if chromium deficiency is responsible for decreased glucose tolerance, then supplementation might improve this. A well-controlled study in elderly adults showed that supplementation of the diet with brewer's yeast improved glucose tolerance compared to no improvement with the use of a yeast that was free of this trace metal.

Measurement of chromium status is difficult. Normal ranges in some laboratories indicate that no detectable chromium may be considered to be normal. Therefore, it is hard to understand at which level deficiency truly occurs. It is said that hair chromium declines in situations associated with deficiency.[36] Some investigators believe that the only way to assess chromium deficiency is to demonstrate normal glucose clearance responding to chromium supplementation.

The true chromium requirements have not been determined. Data is lacking in this area. One adult patient has been reported in the literature to require between 10 to 20 μg/day.[35] In a study of infants, a requirement of 0.14 to 0.2 μg/kg/day was determined. Kien[89] reported a 6-year-old boy who received chromium-free total parenteral nutrition for a period of 16 months but still maintained serum chromium levels that were 4 times the upper limit of normal following a 2-year interval in which he received a 3 mcg/day of chromium in chromic chloride. However, the unsupplemented TPN solution contained 4 mcg/day, leading the investigator to conclude that chromium contamination of standard parenteral nutrition fluid may prevent biochemical evidence of low chromium status.

Only 3 cases of chromium deficiency have been reported in patients receiving TPN, all in women who had undergone massive bowel resection.[35,36,90] All 3 had glucose intolerance that was reversed by daily chromium supplementation. In all three cases, the range of serum chromium concentrations reported as normal was well above the 1 to 0 μg/l (19 nmol/l) now accepted as the upper limit of normal.[89] Since two reports[36,90] did not fully describe the method of measurement, the source of the discrepancy in the normal of chromium ranges cannot be identified.

In 1979, the American Medical Association expert panel on nutrition issued guidelines on essential trace element preparations for parenteral use.[91] The panel suggested the daily administration of 0.14–0.20 μg/kg in children receiving TPN. The committee on clinical practice issues of the American Society for Clinical Nutrition,[92] by extrapolation from the adult parenteral requirements for chromium, has since recommended the same amount for infants and children receiving TPN. In Europe, Ghisolfi[93] has recommended 0.10 to 0.20 μg/kg daily. Since there are few data on the parenteral chromium requirements of infants and children, we recently tested these recommendations by assessing chromium intake, serum chromium concentrations, and renal function in 15 children receiving TPN.[94]

The median duration of TPN use was 9.5 (range 1.3–14) years. The children's glomerular filtration rate (GFR), measured by plasma clearance of indium − 111-DTPA, was lower than that of non-TPN controls (70 vs. 110 ml/min per 1.73m2). The daily chromium intake averaged 0.15 μg/kg daily, but the serum chromium concentration was 20 (4 to 42) times higher than that of the controls (2.1 vs. 0.10 μg/l; $p < 0.0001$). GFR was significantly inversely correlated with serum chromium concentration, daily chromium intake, cumulative parenteral chromium intake, and TPN duration. We discontinued chromium supplementation of TPN solutions and reassessed the children a year later. Contaminating chromium concentrations were 1.0 to 1.8 μg/l in TPN solutions and 0.9 μ/l in fat emulsions. Drinking water contained 4.3 to 5.7 μg/l. Thus, the chromium intake without supplementation was only 0.05 μg/kg daily. The mean serum chromium concentration fell to 0.50 μ/l but was still significantly higher than that in the controls. The GFR did not change significantly (65 ml/min per 1.73 m2). No patient has shown signs of chromium deficiency.

Even though our subjects received less than the recommended parenteral chromium intake, their high serum chromium concentrations suggested that their intake was too high. Thus, the recommendations for parenteral nutrition are excessive and inappropriate. The chromium content of human milk is now known to be less than previously thought (0.3 μg/l),[95,96] which would provide an intake for a breastfed infant of about 0.05 μg/kg daily.[97] Such an intake is presumed to be adequate, at least for an infant born at term. The maximum parenteral requirement should be about this amount, since absorption of chromium from human milk is more efficient than the 0.5% usually accepted as the coefficient of absorption in children and adults.[98] We see no reason for the recommended parenteral intake to be four times the natural enteral intake.

The recommended, parenteral chromium intake for children needs to be revised downwards. A daily dose of 0.05 μg/kg might be more reasonable, but the dose needs further study. Since 1990, we have discontinued chromium supplementation of TPN solutions in all patients receiving long-term TPN, including adult patients since their serum chromium concentrations were also five to twenty times the upper limit of normal. The concentrations of contaminating chromium in TPN solutions seem high enough to prevent chromium deficiency.[99,100] However, changes in purification methods for TPN solutions could lead to insufficient concentrations of chromium.

Chromium contamination of TPN solutions should be monitored as well as chromium intake. Serum chromium concentrations should be measured regularly in patients receiving long-term TPN. If they are high, chromium supplementation needs to be reduced or stopped.

VI. MANGANESE

In the early 1930s, supplementation manganese was recognized to be an essential trace metal in animals. Although knowledge of the metabolism and enzymatic functions of manganese has increased, its significance in humans is still not clearly defined. The adult human body contains 10 to 20 mg of manganese.[101] Manganese is not accumulated in the liver before birth as occurs with copper.[102,103] In animals, regulation of the manganese levels primarily occurs through excretion rather than through regulation of absorption. Absorbed manganese is excreted through the intestine via bile, which is its main regulatory route. It is recognized that manganese retention in very young animals is quite large. It is also recognized that manganese retention may be affected by the addition of iron to the diet. The addition of iron to the diet depresses manganese retention.[104]

Manganese functions biochemically either as a co-factor activating a large number of enzymes which form metal enzyme complexes or as an integral part of metalloenzymes, which metal is built rather firmly into the metalloprotein. A group of enzymes that seem to be involved in manganese deficiency are the glycosal transferases. These enzymes are activated by manganese as a co-factor and are required for synthesis for glucoseaminoglycans.[105] Manganese-containing metalloenzymes are primarily found in the mitochondria. The two manganese metalloenzymes are pyruvate carboxylase and manganese superoxide dismutase.

A deficiency of manganese in animals results in abnormalities of cartilage growth. It also may be necessary for the action of vitamin K in adding the carbohydrate component of prothrombin to the prothrombin protein. A case has been described in which vitamin K could not correct prothrombin levels until the patient was given manganese. A human has been recorded to have manganese deficiency. This followed the inadvertent omission of manganese salt from a vitamin K deficient purified diet that he was receiving as a volunteer under metabolic ward conditions. This patient was known to have hypercholesterolemia, weight loss, transient dermatitis, occasional nausea and vomiting, changes in hair and beard color, and slow growth of beard and hair.[106]

The human body contains 12 to 20 mg of manganese mainly distributed in the mitochondria.[107] As indicated earlier, only 3 to 4% of an oral dose of manganese is absorbed and is very efficiently excreted by the intestine. Lesser amounts are excreted in pancreatic juice and through the intestinal wall. Little manganese is excreted in the urine. Manganese in the circulation is bound to be a beta globulin, transferrin. It is taken up by the mitochondria and more slowly by the nuclei. Oral intake in excess of need is excreted through the gastrointestinal tract.[108]

Dietary requirements are reported to be 2 to 3 mg/day.[109] The amount retained is between 50 to 400 mcg/day. The requirements for patients on parenteral nutrition have still not been well established. Shils,[22] in a study of various commercially produced parenteral nutrition solutions, showed that the amount of manganese in preparations tested varied among manufacturers and lot numbers. He showed that some of the additives in the solutions were high in manganese, particularly potassium phosphate and magnesium sulfate. The calculated manganese content in TPN formulations varied from 8.07 to 21.75 mcg/per total daily volume. These values agreed with those obtained from the analysis of the actual TPN solutions. Although many components of TPN solutions are contaminated with manganese, the contamination is relatively small and represents a small fraction of the lower limit of the suggested daily intravenous intake for stable adults, of 0.15 to 0.8 mg.

The potential daily contribution of infants is unlikely to exceed 0.7 to 1 mcg/kg with the recommended dosage at 2 to 10 mcg/kg of body weight. When there is persistent biliary obstruction, excretion of manganese may be impeded. Therefore, monitoring of blood levels is indicated if supplementary manganese is being contemplated or is being given. Our long-term pediatric patients had normal manganese levels in their plasma, despite the fact that none was added to the solutions. This indicates the natural combination was sufficient to meet their needs.

Although there is no data concerning manganese toxicity in patients treated with TPN, liver damage has been described in self-induced hypermangasemia with a manganese level of liver at 2.17 mg/dL, with a normal value cited at 0.11 mg/dL. Chronic industrial manganese poisoning affects mostly brain, producing extrapyramidal symptoms; manganese levels in urinary excretion were increased to 3.6 and 16.0 mcg/dL, respectively, with normal values given as 2.0 and 0.1 to 0.2 mcg/dL, respectively. Blood manganese four times greater than the highest level of normal values may be accompanied by signs of liver damage.[106]

VII. MOLYBDENUM

Molybdenum is an essential constituent of the enzymes xanthine oxidase, aldehyde oxidase, and sulfite oxidase. The activities of these enzymes decline in experimental deficiency. Xanthine oxidase activity in the blood of humans and in the tissues of animals living in a high molybdenum geochemical province of the Soviet Union was significantly elevated. In that population, uric acid levels in blood and urine were significantly elevated, and the incidence of gout is very high. Human liver and kidney have the highest concentrations of molybdenum.

Molybdenum in the diet is absorbed as molybdate in its hexavalent form and is easily absorbed from salts and vegetables. Excretion is mainly in the urine, but urinary excretion rises as sulfate intake or endogenous sulfate production increases. It has been found that patients with large volume diarrhea such as Crohn's disease may have excessive molybdenum losses in their stool.[111] The minimum requirements for molybdenum are really unknown. Limited balanced studies have shown that between 48 to 96 mcg/day may be required. Larger supplementation may be required in the patient with extra-gastrointestinal losses. Normally, patients with acute stress-like illnesses may have greater requirements. Intravenous data in children is not available. A single case of molybdenum deficiency has been recognized. The patient developed a coma-like syndrome which was reversed with 300 mg of molybdenum per day.[112]

VIII. SELENIUM

Selenium was recognized to prevent dietary liver necrosis in the rat in the early 1950s.[113] Subsequent to this, additional selenium-responsive conditions were recognized in a variety of animal diseases. All of these conditions occurred with low soil selenium content and were prevented by supplementing animal rations with selenium. In the late 1960s, pure selenium deficiency, defined as a pathological clinical condition when vitamin E was adequate, was produced in animals in the laboratory. Pure selenium deficiency in chicks caused pancreatic degeneration.[114] Growth retardation, partial alopecia, aspermatogenesis, and cataracts were manifestations of selenium deficiency in rats.[115–117] The selenium concentration of food depends on the soil in which it is produced and its content of protein. Selenium is typically found in protein fractions of foods; therefore, plants such as fruits and vegetables are poor sources of selenium, and meats are reliable sources of this element. In the United States, the daily dietary selenium intake is 60 to 216 mcg.[118,119] In the areas of China where Keshan disease occurs, selenium intake is less than 30 mcg/day.[109] In the late 1970s, some therapeutic and formula diets were found to contain less than 5 mcg of selenium in the daily intake.[120,121] In some parts of the world, selenium intake is quite high, and intakes have been reported up to 500 mcg/day without any toxicity.[122]

Food selenium is largely in the form of amino acids, such as selenomethionine.[123] Selenomethionine and sodium selenite, the inorganic form, have similar potencies for preventing selenium deficiency states. Both promote tissue glutathione peroxidase when administered to selenium-deficient individuals. Selenomethionine causes a greater rise in blood and tissue selenium levels, most likely because the former is incorporated into the primary structure of tissue proteins in place of methionine. Selenium in this form becomes available to the animal or to humans only after catabolism of the seleno amino acid. This form of selenium may serve as an unregulated storage or buffer pool of the element, providing endogenous selenium when dietary supply is interrupted. There is no other recognized storage form of selenium in the body. The absorption of selenium from the body appears to be under no physiological control. More than 90% of a given dose of selenium is absorbed, even when toxic levels are given. Reports in humans indicate that selenomethionine and selenite are absorbed nearly completely.[124] It appears as if humans regulate their selenium content through excretion. Selenium and vitamin E are interrelated in their actions. A deficiency of one can be partially corrected by giving the other. Glutathione peroxidase is an enzyme made up of 4 subunits, each containing selenocysteine as in integral part of the molecule.[125] In association with superoxide dysmutase, it controls the levels of superoxide and peroxide in the cell. This, in turn, affects lipid peroxidation of polyunsaturated fatty acids in cell membranes. Vitamin E is the second line of defense that controls the formation of hydroperoxides in the fatty acid residues of phospholipids, the process which depends on the antioxidant role of the vitamin and also involves its entering into a structural relation with membrane phospholipids.[126]

Intracellular lipid hydroperoxides may be reduced by glutathione oxidase to hydroxyacids. There are 4 biological reactions which produced superoxide: (1) enzyme reactions, such as those involving xanthine oxidases and galactose oxidase; (2) metabolic pathways, such as the hexose monophosphate shunt and oxidated reactions mediated by cytochrome P450; (3) interaction of dioxygen with the electron transport chain in the mitochondria; and (4) phagocytosis, where a burst of oxidative metabolism is associated with generation of NADPH by the hexosemonophosphate shunt, which in turn is used by the NADPH oxidase to generate superoxide. The excess superoxide is controlled by superoxidedysmutase and glutathione peroxidase. Therefore, it is not expected that bacterial killing is affected by selenium deficiency.

Selenium is bound to albumin, and after being processed by red cells, circulates in association with betalipoprotein.[127] It is taken up by tissues of the body and is incorporated into proteins and glutathione peroxides. Fecal excretion accounts for almost 60% of losses and the urinary excretion the remainder.[128,129]

Plasma selenium and glutathione peroxidase levels are sensitive to selenium intake and can be used to assess the need for this trace element. It has not been clearly determined how selenium affects the need for vitamin E in humans. The exact selenium requirements per day have not been determined. Some metabolic studies in humans suggest a minimum intake of 20 mcg per day; others have estimated the need to be as high as 54 mcg/day.[128]

Patients receiving parenteral nutrition may develop selenium deficiency with associated muscle pain and weakness. This may be manifested with elevations of serum creatine kinase and electromyographic evidence of myositis and nonspecific membrane irritability. Reversal of the myositis and normalization of serum creatine kinase can occur in less than 2 weeks.[129] Patients with selenium deficiency may be asymptomatic but have elevated transaminases and creatine phosphokinase.[130] Pain and muscle weakness have been observed in both children and adults.

Both children and adults have had pseudoalbinism described. A unique characteristic of patients on TPN with selenium deficiency is the presence of macrocytosis without anemia which has been described in adults as well as children. Decreased skin pigmentation has also been seen both in children and adults. Thirty-five percent of children and 75% adults who received long-term parenteral nutrition at UCLA Medical Clinic were found to have manifestations of selenium deficiency prior to the institution of selenium supplementation. Reversal of pseudoalbinism, darkening of skin and pigmentation, and reversal of macrocytosis occurred in all following supplementation of TPN solutions with selenium. Children were repleted with 2 mcg/kg/day and adults with 100 mcg/kg/day for 2 to 4 weeks followed by maintenance selenium levels in children of 0.5–1 mcg/kg/day and 40 mcg/day for adults. Periodic monitoring of serum levels should be done to avoid toxicity. All individuals found to be selenium deficient had normal vitamin E status, and none was shown to have increased hemolysis of red blood cells.[131–136]

IX. ALUMINUM

Today, TPN solutions may be contaminated with small amounts of aluminum, but not nearly as they were a decade ago.[137] Recent experience in our institutions has shown that infants and children receiving long-term parenteral nutrition have serum aluminum which is not any different from age and sex matched controls eating diets appropriate for age.[138] In the past, the major source of aluminum contamination was the use of casein hydrolysate. With the use of crystalline amino acids, sterile water, and dextrose, the contamination of the solutions is relatively low. It is the calcium salts that contributed up to 80% of the total aluminum load in the TPN solutions.[139] One must remember that similar contamination can occur from the ingestion of food. In the past, when aluminum contamination was excessive, there was a question of whether it impaired bone matrix formation and mineralization.[140]

X. IODINE

Iodine is of importance in the cellular oxidative processes associated with thyroid functions. Its principal role in humans is its incorporation into thyroid hormones (triiodothyronine and thyroxine) that regulate cellular metabolism, temperature, and normal growth. Approximately two-thirds of total body iodine is in the thyroid.

The minimum adult requirement[3] for iodine has been estimated at 50 to 75 µg/d; children and pregnant women have a higher requirement.

In 1989, the Committee on Clinical Practice Issues of the American Society for Clinical Nutrition[92] recommended 1 mcg/kg/d parenteral iodine intake for infants and children receiving TPN. In Europe, Ghisolfi[93,141] has recommended 5 mcg/kg/d of parenteral iodine for children. The routine administration of parenteral sodium iodine is recommended for long-term patients who are receiving solely parenteral support who are not absorbing enterally administered nutrients.

Our group has never supplemented iodine in parenteral nutrition solutions.[142] Therefore, we assessed serum iodine and thyroid function in groups of children receiving long-term TPN without iodine supplementation to determine if our patients had developed biochemical evidence of iodine deficiency.

We have found that there is no need to supplement parenteral nutrition solutions with iodine.[143] Typically, the iodine is provided in the water of parenteral nutrition plus as natural contaminants in many of the salts that are added to the solutions. Furthermore, if iodine antiseptic solutions are used to clean the site at which the catheter enters the body, this iodine may be absorbed through the skin and contribute to the normal iodine levels. In over two decades, we have never seen a patient with iodine deficiency who was on parenteral nutrition.

XI. FLUORIDE

The status of fluoride as an essential nutrient has been debated. Fluoride's role, whether physiological or pharmacological, in preventing dental caries has been indisputably documented. Fluoride may also play a physiological role in bone metabolism. Because fluoride is readily absorbed, intravenous requirements for subjects on long-term TPN may approach those recommendations given for oral use.[144]

According to the committee on issues of the American Society for Clinical Nutrition, in infants requiring long-term TPN (3 to 6 mo) without significant enteral feeding, a daily dose of 500 μg fluoride should be considered.[92] In Europe, 30 to 200 μg/kg of fluoride are added routinely to the TPN solutions in pediatrics[93,145] and 1 to 2 mg in adults.[3]

In a recent prospective study,[146] we assessed the trabecular bone mineral content of asymptomatic children receiving long-term parenteral nutrition. In all children, PTH, calcitonin, vitamin D, 25-OH, vitamin D, 125-(OH)2, and serum aluminum levels were within normal range.[138] The trabecular bone mineral content was on average 77% of normal controls. The trabecular bone mineral content was greater than −2SD in 50% of children.

In this population, even though the serum fluoride level was within normal levels, there was a significant positive correlation between the plasma fluoride level and bone density.[146] These findings suggest that fluoride may play a role in TPN-associated osteopenia.

XII. OTHER TRACE ELEMENTS

There is considerable evidence of the essentiality of silicon in mammalian nutrition. Silicon performs an important role in connective tissue, especially in bone and cartilage. Silicon's primary effect in bone and cartilage appears to be on the formation of the organic matrix. Bone and cartilage abnormalities are associated with a reduction in matrix components, resulting in the establishment of a requirement for silicon in collagen and glycosaminoglycan formation. Additional support for silicon's metabolic role in connective tissue is provided by the finding that silicon is a major ion of osteogenic cells, especially high in the metabolically active state of the cell.

Although there is evidence that serum silicone level of subjects receiving long-term parenteral nutrition are lower than that of controls, there is no definitive demonstration of its role in human nutrition.[146]

Arsenic and vanadium are among those essential in some mammalian but probably not in human. Serum nickel levels and rubidium[126] were reported to be unusually low in children receiving long-term TPN. Serum boron levels when studied, were reported either to be low[147] or high.[126]

There is no good indication to supplement TPN with any of these trace elements at this time.

In conclusion, many of the essential trace elements not supplemented in the solutions will become depleted in patients receiving long-term parenteral nutrition. Although many are contained as contaminants in infused solutions, proper knowledge of the amount of contamination is mandatory if TPN is expected to be for long term, as deficiency[131] or toxicity[137] may occur.

REFERENCES

1. Ulmer, D. D., Trace elements, *N. Engl. J. Med.*, 297, 318–321, 1977.
2. Underwood, E. J., *Trace Elements in Human and Animal Nutrition*, 4th ed., New York, Academic Press, 1977.
3. American Medical Association, Department of Foods and Nutrition, Guidelines for essential trace element preparations for parenteral use, *JAMA*, 241, 2051–2054, 1979.
4. Anspaugh, L. R. and Robison, W. L., Trace elements in biology and medicine, *Prog. Atom. Med.*, 3, 63–138, 1971.
5. Mertz, W., Some aspects of nutritional trace element research, *Fed. Proc.*, 29, 1482–1488, 1970.
6. Mendel, L. B. and Bradley, H. C., Experiment studies on the physiology of the molluscs: Second paper, *Am. J. Physiol.*, 14, 313–327, 1905.
7. Beisel, W. R., Trace elements in infectious processes, *Med. Clin. N. Am.*, 60, 831–849, 1976.
8. Biesel, W. R. and Pekarek, R. S., Acute stress and trace element metabolism, *Int. Rev. Neurobiol.*, 1, (suppl), 53–82, 1972.
9. Schwartz, M. K., Role of trace elements in cancer, *Cancer Res.*, 35, 3481–3487, 1975.
10. Schrauzer, G. N., Inorganic and nutritional aspects of cancer, *Adv. Exp. Med. Biol.*, 1978, Vol. 91, *Med. Clin. N. Am.*, 60, 779–797, 1976.
11. Revelant, V., Trace elements in total parenteral nutrition, *Acta Client. Venez.*, 41, 171–176, 1990.
12. Becking, G. C., Trace elements and drug metabolism, *Med. Clin. N. Am.*, 60, 813–830, 1976.

13. Golden, M. H. N. and Bolden, B. E., Trace elements, *Br. Med. Bull.*, 37, 31–36, 1981.
14. Foy, H. and Kondi, A., Anemias of the tropics; relation to iron intake, absorption and losses during growth, pregnancy and lactation, *J. Trop. Med. Hyg.*, 60, 105–118, 1957.
15. Bush, J. A., Mahoney, J. P., Gubler, C. J., et al., Studies on copper metabolism; transfer of radio-copper between erythrocytes and plasma, *J. Lab. Clin. Med.*, 47, 898–906, 1956.
16. Sternlieb, I., Morell, A. G., Tucker, W. D., et al., The incorporation of copper into ceruloplasmin in vivo: studies with copper-64 and copper-67, *J. Clin. Invest.*, 40, 1834–1840, 1961.
17. Wan, K. K. and Tsallas, G., Dilute iron dextran formulation for addition to parenteral nutrient solutions, *Am. J. Hosp. Pharm.*, 37, 206–210, 1980.
18. Finch, C. A. and Hubers, H., Perspectives in iron metabolism, *N. Engl. J. Med.*, 306, 1520–1528, 1982.
19. Shoden, A., Gabrio, B. W., and Finch, C. A., The relationship between ferritin and hemosiderin in rabbits and man, *J. Biol. Chem.*, 204, 823–830, 1953.
20. Dubach, R., Moore, C. V., and Callender, S., Studies in iron transportation and metabolism. IX. The excretion of iron as measured by the isotope technique, *J. Lab. Clin. Med.*, 45, 599–615, 1955.
21. Ellis, B. W., Zinc content of amino acid solutions, *Lancet,* 2(8085), 380, 1978.
22. Shils, M. E., More on trace elements and total parenteral nutrition solutions, *Am. J. Hosp. Pharm.*, 32, 141–142, 1975.
23. Hinkins, D. A., Whole blood trace element concentrations during total parenteral nutrition, *Surgery,* 79, 674–677, 1976.
24. Solomons, N. W., Plasma trace metals during total parenteral alimentation, *Gastroenterology,* 70, 1022–1025, 1976.
25. Hoffman, R. P. and Ashby, D. S., Trace element concentrations in commercially available solutions, *Drug Intell. Clin. Pharm.*, 10, 74–76, 1976.
26. Gettin, M. M., Trace element contamination of intravenous solutions, *Arch. Intern. Med.*, 136, 782–784, 1976.
27. Davis, N. M., Turco, S. J., and Sivelly, E., The study of particulate matter in IV infusion fluids, *Am. J. Hosp. Pharm.*, 27, 822–826, 1970.
28. Turco, S. J. and Davis, N., Particulate matter in intravenous infusion fluids: phase VII, *Am. J. Hosp. Pharm.*, 28, 620–623, 1971.
29. Turco, S. J. and Davis, N., Particulate matter in intravenous infusion fluids: phase VII, *Am. J. Hosp. Pharm.*, 30, 611–613, 1973.
30. Shear, E. R. and Bozian, R. C., Availability of trace elements in intravenous hyperalimentation solutions, *Drug Intell Clin. Pharm.*, 11, 465–469, 1977.
31. Hauer, E. C. and Kaminski, M. B., Trace metal profile of parenteral nutrition solutions, *Am. J. Clin. Nutr.*, 31, 264–268, 1978.
32. Odne, M. A., Rationale for adding trace elements to total parenteral solutions: a brief review, *Am. J. Hosp. Pharm.*, 35, 1057–1059, 1978.
33. Fliss, D. M. and Lamy, P. P., Trace elements and total parenteral nutrition, *Hosp. Formul.*, 36, 698–705, 1979.
34. Strobel, C. T., A zinc deficiency dermatitis in patients on total parenteral nutrition, *Int. J. Dermatol.*, 17, 575–581, 1978.
35. Freund, H., Atamin, S., and Fischer, J. E., Chromium deficiency during total parenteral nutrition, *JAMA,* 241, 496–498, 1979.
36. Jeejeebhoy, K. N., Chu, R. C., Marliss, E. B., et al., Chromium deficiency, glucose intolerance, and neuropathy reversed by chromium supplementation in a patient receiving long-term total parenteral nutrition. *Am. J. Clin. Nutr.*, 30, 531–538, 1977.
37. Kay, R. G., A syndrome of acute zinc deficiency during total parenteral alimentation in man, *Ann. Surg.*, 183, 331–340, 1976.
38. Shike, M., Copper metabolism and requirements in total parenteral nutrition, *Gastroenterology,* 81, 290–297, 1981.

39. Wolman, S. L., Zinc and total parenteral nutrition: requirements and metabolic effects, *Gastroenterology,* 76, 458–467, 1979.
40. Hull, R. L., Use of trace elements in intravenous hyper-alimentation solutions, *Am. J. Hosp. Pharm.,* 31, 759–761, 1974.
41. Prasad, A. S., Schulert, A. R., Sandstead, H. H., et al., Zinc, iron and nitrogen content of sweat in normal and deficient subjects, *J. Lab. Clin. Med.,* 62, 84–89, 1963.
42. Main, A. N., et al., Clinical experience of zinc supplementation during intravenous nutrition in Crohn's disease. Value of serum and urine zinc measurement, *Gut,* 23, 984–991, 1982.
43. Takagi, Y., Okada, A., Itakura, T., and Kawashima, Y., Clinical studies on Zinc metabolism during total parenteral nutrition as related to zinc deficiency, *JPEN,* 10, 195–202, 1986.
44. Widdowson, E. M., Dauncey, J., and Shaw, J. C. L., Trace elements in fetal and early postnatal development, *Proc. Nutr. Soc.,* 33, 275–284, 1974.
45. James, B. E. and MacMahon, R. A., Balance studies of 9 elements during complete intravenous feeding of small premature infants, *Aust. Ped. J.,* 12, 154–162, 1976.
46. Shaw, J. C. L., Trace elements in the fetus and young infant. I. Zinc. *Am. J. Dis. Chil.,* 133, 1260–1268, 1979.
47. Arakawa, T., Tamara, T., Igarashi, Y., et al., Zinc deficiency in two infants during total parenteral alimentation for diarrhea, *Am. J. Clin. Nutr.,* 29, 197–204, 1976.
48. Hambridge, K. M. and Walravens, P. A., Trace elements in nutrition, *Pract. Pediatr.,* 1, 17, 1975.
49. Mason, K. E., A conspectus of research on copper metabolism and requirements in man, *J. Nutr.,* 109, 1979–2066, 1979.
50. Evans, J. L. and Abraham, P. A., Anemia, iron storage and ceruloplasmin in copper nutrition in the growing rat, *J. Nutr.,* 103, 196–201, 1973.
51. O'Dell, B. L., Roles for iron and copper in connective tissue biosynthesis, *Phil. Trans. R. Soc. Lond. B.,* 294, 91–104, 1981.
52. Osaki, S., Johnson, D. A., and Freiden, E., The possible significance of the ferrous oxidase activity of ceruloplasmin in normal human serum, *J. Biol. Chem.,* 241, 2746–2751, 1966.
53. Golden, M. H. N., Trace elements in human nutrition, *Hum. Nutr. Clin. Nutr.,* 36C, 185–202, 1982.
54. Cordano, A., Baertl, J. M., and Graham, G. G., Copper deficiency in infancy, *Pediatrics,* 34, 324–336, 1964.
55. Hamilton, E. I., Minsky, M. J., and Clearly, J. J., The concentration and distribution of some stable elements in healthy human tissues from the United Kingdom, *Sci. Total Environ.,* 1, 341–374, 1972.
56. Gubler, C. J., Lahey, M. E., Cartwright, G. E., et al., Studies on copper metabolism; transportation of copper in blood, *J. Clin. Invest.,* 32, 405–414, 1953.
57. Bremner, I., Absorption, transport and storage of copper, in *Biological Roles of Copper,* CIBA Foundation Symposium, Amsterdam, Excerpta Medica, 1980, 79, 23–48.
58. Own, C. A., Absorption and excretion of^{64} Cu-labelled copper by the rat, *Am. J. Physiol.,* 207, 1203–1206, 1964.
59. Hall, A. C., Young, B. W., and Bremner, I., Intestinal metallothionein and the mutual antagonism between copper and zinc, *J. Inorg. Biochem.,* 11, 57–66, 1979.
60. Cartwright, G. E. and Wintrobe, M. M., Copper metabolism in normal subjects, *Am. J. Clin. Nutr.,* 14, 224–232, 1964.
61. Shike, M., Roulet, M., Kurian, R., et al., Copper metabolism and requirements in total parenteral nutrition, *Gastroenterology,* 81, 290–297, 1981.
62. Jacob, R. A., Sandstead, H. H., Munoz, J. M., et al., Whole body surface loss of trace metals in normal males, *Am. J. Clin. Nutr.,* 34, 1379–1383, 1981.
63. Henkin, R. I., On the role of adrenocorticosteroid in the control of zinc and copper metabolism, in *Trace Element Metabolism in Animals,* Vol. 2, Hoekstra, W. G., Suttie, J. W., Ganther, H. E., et al., Eds., Baltimore, University Park Press, 1974, 647–651.

64. Askari, A., Long, C. L., Murray, R. R. L., et al., Zinc and copper balance in the severely injured patient, (Abstr) *Fed. Proc.*, 38, 707, 1979.
65. Bremner, I. and Mills, C. F., Absorption, transport and tissue storage of essential trace elements, *Phil. Trans. R. Soc. Long. B.*, 294, 75–89, 1981.
66. Kovalsky, V. V., The geochemical ecology of organisms under conditions of varying contents of trace elements in the environment, in *Trace Element Metabolism in Animals*, Vol. 1, Mills, C. F., Ed., Edinburgh, Livingstone, 1970, 385–397.
67. Hambridge, K. M., Increase in hair copper concentration with increasing distance from the scalp, *Am. J. Clin. Nutr.*, 26, 1212–1215, 1973.
68. Jacobson, S. and Western, P. O., Balance study of twenty trace elements during total parenteral nutrition in man, *Br. J. Nutr.*, 37, 107–126, 1977.
69. Phillips, G. D. and Garnys, V. P., Parenteral administration of trace elements to critically ill patients, *Ann. Intens. Care.*, 9, 221–225, 1981.
70. Mertz, W., Roginski, E. E., and Schwartz, K., Effects of trivalent chromium complexes on glucose uptake by epididymal fat tissue of rats, *J. Biol. Chem.*, 236, 318–322, 1961.
71. Sriram, K., O'Gara, J. A., Strunk, J. R., and Peterson, J. K., Neutropenia due to copper deficiency in total parenteral nutrition, *JPEN*, 10, 530–532, 1986.
72. Tokuda, Y., Yokoyama, S., Tsuti, M., Sugital, T., Tajima, T., and Mitomi, T., Copper deficiency in an infant on prolonged total parenteral nutrition, *JPEN*, 10, 242–244, 1986.
73. Rollinson, C. L. and Rosenbloom, W. E., in *Coordination Chemistry*, Bailar, J. C., Jr., Ed., New York, Kirschner, 1969, 103–124.
74. Toepfer, W. W., Mertz, W., Polansky, M. M., Roginski, E. E., and Wolf, W. R., Synthetic organic chromium complexes and glucose tolerance, *J. Agr. Food. Chem.*, 25, 162–165, 1977.
75. Schwartz, K. and Mertz, W., Chromium (III) and the glucose tolerance factor, *Arch. Biochem. Biophys.*, 85, 292–295, 1959.
76. Evans, G. W., Roginski, E. E., and Mertz, W., Interaction of the glucose tolerance factor (GTF) with insulin, *Biochem. Biophys. Res. Comm.*, 50, 718–722, 1973.
77. Mertz, W., Chromium occurrence and function in biological systems, *Physiol. Rev.*, 49, 163–203, 1969.
78. Sargent, R., Lim, T. H., and Gensen, R. L. Reduced chromium retention in patients with hemochromatosis, a possible basis for hemochromatotic diabetes, *Metabolism*, 28, 70–79, 1979.
79. Glinsmann, W. H., Feldman, F. J., and Mertz, W., Plasma chromium after glucose administration, *Science*, 152, 1243–1245, 1966.
80. Guthrie, B. E., Wolf, W. R., Veillon, C., and Mertz, W., in *Trace Substances in Environmental Health XII.*, Hemphill, D. C., Ed., University of Missouri Press, Columbia, 1978.
81. Veillon, C. and Wolf Buthrie, B. E., Determination of chromium in biological materials by stable isotope dilution, *Anal. Chem.*, 51, 1022–1024, 1979.
82. Hopkins, L. L. Jr., Ransome-Kuti, O., and Majaj, A. S., Improvement of impaired carbohydrate metabolism in malnourished infants, *Am. J. Clin. Nutr.*, 21, 203–211, 1968.
83. Schroeder, H. A., Balassa, J. J., and Tipton, I. H., Abnormal trace metals in man chromium, *J. Chron. Dis.*, 15, 941–964, 1962.
84. Schroeder, H. A., The role of chromium in mammalian nutrition, *Am. J. Clin. Nutr.*, 21, 230–244, 1968.
85. Hambridge, K. M., Chromium nutrition in the mother and the growing child, in Mertz, W. and Cornatzer, W. E., Eds., *Newer Trace Elements in Nutrition*, New York, Marcel Dekker, Inc., 1971, 169–194.
86. Freed, B. A., Pinchofsky, G., Nasr, N., et al., Normalization of serum glucose levels and decreasing insulin requirements by the addition of chromium to TPN, (Abstr) *JPEN*, 5, 568, 1981.
87. Kien, C. L., Veillon, C., Patterson, K. Y., and Farrell, P. M., Mild, peripheral neuropathy but biochemical chromium deficiency during 16 months of "chromium-free" total parenteral nutrition, *JPEN*, 10, 662–664, 1986.
88. Brown, R. O., Forlonies-Lynn, S., Gross, R. E., et al., Chromium deficiency after long term total parenteral nutrition, *Dig. Dis. Sci.*, 31, 661–664, 1986.

89. Guidelines for essential trace element preparation for parenteral use: a statement by an expert, *JAMA,* 241, 2051–2054, 1979.
90. Green, H. L., Hambridge, K. M., Schanler, R., et al., Guidelines for the use of vitamins, trace elements, calcium, magnesium, and phosphorus in infants and children receiving total parenteral nutrition: Report of the subcommittee on pediatric parenteral nutrient requirements from the committee on clinical practice issues of the American Society for Clinical Nutrition, *Am. J. Clin. Nutr.,* 48, 1324–1342, 1988.
91. Ghisolfi, J., Apports nutritionnels et alimentation artificielle, in *Journees Parisiennes de Pediatrie,* Flammarion Medecine, Ed., Sciences, Paris, 1984, 97–107.
92. Moukarzel, A. A., Song, M. K., Buchman, A. I., Vargas, J. H., Guss, W., McDiarmid, S., Reyen, L., and Ament, M. E., Excessive chromium intake in children receiving total parenteral nutrition, *Lancet,* 339, 385–388, 1992.
93. Kumpulainen, J. and Vuori, E., Longitudinal study of chromium in human milk, *Am. J. Clin. Nutr.,* 33, 2299–2302, 1980.
94. Casey, C. E. and Hambridge, K. M., Chromium in human milk from American mothers, *Br. J. Nutr.,* 52, 73–77, 1984.
95. Hambridge, K. M., Trace element requirements in premature infants, in *Textbook of gastroenterology and nutrition in infancy,* Lebenthal, E., Ed., New York, Raven Press, 1989, 398.
96. Subcommittee on the Tenth Edition of RDAs, Food and Nutrition Board Commission on Life Sciences, Recommended dietary allowances, National Academy Press, Washington, D.C., 1989, 242.
97. Ito, Y., Alcock, N. W., and Shills, M. E., Chromium content of total parenteral nutrition solutions, *JPEN,* 14, 610–614, 1990.
98. Kien, C. L., Veilon, C., Patterson, K. Y., and Farrel, P. M., Mild peripheral neuropathy but biochemical chromium sufficiency during 16 months of ''chromium-free'' total parenteral nutrition, *JPEN,* 10, 662–664, 1986.
99. Schroeder, H. A., Balassa, J. J., and Tipton, I. H., Essential trace metals in man: manganese, a study in homeostasis, *J. Chron. Dis.,* 19, 545–571, 1966.
100. Widdowson, E. M., Chan, H., Harrison, G. E., and Milner, R. D. G., Accumulation of Cu, Zn, Mn, Cr, and Co in the human liver before birth, *Biol. Neonat.,* 20, 360–367, 1972.
101. Casey, D. E. and Robinson, M. F., Copper, manganese, zinc, nickel, cadmium and lead in human fetal tissue, *Br. J. Nutr.,* 39, 639–646, 1978.
102. Gruden, N., Suppression of transduodenal manganese transport by milk diet supplemented with iron, *Nutr. Metab.,* 21, 305–309, 1977.
103. Leach, R. M., Jr., in *Trace Elements in Human Health and Disease,* Vol. 2, Prasad, A., Ed., Academic Press, New York, 1976, 235–247.
104. Lustig, S., Pitlik, S. D., and Rosenfeld, J. B., Liver damage in acute self-induced hypermangasemia, *Arch. Intern. Med.,* 142, 405–406, 1982.
105. Cotzias, G. C., Manganese in health and disease, *Physiol. Rev.,* 38, 503–532, 1958.
106. Greenberg, D. M., Copp, D. H., and Cuthbertson, E. M., Studies in mineral metabolism with the aid of artificial radioactive isotopes, *J. Biol. Chem.,* 147, 749–757, 1943.
107. Kurkus, J., Alcock, N. W., and Shils, M. E., Manganese content of large-volume parenteral solutions and nutrient additives, *JPEN,* 8, 254–257, 1984.
108. Underwood, E. J., *Trace Elements in Human and Animal Nutrition,* 4th Ed., Academic Press, New York, 1977.
109. Abumrad, N. N., Molybdenum—is it an essential trace metal? *Bull. NY Acad. Med.,* 60, 163–171, 1984.
110. Abumrad, N. N., Schneider, A. J., Steel, D., et al., Amino acid intolerance during prolonged total parenteral nutrition reversed by molybdate therapy, *Am. J. Clin. Nutr.,* 34, 2351–2359, 1981.
111. Schwartz, K. and Foltz, C. M., Selenium as an integral part of Factor 3 against dietary necrotic liver degeneration, *J. Am. Chem. Soc.,* 79, 3292–3293, 1957.

112. Thompson, J. N. and Scott, M. L., Impaired lipid and vitamin E absorption related to atrophy of the pancrease in selenium-deficient chicks, *J. Nutr.*, 100, 797–809, 1970.
113. Sprinker, L. H., Harr, J. R., Newberne, P. M., Whanger, P. D., and Weswig, P. H., Selenium deficiency lesions in rats fed vitamin E-supplemented rations, *Nutr. Rep. Int.*, 4, 335–340, 1971.
114. Keshan Disease Research Group, Observations on effect of sodium selenite in prevention of Keshan disease, *Chinese Med. J.*, 92, 471–476, 1979.
115. Keshan Disease Research Group, Epidemiologic studies on the etiologic relationship of selenium and Keshan disease, *Chinese Med. J.*, 92, 477–482, 1979.
116. Levander, O. A., in *Proceedings of the Symposium on Selenium-Tellurium in the Environment*, Industrial Health Foundation, Pittsburgh, 1976, 26–53.
117. Olson, O. E. and Palmer, I. S., Selenium in foods consumed by South Dakotans, *Proc. S. D. Acad. Sci.*, 57, 113–121, 1978.
118. Lombeck, I., Kasperek, K., Harcisch, H. D., et al., The selenium state of children, *Eur. J. Pediatr.*, 129, 213–234, 1978.
119. Zabel, N. L., Harland, J., Gormican, A. T., and Ganther, H. E., Selenium content of commercial formula diets, *Am. J. Clin. Nutr.*, 31, 850–858, 1978.
120. Sakuari, H. and Tsuchiya, K., A tentative recommendation for the maximum daily intake of selenium, *Environ. Physiol. Biochem.*, 5, 1207–1218, 1975.
121. Olson, O. E., Novacek, E. J., Whitehead, E. I., and Palmer, I. S., Investigations on selenium in wheat, *Phytochemistry*, 9, 1181–1188, 1970.
122. Thomson, C. D., Burton, C. E., and Robinson, M. F., On supplementing the selenium intake of New Zealanders. I. Short experiments with large doses of selenite or selenomethionine, *Br. J. Nutr.*, 39, 579–587, 1978.
123. Rotruck, J. T., Pope, A. L., Gather, H. E., et al., Selenium: biochemical role as a component of glutathione peroxidase, *Science*, 179, 588–590, 1973.
124. Diplock, A. T. and Lucy, J. A., The biochemical modes of action of vitamin E and selenium: a hypothesis, *FEBS lett.*, 29, 205–210, 1973.
125. Underwood, E. J., Selenium, in *Trace Elements in Human and Animal Nutrition*, 4th Ed., New York, Academic Press, 1977, 246–302.
126. Levander, O. A., Sutherland, V., Morris, V. C., et al., Selenium balance in young men during selenium depletion and repletion, *Am. J. Clin. Nutr.*, 23, 2662–2669, 1981.
127. Kelly, D. A., Coe, A. W., Shenkin, A., Lake, B. D., and Walker-Smith, J. A., Symptomatic selenium deficiency in a child on home parenteral nutrition, *J. Pediatr. Gastroenterol. Nutr.*, 7, 783–785, 1988.
128. Vinton, N. E., Ament, M. E., Dahlstrom, K. A., and Strobel, C. T., Treatment of selenium deficiency in stable long term total parenteral nutrition patients, *JPEN*, in press.
129. Lane, H. W., Barroso, A. O., Englert, D. A., Dudrick, S. T., and MacFadyen, B. S., Selenium status of seven chronic intravenous hyperalimentation patients, *JPEN*, 6, 426–431, 1982.
130. Baptista, R. J., Bistrian, B. R., Blackburn, F. L., Miller, D. G., Champagne, C. D., and Buchanan, L., Suboptimal selenium status in home parenteral nutrition patients with small bowel resections, *JPEN*, 8, 542–545, 1984.
131. Fleming, C. R., McCall, J. J., O'Brien, J. F., Forsman, R. W., Ilstrup, D. M., and Petz, J., Selenium status in patients receiving parenteral nutrition, *JPEN*, 8, 258–262, 1984.
132. Watson, R. D., Cannon, R. A., Kurland, G. S., Cox, K. L., and Frates, C., Selenium responsive myositis during prolonged home total parenteral nutrition in cystic fibrosis, *JPEN*, 9, 58–60, 1985.
133. Dahlstrom, K. A., Ament, M. E., Medhin, M. G., and Meurling, J., Serum trace elements in children receiving long term parenteral nutrition, *J. Pediatr.*, 109, 625–630, 1986.
134. Vinton, N. E., Dahlstrom, K. A., Strobel, C. T., and Ament, M. E., Macrocytosis and pseudoalbinism; manifestations of selenium deficiency, *J. Pediatr.*, 111, 711–719, 1987.

135. Vargas, J., Klein, G., Ament, M. E., Ott, S. M., Sherrard, D. J., Horst, R. L., Berquist, W. E., Alfrey, A. C., Slatopolsky, E., and Coburn, J. W., Metabolic bone disease of total parenteral nutrition: Course after changing from casein to amino acids in parenteral solutions with reduced aluminum content, *Am. J. Clin. Nutr.,* 48, 1–9, 1988.
136. Moukarzel, A., Ament, M. E., Vargas, J., McDiarmid, S., Reyen, L., Najm, I., and Guss, W., Non-aluminum dependent osteopathy in children on long-term parenteral nutrition, *Am. J. Clin. Nutr.,* 51, 520, 1990.
137. Moukarzel, A., Ament, M. E., Vargas, J., Reyen, L., and Guss, W., (Abstract) Parenteral nutrition bone disease in children, *Clinical Research,* 38, 190A, 1990.
138. Koo, W. W. and Kaplan, L. A., Aluminum and bone disorders: with specific reference to contamination of infant nutrients, *J. Am. Coll. Nutr.,* 7, 199–214, 1988.
139. Ricour, C., Duhamel, J. F., Gros, J., et al., Estimates of trace element requirements of children receiving total parenteral nutrition, *Arch. Fr. Pediatr.,* 34 (supp 7), 92–100, 1977.
140. Moukarzel, A., Vargas, J. H., McDiarmid, S. V., Reyen, L., Guss, W., and Ament, M. E., Parenteral nutrition (TPN) in children: Intravenous iodide supplements are not necessary, *Pediatr. Res.,* 27(4), 112A, 1990.
141. Moukarzel, A. A., Buchman, A. L., Salas, J. S., et al., Iodine supplementation is not necessary in children receiving long term parenteral nutrition, *J. Peds.,* 121, 252–254, 1992.
142. Committee on Nutrition, American Academy of Pediatrics, Fluoride supplementation: revised dosage schedule, *Pediatrics,* 63, 150–152, 1979.
143. Ricour, C. and Nihoul-Fekete, C., Nutrition parenterale prolongee chez l'enfant, *Arch. Fr. Pediatr.,* 30, 469–490, 1973.
144. Moukarzel, A. A., Ament, M. E., Vargas, J., Reyen, L., and Guss, W., Is fluoride deficiency related to the bone disease of parenteral nutrition? *Clin. Nutr.,* 9, 65, 1990.
145. Moukarzel, A., Song, M. K., Haddad, I., Buchman, A. L., Baron, H., Vargas, J., Reyen, L., Guss, W., and Ament, M. E., Is silicon deficiency involved in the pathogenesis of metabolic bone disease of children receiving parenteral nutrition? *JPEN,* 16(1), 31S, 1992.
146. Moukarzel, A. A., Buchman, A. L., Song, M., Vargas, J., Baron, H. I., Reyen, L., and Ament, M. E., Osteopenia of parenteral nutrition bone disease in children: Boron deficiency is not an etiological factor, *Clin. Res.,* 40, 60A, 1992.

Chapter 16

HOME PARENTERAL NUTRITION*

Adib A. Moukarzel and Marvin E. Ament

TABLE OF CONTENTS

I. Introduction 184

II. Indications and Contraindications 184

III. Implementation of HPN 186
A. Patient Preparation 186

IV. Parenteral Nutrition Access 186

V. HPN Formula 187

VI. Discharge Planning 188

VII. Complications of HPN 188
A. Technical Problems 188
B. Infectious Problems 189

VIII. Metabolic Problems 190
A. Lipid and Platelets 190
B. Trace Elements and Vitamins 190

IX. Cholelithiasis, Cholecystitis, and Chronic Liver Disease 191

X. Bone Disease 192

XI. Renal Disease 192

XII. Visual Function 192

XIII. Mortality 192

XIV. Conclusions 192

References 194

*Partially funded by Maimonides Research Foundation.

0-8493-2764-4/95

I. INTRODUCTION

Home parenteral nutrition (HPN) is the provision of parenteral nutrition (PN) in infants, children, and adults in a non-hospital setting. The goal of HPN is to achieve positive nitrogen balance, weight gain, and to improve clinical outcome in patients who do not have adequate gastrointestinal tract function. HPN has saved the lives of many patients who previously might have died with catastrophic gastrointestinal conditions. Since its inception almost 25 years ago, the use of HPN has increased tremendously, and it has become accepted as a useful supportive and therapeutic technique for a variety of gastrointestinal diseases and conditions. In the United States, it is estimated that there are over 19,000 patients on home parenteral nutrition. We have treated more than 1000 patients for 18 years with HPN for over 100,000 days of therapy. The mean patient days on HPN has been nearly 1000 days, and the longest period for any single patient has been well over 16 years. Half of the patients we are currently following have received parenteral support at home for 8 or more years. Many of our patients are infants and children.

II. INDICATIONS AND CONTRAINDICATIONS

HPN is generally limited to patients who have been refractory to all other forms of treatment (e.g., enteral nutrition) or for whom other forms of treatment (e.g., surgery) are inadvisable. The most common indications and conditions for HPN are listed in Table 1.[3–14]

Short bowel syndrome is the most frequent condition associated with the need for HPN in pediatric age. In the last years, we have seen a decline in the incidence of this condition, because of increasing sophistication in the management of infants with prematurity and necrotizing enterocolitis. Therefore, fewer of these infants are having massive intestinal resections resulting in the incidence of this condition. Chronic intestinal pseudo-obstruction syndrome is another condition in which HPN has made the difference between life and death. These chronic motility disorders can present in the neonatal period, during infancy, or later in childhood. Many children are operated two or three times before the condition is recognized. Some children with this condition who require HPN do not become symptomatic until later in life.

Crohn's disease in the pediatric patient rarely leads to multiple and massive intestinal resections; however, there are some occasional patients in whom this occurs. In recent years, because of increasing sophistication in the use of medication for management of Crohn's disease, it is hoped that this will reduce the number of pediatric patients who undergo major intestinal resection. HPN for oncology patients is often indicated for mucositis in graft vs. host disease and in

TABLE 1
Common HPN Indications

Short bowel syndrome
Intestinal motility disorders
Intractable diarrhea
Acquired immune deficiency syndrome
Cancer-related disorders
Other

individuals with anorexia or vomiting associated with chemotherapy and/or radiation therapy. In recent years, we have had an increasing number of youngsters with intra-abdominal malignancies who require HPN because of the damage to the intestine from radiotherapy and intra-abdominal operative procedures.

Some patients require complete PN, while others receive supplemental or intermittent support. HPN is also utilized with patients with AIDS who have intractable diarrhea and/or pancreatitis. Although PN is beneficial in slowing or reversing the malnutrition in AIDS patients, it does not improve their measurable immune functions. Experience indicates that in these patients, HPN may be worthwhile provided the patient has not had serious and irreversible central nervous system damage. Patients with immunodeficiency disorders of the combined type may require prolonged parenteral support because of intractable diarrhea. In some of these patients, a severe mucosal lesion develops for which no specific cause can be found. Occasionally HPN is used in these patients to support them until bone marrow transplantation becomes feasible. Cystic fibrosis is another condition in which the placement of a catheter may be used at home to support nutrition and provide antibiotics. Intestinal lymphangectasia is typically thought of as a condition that can be managed by dietary means. However, a substantial number of affected patients do not benefit from low fat diets with medium-chain triglycerides and may benefit from parenteral support on a long-term basis at home.

Contraindications for HPN include a functional, available gastrointestinal tract, or other extenuating circumstances precluding vascular access and the use of PN. Whenever enteral feeding is possible, it should be utilized. HPN patients who have some intestinal function should be encouraged to feed as much as possible. Even small amounts of liquid should be given by mouth, as soon as possible, to ensure maximal stimulation of the gastrointestinal tract for its adaptation and to diminish bacterial translocation.[23] This also applies to infants who have minimal chance of surviving without PN. They should receive at least 5% of their caloric requirements orally. Failure to initiate oral feeding in infants can later result in sucking or swallowing problems.[24] Patients with anorexia nervosa are not candidates for HPN. HPN should not be utilized when there is not a family member dedicated to learn and perform the daily techniques required for a successful program or when the family lacks sufficient intelligence and motivation to learn.

TABLE 2
Technical Complications

Mechanical
  Broken catheter requiring repair
Catheter occlusion or slow infusion
  Clear blockage with urokinase
Drug precipitate occlusion
  Clear blockage with HCI
Thrombosis at vascular access
  Early diagnosis and treatment essential
  Echocardiography for early detection
  Prophylactic warfarin therapy in adults may be of benefit

Data from Berm, M. M., Bohe, A., Jr., Bistrian, B., et al., *Surgery,* 99, 216, 1986, and Berm, M. M., Lokich, J. J., Vallach, S. R., et al., *Ann. Intern. Med.,* 112, 423, 1990.

III. IMPLEMENTATION OF HPN

A. PATIENT PREPARATION

Preparing a patient for HPN involves an experienced and knowledgeable multidisciplinary health care team, with the physician assuming primary responsibility. General information outlining benefits and risks is the first aspect of HPN discussed with the patient and family. Potential complications (Table 2) including catheter infection, sepsis, thrombosis, bleeding from inadvertent tubing disconnection, hyperglycemia, hypoglycemia, and unknown metabolic complications are presented. The discussion should include expected outcome of the TPN therapy and the predicted degree of bowel adaptation. Patients who have 15 to 20 cm of small bowel and an intact ileocecal valve may eventually have partial or complete bowel adaptation.[3,4] It is also necessary to talk openly about the anticipated effect of HPN on the patient and family lifestyle. In all these areas, it is important to have as straightforward a discussion as possible. If not already in place, a vascular access device (catheter or port) is inserted, and an infusion pump is selected. Both should be chosen based upon patient needs, capability, lifestyle, preference, and the HPN team's experience and knowledge of available products.

IV. PARENTERAL NUTRITION ACCESS

For long-term parenteral nutrition, cuffed, silastic tunneled central venous catheters such as Hickman or Broviac catheters have been commonly used. They may be placed by either a cutdown or a percutaneous method in a central vein in the operating room or in special patient care areas where sterile technique can be provided. These catheters provide stability and decrease the risk of infection by subcutaneously tunneling the catheter to a distant exit site. Once the catheter has been inserted, it is mandatory to confirm proper placement to rule out mechanical

TABLE 3
Recommended Energy and Protein Intakes

Age	Energy* (mean) (Kcal/kg)	Kcal/day (mean)	Protein (g/kg) or (g/day)
Infant & Children			
Birth to 6 months	115	300	2.2 g/kg
6 months to 1 year	105	600	2.0 g/kg
1 to 3 years	100	1300	1.8 g/kg
4 to 6 years	85	1700	1.5 g/kg
7 to 10 years	85	2400	
Adolescents			
Boys			
11 to 14 years	60 to 64	2800	45 g/day
15 to 18 years	43 to 49	3000	56 g/day
Girls			
11 to 16 years	38 to 55	2400	46 g/day
Adults			
Men & women	25 to 35	2000	1 to 1.5 g/kg

*Best assessed by estimation of resting energy expenditure (indirect calorimetry).

TABLE 4
Typical HPN Solutions

Final Concentrations	Infants	Children & Adults
Neonatal Amino Acids	2%	—
Balanced Amino Acids	—	3.5 to 4.25%
Dextrose	10 to 25%	10 to 25%
Sodium	30 mEq/L	35 mEg/L
Potassium	25 mEq/L	30 mEq/L
Calcium	10 mEq/L	5 mEq/L
Magnesium	10 mEq/L	10 mEq/L
Phosphate	10 mmol/L	7.5 mmol/L
Chloride	30 mEq/L	35 mEq/L
Acetate	27 mEq/L	64 mEq/L
Zinc	2 mg/L	2 mg/L
Copper	1 mg/l	1 mg/L
Selenium	20 mcg/l	20 mcg/L

complications. Dual and triple lumen catheters are used in pediatric patients specifically for oncologic and transplantation patients. They require more care, and the incidence of infection is greater than with single lumen catheters.

V. HPN FORMULA

The patient's home nutritional requirements are based on the expected weight and height of the patient according to age and sex (Table 3). Fluid needs are determined by the patient's clinical condition as well as by anthropometric measurements, and then a PN solution is prescribed (Table 4). Patients are generally established on a 24-hour infusion in the hospital, and once they achieve an optimal formula, the infusion time is decreased. Schedules for HPN infusions vary from 24 hours a day to an 8-hour infusion. The majority of patients receive their HPN over 10–12 hours at night while they sleep and are disconnected from the PN during the day. Infusion times vary with the patient's individual TPN formula, volume, nutritional needs, metabolic response, and personal preference.

VI. DISCHARGE PLANNING

Discharge planning and coordination of home care activities should be initiated as soon as it is determined the patient will receive HPN. The amount of time needed to prepare for HPN varies from several days to a month. The patient and family home situation should be evaluated to assure its appropriateness and determine if additional resources may be needed. The patient and primary caregiver are taught as much of the care as they are capable of doing. It is optimal for them to be self-sufficient and perform all tasks. Various techniques may be utilized to teach HPN, including written instructions outlining the procedures, demonstrations of care, videotapes, and use of practice catheters. Patients should be instructed to recognize and notify the HPN team of any fever, infection, metabolic complication, or mechanical problem.

The patient should be referred to an experienced, accredited, and reliable home care and HPN provider. The provider pharmacy should meet or exceed standards of practice. It may be beneficial for a Registered Nurse with HPN expertise to provide follow-up home visits and intermittent care at home, particularly during the initial phase of HPN. More extensive nursing is occasionally needed if the patient needs are complicated and beyond the capability of the family. The provider should monitor the patient's progress and compliance with procedure and report all pertinent patient information to the physician.

VII. COMPLICATIONS OF HPN

Complications of HPN are fewer when PN protocols are administered by a team with substantial experience and familiarity with the techniques. The three types of complications which can occur are technical, infectious, and metabolic.

A. TECHNICAL PROBLEMS

The central venous catheters may become damaged at their external segment from repeated clamping and unclamping. Broken or disconnected catheters should be repaired emergently utilizing catheter-specific repair kits available from each catheter manufacturer. To prevent the catheter occlusion, heparin flushes are used, and urokinase has been reported for clearing the occluded catheter.[36] Patency in long-term catheters is generally maintained with intermittent flushes of heparinized solution into the catheter. Heparin concentrations ranging from 10 units per millimeter to 1000 units per millimeter and volumes of flush solutions ranging from 1.5 to 10 ml used twice daily to once weekly have been reported.[37] If drug practice is a suspected cause of occlusion, irrigation of the catheter with hydrochloric acid is attempted.[38] All catheters have the potential to form thrombus around and within the lumen of the vein. The diagnosis of thrombosis can usually be made by two-dimensional and M-mode echocardiography and confirmed, if necessary, by angiography. We found that the sensitivity of cardiac thrombus detection by echocardiography was 100% and the specificity 93%.[39]

B. INFECTIOUS PROBLEMS

Despite the widespread use of PN for over two decades, infectious complications are the most frequent PN-related problem, resulting in increased morbidity, mortality, and health care costs. Sepsis rates vary and depend on the definition, methodology, and institution. It is not uncommon for patients receiving PN at home to have a lower sepsis rate than hospitalized patients. A majority of patients never have a catheter infection. Excellent pre-discharge teaching, meticulous catheter care, and strict PN technique all contribute to low infection rates.

PN infections are always due to some known or unsuspected break in technique. With each fever, our patients come in for a careful examination. A history is taken to determine if there is a known cause for the fever. The care provider is questioned concerning technique and routine. If there is no recognizable source, the most likely possibility is either a catheter infection or viral infection. The patient has both central line and peripheral blood cultures done for aerobes, anaerobes, and fungi. A CBC with differential, a urinalysis, and a chest x-ray are also done. Other tests depend on the clinical findings and suspicions of the examining physician. If a specific source of infection is identified, the patient should be treated appropriately. Since most patients receiving HPN cannot effectively absorb oral antibiotics, they are usually given intravenously. Often patients receive their IV antibiotics at home. When given at home, we attempt to avoid those that are administered more often than every 8 to 12 hours.

If no obvious source of infection is found, antibiotic coverage may be started to treat the suspected infection in the catheter. Vancomycin is almost always the initial therapy of choice in suspected catheter infection.[5] Typically, gentamicin is added to the initial regime pending final culture report and sensitivities. Catheter infections in HPN patients are usually treated with IV antibiotics for 4 weeks. We avoid catheter removal because of the patient's dependence on HPN and the long-term need for vascular access. IV antibiotics are most effective in case of gram positive bacteria, but success in gram negative is not uncommon. The most common cause of infection is gram positive bacteria (including oxacillin-resistant coagulase negative staphylococci), followed by gram negative bacteria, fungi, and mycobacteria (Table 5). Catheter removal is required in the presence of fungi infection, septic shock, endocarditis, embolism, persistent fever, or disseminated intravascular coagulation. Infections of the tunnel tract usually require catheter

TABLE 5
Organism Cultured at Catheter Insertion Site

Pityrosporum orbicular
Streptococcus viridans
Klebsiella species
Escherichia coli
Pseudomonas aerugenosa
Candida albicans
Staphylococcus epidermidis
Staphylococcus aureus

removal. The catheter is removed, antimicrobial therapy is continued for 5 to 7 days, and a new catheter inserted after the patient is afebrile for 72 hours and blood cultures no longer contain the infectious organism.

In our population,[2] the typical catheter remains in place for more than 700 days. We have patients who have had the same catheter for more than 10 years. Experience suggests a longer life span of the second catheter, as well as the higher incidence of catheter-related complications in the first 2 years of HPN vs. later years.

VIII. METABOLIC PROBLEMS

A. LIPID AND PLATELETS

Fat overload syndrome occurs when patients receive excessive quantities of lipids, usually more than 3 gm/kg. Treatment of fat overload syndrome consists of removing or decreasing the lipids and giving supportive care as indicated. Occasionally, parenteral steroids are used. Hyperlipidemia may be due to overinfusion of lipids or reduced utilization of fat. Before HPN is initiated, serum triglyceride and cholesterol levels should be measured 6 to 8 hours after the infusion of lipids to verify effective clearance.

Patients on long-term HPN who infuse lipids on a regular basis should have triglyceride and cholesterol determinations quarterly. If either is elevated, the intravenous fat dosage should be adjusted. Serum triglyceride values should be obtained 24 hours after any increase in dose of lipids to be sure the patient can tolerate the new dosage. A sudden increase in the triglyceride level may be indicative of sepsis.

B. TRACE ELEMENTS AND VITAMINS

A variety of trace element deficiencies have been described in patients receiving long-term PN, and recent recommendations for trace elements have been published.[6] Zinc and copper deficiencies are very uncommon. Zinc supplementation is required in patients with massive diarrhea and malabsorption. Selenium deficiency can occur in patients not supplemented with selenium. Intravenous supplementation with 2 mcg/kg/day of selenium can totally reverse the deficiency. Selenium should always be included as one of the trace metals necessary in patients with chronic diarrhea and large ostomy output.

Patients absorb sufficient quantities of chromium from drinking water and PN solutions. Similarly, iodide is in the water of the PN solution, and patients may absorb it through the skin if they use iodine skin preparations. Neither chromium or iodide are routinely added to the parenteral nutrition solutions.

Intravenous iron is provided for patients who are unable to absorb it through the enteral route. Since iron is absorbed proximally in the duodenum and jejunum, a trial of oral iron is given first. Dilute forms of iron dextran have been added to PN solutions (dextrose and amino acid solution) without any serious consequences. If one limits the amount of iron in a single day to what the normal requirements are, there is unlikely to be any problem with its administration. No consequences have been observed at our institution with providing 0.5 to 1 mg of a daily iron dextran preparation daily. There are a variety of other trace metals, but it is not clear whether they should be supplemented. These included fluoride, manganese, and cobalt. There is no question that other trace metals may need to be added to the parenteral nutrition solutions, but the documentation is not available. Vitamin deficiency has not been apparent in most individuals reported on HPN. It is quite uncommon and should not occur if patients are provided with the appropriate supplementation. The multiple vitamin preparations should be added to the PN solutions just before administration to minimize losses while the solution is hanging.

IX. CHOLELITHIASIS, CHOLECYSTITIS, AND CHRONIC LIVER DISEASE

The chance of developing gallbladder sludge as a result of PN is almost 100% after patients have been receiving PN for 6 full weeks. Any patient who is on PN for more than thirty days and develops abdominal pain should be evaluated for cholecystitis. It is possible that stimulating the gallbladder by more frequent feeding would reduce the incidence of biliary sludge and stones. There has been a dramatic reduction in the incidence of PN-induced liver disease. The complication occurs less frequently because of a change in most protocols in initiating earlier enteral nutrition. The decrease of PN-associated liver disease in infants is due to the use of a balanced amino acid solution specifically designed for infants. There are multiple factors which explain the occurrence of liver disease. Monitoring patients for evidence of early cholestasis can be done by measuring the GGT, 5′-nucleotidase, serum bile acids, and direct bilirubin. Baseline and repeat ultrasound examinations of gallbladder for sludge may be appropriate for long-term PN.

X. BONE DISEASE

Osteopenia is a characteristic of patients who receive long-term parenteral nutrition. Patients may be osteopenic with a mean loss of 25% of the calcium in their trabecular bone.[8] A number of factors have been considered that may contribute to the decreased mineralization. Deficiencies in manganese, fluoride, boron, and silicone have been hypothesized as potential factors. In a recent study,[9] we found that serum silicone levels in those receiving HPN were 50% lower than those in non-PN controls. Furthermore, the significant correlation between silicone intake and degree of demineralization suggests an involvement of silicone in the pathogenesis of the bone disease.

XI. RENAL DISEASE

Glomerular filtration rates may be reduced in patients receiving long-term HPN. No nephrocalcinosis or tubular dysfunction was identified in a group of patients studied.[10,11] The decrease in glomerular filtration rate was not related to the underlying disease, infectious episodes, or to the nephrotoxic drugs used. Although high amino acid intake has been implicated, the cause remains uncertain, but longer duration of PN is associated with lower GFR. The significance of this for long-term HPN is of concern.

XII. VISUAL FUNCTION

In a recent study, despite normal visual acuity, one-half of the children receiving long-term HPN had at least one and usually two abnormalities in their electroretinogram. Although deficiency of vitamins A and E, zinc, selenium, linolenic acid, and taurine have each been implicated as causes of retinal dysfunction, the etiology and the progression of the visual dysfunction remain unclear.

XIII. MORTALITY

HPN has associated risks. Sepsis, secondary to infection of the central venous catheter, is the number one cause of death in patients receiving HPN. This is a preventable complication, and in most instances, there is delay in seeking medical attention. However, on occasion, failure of the physician to recognize the early signs of sepsis or a fulminant septic course may be responsible. Liver failure as the primary cause of death is uncommon after the first year of PN.

XIV. CONCLUSIONS

Patients receiving HPN live long enough to encounter medical problems and should be supervised by a multi-disciplinary team cognizant of these syndromes. Standards of Practice for Home Nutrition Support have been developed by ASPEN.[12] The primary provider of specialized home nutrition support is the

TABLE 6
Ongoing Nutritional Assessment

Minimum monitoring includes

With each follow-up visit
- Anthropometric measurements
- Labs: CBC, electrolytes, calcium, magnesium, total protein, albumin, prealbumin
- Recalculation of TPN formula
- General patient evaluation and interview
- In pediatric patients, an emphasis on developmental aspects

Every 3 months
- Liver function tests
- Triglyceride and cholesterol
- Serum iron, TIBC, % saturation

Every 6 months
- Fat-soluble vitamin deficiencies

Yearly
- Trace element determinations
- Renal clearance
- Trabecular bone density
- Gallbladder ultrasound

TABLE 7
Frequency of Follow-up Visits

Within 1 week of start of HPN
Every 2 weeks for 1 month
Every month for 1 year
Every 2 to 3 months for second year
Every 3 months thereafter

physician who is responsible for the patient's nutrition care. He should be assisted by a registered nurse, a registered dietitian, and a registered pharmacist, each having appropriate education, specialized training, and experience in the discipline of specialized nutrition support. The specialized home nutrition support service is initiated and coordinated by the physician. The team is responsible for performing and monitoring ongoing nutritional assessments (Table 6, Table 7). A company that provides HPN products and services to these patients should also have the requisite expertise and personnel to provide monitoring assessment, reporting, and follow-up as needed and requested by the physicians.

Theoretically, HPN should be possible to provide nutritional support, if necessary, for an entire life span. The techniques for this support can be mastered by adults and teenagers for themselves and by most parents for their infants and children. With astute monitoring and ongoing nutritional assessments, complete nutritional support may be carried out without serious complications in most cases. It has been well established that patients on HPN can lead productive lives and do most things that other adults and children do.

With the coming of small intestinal transplantation (with or without liver transplantation), it may be possible in the future that those who formerly would have anticipated living their life on HPN will be able to return to an existence which is more normal.

REFERENCES

1. Howard, L., Heaphey, L., Fleming, R., et al., Four years of North American Registry Home parenteral nutrition outcome data and the implications for patient management, *JPEN,* 15, 384–393, 1991.
2. Moukarzel, A. A., Haddad, I., Ament, M. E., et al., 230 patient-years of experience with home long-term parenteral nutrition in childhood: Natural history and life of central venous catheters, *Journal Peds Surg.,* 29, 1–5, 1994.
3. Vargas, J. H., Ament, M. E., and Berquist, W. E. Long-term home parenteral nutrition in pediatrics: ten years of experience in 102 patients, *J. Pediatr. Gastr. Nutr.,* 6, 24–32, 1987.
4. Mughal, M. and Irving, M., Home parenteral nutrition in the United Kingdom and Ireland, *Lancet,* 16, 383–387, 1986.
5. Wolfe, B. M., Beer, W. H., Hayashi, J. R., et al., Experience with home parenteral nutrition, *Am. J. Surg.,* 146, 7–14, 1983.
6. Grosfeld, J. L., Rsecorla, F. J., and West, K. W., Short bowel syndrome in infancy and childhood. Analysis of survival in 60 patients, *J. Surg.,* 151, 41–46, 1986.
7. Cochran, W. J., Klish, W. J., Brown, M. R., Lyons, J. M., and Curtis, T., Chylous ascites in infants and children: a case report and literature review, *J. Pediatr. Gastr. Nutr.,* 4, 668–673, 1985.
8. Postuma, R. and Moroz, S. P., Pediatric Crohn's disease, *J. Pediatr. Surg.,* 20, 478–482, 1986.
9. Stokes, M. A. and Irving, M. H., How do patients with Crohn's disease fare on home parenteral nutrition? *Dis. Colon Rectum,* 31, 454–458, 1988.
10. Pitt, H. A., Mann, L. L., Berquist, W. E., Ament, M. E., Fonkalsrud, E. W., and DenBesten, L., Chronic intestinal pseudo-obstruction. Management with total parenteral nutrition and a venting enterostomy, *Arch. Surg.,* 120, 614–618, 1985.
11. Wesley, J. R., Home parenteral nutrition: indication, principles, and cost-effectiveness. *Compr. Ther.,* 9, 29–36, 1983.
12. Skehan, A. M. and Fitzgerald, R. J., Home parenteral nutrition in childhood, *Ir. Me. J.,* 79, 253–254, 1986.
13. Taylor, C. J. and Manning, D., Home parenteral nutrition in infantile short bowel syndrome, *Acta Pediatr. Scand.,* 75, 866–867, 1986.
14. A.S.P.E.N. Board of Directors, Guidelines for use of home total parenteral nutrition, *JPEN,* 11, 342–344, 1987.
15. Richard, K. A., Kirksey, A., Baehner, R. L., Grosfeld, J. L., Provisor, A., Weetman, R. M., Boxer, L. A., and Ballantine, T. V. N., Effectiveness of enteral and parenteral nutrition in the nutritional management of children with wilms tumors, *Am. J. Clin. Nutr.,* 33, 2622–2629, 1980.
16. Weich, D., Nutritional compromise in radiation therapy patients experiencing treatment-related emesis, *JPEN,* 5, 57–60, 1980.
17. Van Eys, J., Nutritional therapy in children with cancer, *Cancer Res.,* 37, 2457–2461, 1977.
18. Kirschner, B. S., Enteral nutrition in enteral bowel disease, in *Report of the 94th Ross Conference on Pediatric Research: Enteral feeding: scientific basis and clinical applications,* 1988, 103–109.

19. Biller, J. A., Short small-bowel syndrome, in *Pediatric nutrition theory and practice,* Grand, R. H., Sutphen, J. L., and Dietz, W. H., Eds. Boston, Butterworths, 1987, 481.
20. Culpepper-Morgan, J. A. and Floch, M. H., Bowel rest or bowel starvation: defining the role of nutritional support in the treatment of inflammatory bowel diseases, *Am. J. Gastroenterol.,* 86, 269, 1991.
21. Navarro, J., Goulet, J., Charritat, J., et al., Constant rate enteral nutrition at home in gastrointestinal pediatric practice, *Nutr. Sup. Serv.,* 1(2), 21–22, 1981.
22. Halmi, K. A., Powers, P., and Cunningham, S., Treatment of anorexia nervosa with behavior modification. Effectiveness of formula feeding and isolation, *Arch. Gen. Psychiatry,* 32, 93–96, 1975.
23. Alverdy, J. C., Chi, H. S., and Sheldon, G. F., The effect of parenteral nutrition on gastrointestinal immunity: the importance of enteral stimulation, *Ann. Sur.,* 202, 6981, 1985.
24. Blackman, J. A. and Nelson, C. L. A., Reinstitution of oral feedings in children fed by gastrostomy tube, *Clin. Pediatr.,* 24, 434–438, 1985.
25. Dorey, S. F., Ament, M. E., Berquist, W. E., Vargas, J. H., and Hassal, E., Improved survival in very short small bowel of infancy with use of long-term parenteral nutrition, *J. Pediatr.,* 107, 521–525, 1985.
26. Postuma, R., Moroz, S., and Friesen, F., Extreme short-bowel syndrome in an infant, *J. Pediatr. Surg.,* 18, 264–268, 1983.
27. Levitt, M. B., *Families at Risk: Primary Prevention in Nursing Practice,* Boston, Little, Brown, 1982.
28. Schreiner, R. L., Eitzen, H., Gfell, M. A., et al., Environmental contamination of continuous drip feedings, *Pediatrics,* 63, 232, 1979.
29. Nicholson, L. J., Declogging, small-bore feeding tubes, *JPEN,* 11, 594–597, 1987.
30. Broadwell, D. C. and Jackson, B. S., Eds., *Principles of ostomy care,* St. Louis, CV Mosby, 1982, 257.
31. Nixon, H. H., Paediatric problems associated with stomas. Intestinal surgical procedures, in *Stomas, Clinics in Gastroenterology,* Brooke, B. N., Jeter, K. F., and Todd, I. P., Eds., London, WGB Saunders, 1982, 351.
32. Faith, S., Streiger, E., Gulledge, A. D., et al., Psychological issues of nutrition support, *Nursing Clin. N. Am.,* 24, 447, 1989.
33. Gulledge, A. D., SR, F., Sharp, J. W., et al., Subjective distresses of nasogastric tube feeding, *J. Parenter. Enter. Nutr.,* 3, 53–57, 1979.
34. Gulledge, A. D., Common psychiatric concerns in home parenteral nutrition, *Clev. Clin. Guart.,* 53, 329–332, 1985.
35. Hanson, R. L., Walike, B. C., Grant, M., et al., Patient responses and problems associated with tube feedings, *Was. St. J. Nurs.,* 12, 9–13, 1975.
36. Winthrop, I. L. and Wesson, D. E., Urokinase in the treatment of occluded central venous catheters in children, *J. Pediatr. Surg.,* 19, 536–538, 1984.
37. Glynn, M. F., Langer, B., and Jeejeebhoy, K. N., Therapy for thrombotic occlusion in long term intravenous alimentation catheter, *JPEN,* 4, 387–390, 1980.
38. Duffy, L. F., Kerzner, B., and Gebus Vand Dice, J., Treatment of central venous catheter occlusion with hydrochloric acid, *J. Pediatr.,* 114, 1002–1004, 1989.
39. Moukarzel, A., Azancot, A., Brun, P., Vitoux, C., Cezard, J. P., and Navarro, J., M-mode and two dimensional echocardiography in the routine follow up of central venous catheter in children receiving total parenteral nutrition, *JPEN,* 15, 551–555, 1991.
40. King, D. R., Komer, M., Hoffman, J., et al., Broviac catheter sepsis: The natural history of an iatrogenic infection, *J. Pediatr. Surg.,* 20, 728–733, 1985.
41. Schropp, K. P., Ginn-Pease, M. E., and King, D. R., Catheter-related sepsis: a review of the experience with Broviac and Hickman catheters, *Nutrition,* 4, 195–200, 1988.
42. Wesson, D. E., Ric, R. H., Zlotkin, S. H., and Pencharz, P. B., Fat overload syndrome causing respiratory insufficiency, *J. Pediatr. Surg.,* 19, 777–778, 1984.

43. Heyman, M. B., Storch, S., and Ament, M. E., The fat overload syndrome, *Am. Dis. Child.*, 135, 628–630, 1981.
44. Dahlstrom, K. A., Goulet, O. J., Roberts, R. I., Rlcour, C., and Ament, M. E., Lipid tolerance in children receiving long-term parenteral nutrition: a biochemical and immunologic study, *J. Pediatr.*, 113, 985–990, 1988.
45. Greene, H. L., Hambridge, K. M., Schanler, R., and Tsang, R. T., Guidelines for the use of vitamins, trace elements, calcium, magnesium, and phosphorus in infants and children receiving total parenteral nutrition: report of the subcommittee on pediatric parenteral nutrition requirements from the committee on clinical practice issues of the American Society of Clinical Nutrition, *Am. J. Cin Nutr.*, 48, 1324–1342, 1988.
46. Vinton, N. E., Dahlstrom, K. A., Strobel, C. T., and Ament, M. E., Macrocytosis and pseudoalbinism: manifestations of selenium deficiency. *J. Pediatr.*, 111, 711–717, 1987.
47. Kelly, D. A., Coe, A. W., Shenkin, A., Lake, B. D., and Walker-Smith, J. A., Symptomatic selenium deficiency in a child on home parenteral nutrition, *J. Pediatr. Gastroenterol. Nutr.*, 7, 783–786, 1988.
48. Moukarzel, A. A., Song, M. K., Buchman, A. L., et al., Excessive chromium intake in children receiving total parenteral nutrition, *Lancet*, 1, 385–388, 1991.
49. Moukarzel, A., Ament, M. E., Vargas, J. H., et al., Iodide supplementation is not necessary in children receiving long term parenteral nutrition, *Clin. Nutr.*, (Abstr.), 61, 1990.
50. Wan, K. K. and Tsalas, G., Dilute iron dextran formulation for addition to parenteral nutriment solutions, *Am. J. Hosp. Pharm.*, 37, 206–210, 1980.
51. Pitt, H. A., King, W., Mann, L. L., Roslyn, J. J., Bequist, W. E., and Ament, M. E., Increased the risk of cholelithiasis with prolonged total parenteral nutrition, *Am. J. Surg.*, 145, 106–112, 1983.
52. Roslyn, J. J., Berquist, W. E., Pitt, H. A., Mann, L. L., Kangharloo, H., DenBesten, L., and Ament, M. E., Increasing risk of gallstones in children receiving total parenteral nutrition, *Pediatrics*, 9, 787–784, 1983.
53. Hodes, J. E., Grosfeld, J. L., Weber, T. R., Schreiner, R. I., Fitzgerald, J. F., and Mirkin, L. D., Hepatic failure in infants on total parenteral nutrition: clinical and histopathologic observations, *J. Pediatr. Surg.*, 17, 463–468, 1982.
54. Suite, S., Ikeda, K., Nagasaki, A., et al., Follow-up studies of the children treated with long term intravenous nutrition during the neonatal period, *J. Pediatr. Surg.*, 17, 37–42, 1982.
55. Abad-Lacruz, A., Gonzalez-Huiz, F., Esteve, M., et al., Liver function tests abnormalities in patients with inflammatory bowel disease receiving artificial nutrition: a prospective randomized study of total enteral nutrition vs. total parenteral nutrition, *JPEN*, 14, 618, 1990.
56. Roslyn, J. J., Pitt, H. A., Mann, L. L., Ament, M. E., and DenBesten, L., Gallbladder disease in patients on long-term parenteral nutrition, *Gastroenterology*, 894, 148–154, 1983.
57. Moukarzel, A., Ament, M. E., Vargas, J., et al., Parenteral nutrition bone disease in children, *JPEN*, (Abstr.), 16, 3S, 1992.
58. Moukarzel, A., Song, M. K., Haddad, I., et al., Is silicon deficiency involved in the pathogenesis of metabolic bone disease of children receiving parenteral nutrition? *JPEN*, (Abstr.), 16–3S, 1992.
59. Moukarzel, A., Ament, M., Vargas, J., et al., True glomerular filtration rate of children on long term parenteral nutrition, *Clin. Res*, (Abstr.), 38–190A, 1990.
60. Moukarzel, A., Ament, M., Buchman, A., et al., Renal function of children receiving long term parenteral nutrition, *J. Pediatr.*, 119, 864–868, 1991.
61. Geggel, H. S., Ament, M. E., Heckenlively, J. R., Martin, D. A., and Kopple, J. D., Nutritional requirement for taurine in patients receiving long term parenteral nutrition, *New. Engl. J. Med.*, 312, 142–146, 1985.
62. Vinton, N. E., Laidlaw, S. A., Ament, M. E., and Kopple, J. D., Taurine concentrations in plasma, blood cells and urine of children undergoing long term TPN, *Pediatr. Res.*, 21, 399–403, 1987.

Chapter 17

PRENATAL AND GENETIC MAGNESIUM DEFICIENCY IN CARDIOMYOPATHY: POSSIBLE VITAMIN AND TRACE MINERAL INTERACTIONS

Mildred S. Seelig

TABLE OF CONTENTS

I. Introduction 197

II. Heritable Magnesium Absorption, Renal Excretion, and Tissue Levels 198
- A. Isolated Intestinal Magnesium Malabsorption 198
- B. Renal Tubular Magnesium Wasting 199

III. Magnesium and Cardiomyopathy: Congenital and Genetic 202
- A. Fetal and Infant Cardiovascular Deaths; Maternal Magnesium Deficiency 202
- B. Cardiomyopathies in Diseases with Genetic Components Involving Magnesium 204

IV. Concluding Comments 211

References 214

I. INTRODUCTION

Infantile cardiovascular lesions that have caused early serious disease or death and that can contribute to the common arterial and cardiac diseases of later life have been studied for roles of nutritional or genetic factors. It is proposed here that absolute or relative dietary magnesium deficiency, which is common,[1–4] might contribute to the complications of these diseases, particularly in mothers and infants with genetic variations in handling magnesium. Reviewed elsewhere is substantial experimental evidence that diets low in magnesium, but otherwise adequate, produce microangiopathy and resultant cardiomyopathy (CMP), and that magnesium deficiency intensifies the thrombogenicity and macroangiopathy caused by atherogenic diets.[4–7] Analysis of reports on infants with cardiovascular lesions resembling those of ischemic heart disease of later life has disclosed abnormalities resembling those of induced magnesium deficiency.[4,8] Data on maternal magnesium inadequacy, as contributory to poor outcomes of pregnancy,

0-8493-2764-4/95

have been considered elsewhere.[8–12] Familial patterns of the common cardiovascular diseases suggest that attention be paid to genetic factors in magnesium utilization in afflicted families. Severe metabolic errors, such as isolated malabsorption,[13–17] and renal tubular wasting of magnesium,[18–20] that are associated with hypomagnesemia usually requiring parenteral magnesium therapy to control convulsive and tetanic manifestations, have been shown to be familial.[15,17–20] That such abnormalities may not be an "all or none" phenomenon is suggested by genetic differences in plasma and cellular magnesium levels that have been elicited in different ethnic groups, in Type A and B subjects, in twin studies, and in genetically selected mice.[21–23] Not studied is whether these differences might be contributed to by malabsorption, renal wasting, or differences in membrane permeability to magnesium.

Since magnesium acts like a trace mineral as a cofactor for over 300 enzymes,[24,25] its possible contributory role in genetic diseases involving trace minerals and vitamins, in which arterial disease develops, should be explored. Suggested here is the possibility that the microangiopathy of the common genetic disease, diabetes mellitus, and the cardiomyopathies seen in several inborn errors of metabolism that entail vitamin dependencies and abnormal mineral metabolism, might also involve function of enzymes in which magnesium is a cofactor.

II. HERITABLE MAGNESIUM ABSORPTION, RENAL EXCRETION, AND TISSUE LEVELS

A. ISOLATED INTESTINAL MAGNESIUM MALABSORPTION

Magnesium deficiency that results from generalized intestinal malabsorption—whether caused by chronic diarrhea, Crohn's disease, steatorrhea, celiac disease, or sprue—has long been recognized,[26] but is not the subject of this report. The metabolic error, isolated defect of intestinal absorption of magnesium, was first identified in a French Canadian baby.[13,27,28] Soon thereafter, it was recognized in American,[29] French,[14,20] Swedish,[16,17] and Italian[30] infants and then found to be familial when siblings developed the same syndrome.[15,17] As a result of efforts to repair the hypocalcemia by calcium, high-dose vitamin D, and parathyroid hormone therapy, in the first identified case of magnesium malabsorption, glomerular and renal interstitial fibrosis developed, with calcification, which was described as a major complication of infantile hypomagnesemia.[13,27,28]

The risk of producing hypercalcemia by calcemic treatment of hypocalcemia in the presence of hypomagnesemia was cautioned against in an early study of magnesium malabsorption.[26] That low serum magnesium levels could not be relied upon for diagnosis of severe magnesium deficiency of intestinal malabsorption was commented on in a 1981 study of 17 hospitalized patients with severe Crohn's disease, only six of whom had overt hypomagnesemia, but 15 of whom had low urine magnesium.[31]

B. RENAL TUBULAR MAGNESIUM WASTING

More frequently diagnosed (generally in older children and in adults) is renal wastage of magnesium,[18–20,32–48] usually but not always in association with hypokalemia or hypocalcemia or both. Two families with members suffering from hypomagnesemia caused by renal magnesium loss were reported in 1966.[18,19] Each renal magnesium wasting syndrome has been reported in more than one family member.[18–20,33,41,43,45,47]

Hypochloremic Hypokalemic Alkalosis with Magnesium Wasting—Bartter's syndrome was diagnosed in a three-month-old febrile, sweating baby with apathy, weakness and growth failure, and laboratory findings characteristic of that syndrome: hypokalemic hypochloremic alkalosis, hyper-reninism, aldosteronism, and urinary wasting of sodium, as well as of potassium and chloride.[34] Despite only marginally low serum magnesium, mean of 1.7 mEq/>L (range: 1.5–1.9), oral magnesium supplementation (6 mg of Mg/kg/day) was prescribed because a muscle biopsy disclosed subnormal magnesium levels (11.6 mEq/kg fat-free weight; normal = 20 +/− 1.8). A renal biopsy disclosed hyperplasia of the juxtaglomerular apparatus and cloudy swelling of the proximal tubules. Sustained magnesium supplements lowered his high renin and aldosterone levels and corrected all of the electrolyte and clinical abnormalities, except for slow growth, by 16 months of age. Since this baby's magnesium deficit was not detectable by serum values, the authors suggested that low intracellular magnesium might be more common than suspected in Bartter's syndrome. In another family,[47] three siblings had Bartter's syndrome with pronounced hypomagnesemia; necropsy findings of juxtaglomerular hyperplasia were similar to that reported from a renal biopsy from the infant with low muscle magnesium.[34] Renal magnesium wasting has been reported by others in patients with hypokalemic alkalosis as part of Bartter's syndrome.[20,40] In three siblings (two girls and a boy) with familial hypokalemic alkalosis with tubulopathy, rather than the glomerular abnormality generally considered part of Bartter's syndrome, oral magnesium supplements of 40–60 mEq/day as MgC12 corrected the potassium loss, but had no effect on the elevated renin, and raised the aldosterone level.[41] The hypomagnesemia of two adult sisters in another family was associated with hypokalemic alkalosis, normotensive hyper-reninemism, and marginally high mineralocorticoids.[18]

Renal Tubular Acidosis with Magnesium Wasting—Renal magnesium leakage has also been associated with renal tubular acidosis, usually with hypocalcemia, hypercalciuria, and nephrocalcinosis[33,39,40,45] and/or chondrocalcinosis.[32,36,37,39,42,46] Of two siblings with renal tubular acidosis and nephrocalcinosis, the sister exhibited weakness and trembling and developed active rickets at ten years of age—manifestations of hypomagnesemia from the renal loss. Suggestive of the possibility that they might have had magnesium inadequacy from birth is the fact that their mother had a history of seven spontaneous abortions;[33] magnesium deficiency is contributory to abortions and preterm births.[4,8,9,11,12] A young boy with renal magnesium wasting associated with aminoaciduria, glycosuria (at normal to low blood glucose levels), growth retardation, and osteoporosis, without intestinal magnesium malabsorption, has been reported to respond well to high-dosage oral magnesium supplements;[33] most patients with renal magnesium wasting have required parenteral magnesium treatment.

Renal Damage from Calcemic Therapy of Magnesium-Low Hypocalcemia—Tetany and convulsive seizures characterize both hypomagnesemia and hypocalcemia. The usual clinical pathological testing procedures disclose the calcium deficit, whereas the magnesium deficit is more difficult to detect;[49] an underlying magnesium inadequacy is apt to be missed, and treatment is directed to correction of the hypocalcemia. Since animal studies have shown that magnesium-deficient rats, on high calcium intakes, develop calcific lesions in the corticomedullary area of the kidneys and calcium microliths in the loop and the ascending limb of the loop of Henle (ALLH), the major site of magnesium reabsorption,[5,50–54] it is possible that calcemic therapy, given to hypocalcemic infants and children, who are magnesium deficient, might cause renal tubular lesions in such areas. The laboratory findings and several clinical data support the premise that calcemic rather than magnesium treatment of hypocalcemia with underlying magnesium deficiency might play a role in long-lasting renal magnesium leakage.

Two of the renal magnesium-wasting young men, with hypocalcemia, hypercalciuria, and bilateral nephrocalcinosis, as well as chondrocalcinosis, also malabsorbed magnesium.[37] Another young man with chondrocalcinosis and renal magnesium wasting had his disease first diagnosed at the age of six, when hypocalcemic convulsions and hypokalemia failed to improve with calcium and potassium treatment.[36] At that time, a renal biopsy showed intraluminal tubular microliths in the renal corticomedullary area; possibly, the ALLH was involved. He, too, was found to be a poor absorber of magnesium. The first reported case, in 1944, of hypomagnesemic osteochondrosis was a six-year-old boy who had been treated unsuccessfully with high-dosage vitamin D for neonatal hypocalcemic tetany and convulsions.[55] His magnesium deficiency was not identified until he was six months of age. Parenteral magnesium then controlled the neuromuscular manifestations, but magnesium supplementation was not sustained. Perhaps magnesium malabsorption caused his early hypomagnesemic hypocalcemia; his later persistent magnesium deficit might have been caused by renal damage-induced magnesium wastage. Another young boy's renal magnesium wasting that could not be compensated for by oral or parenteral magnesium supplements, developed osteochondritis and later died of CMP.[32] A child with a comparable history, but without osteochondritis, died with hypertrophic CMP (at 13 years of age).[48] Autopsy of a five-month-old baby girl, whose convulsions had been unsuccessfully treated with high-dose vitamin D and intravenous calcium, and who was found to have profound hypomagnesemia the last day of his life,[56] disclosed renal calcinosis, with calcium deposits in proximal tubules and Henle's loop, as well as focal myocardial necrosis and calcium deposits and coronary arterial calcification. This case report seems supportive of the premise that early calcemic therapy, in the presence of magnesium deficiency (possibly caused by malabsorption), can cause renal tubular damage that diminishes renal magnesium reabsorptive capability. In view of the history of infantile convulsions in four of her siblings, and death of another sibling from cerebral arterial calcification at three months, the metabolic disorder seems to be familial.

Patterns of Hereditary Renal Magnesium Wasting—Three types of hereditary renal hypomagnesemia have been suggested:[43] an autosomal dominant pattern of inheritance is believed to be present in patients with isolated familial hypomagnesemia; an autosomal recessive trait has been suggested in familial hypomagnesemia caused by a low renal magnesium threshold that is associated with hypokalemia, metabolic alkalosis, hypocalciuria, and moderate salt wasting; and familial hypomagnesemic hypercalciuria, caused by a defect in absorption of both magnesium and calcium at the ALLH, with renal calcinosis, also may be inherited as an autosomal recessive trait. This condition has not been clearly separated from hereditary distal renal tubular acidosis. A form of hypomagnesemic hypokalemia that may be transmitted by an autosomal recessive gene has been described in a young son of a woman suffering from chronic hypomagnesemia.[57] He had no symptoms of his subnormal magnesium and potassium serum levels, other than carpopedal spasms, that were not relieved by magnesium or potassium therapy. This defect was postulated to be caused by inability to maintain a normal gradient between intra- and extracellular magnesium and potassium.

Genetic Factors in Control of Magnesium Levels in Plasma and Erythrocytes—Studies of identical and fraternal twins have shown that red cell magnesium is significantly more similar in monozygotic than in dizygotic twins and among family members than in unrelated subjects.[21–23,58] Association of major histocompatibility complex human leukocyte antigens (HLA) alleles have been shown to regulate red blood cell magnesium in humans; H–2 alleles are involved in mice genetically selected for low red cell magnesium.[22,23] HLA-B38 carriers, among blood donors, have lower plasma and red cell magnesium levels than do other genetically related groups.[23,59]

The latent tetany syndrome of magnesium deficiency[60] has been associated with the HLA-B35 allele;[61] chondrocalcinosis, which has been found in renal wasters of magnesium, has been associated with the B15 antigen.[62] Type A behavior students, many of whom were in the HLA-Bw35 group,[63,64] had a lowered red cell level of Mg under stress, as well as lower plasma magnesium levels. Interestingly, they had higher red cell zinc levels (possibly linked with the GLO1 locus[63]) and excreted more urinary zinc under stress.[21] Mitral valve prolapse, which has also been associated with low magnesium levels, occurring more frequently in patients with the latent tetany syndrome,[65–67] also is associated with HLA-Bw35.[65,67] This allele is associated with lower red cell Mg and much greater antibody response to anti-influenza vaccination,[68] which might relate to the role of magnesium in immunologic disorders, in which the major histocompatability complex plays an important role.[69]

TABLE 1
Arterial Abnormalities in Stillborn Infants or in Those Dead in First Month of Life*

Affected arteries	
Small, Medium Coronaries*	16
Large Coronaries	25
Aorta	27
Pulmonary	11
Cerebral	1
Visceral (renal, pancreatic, etc.)	12
Generalized	13
Pathologic changes	
Intimomedial thickening	15
Subintimal, medial elastica degeneration	13
Subintimal, medial elastica calcification	14
Thrombi	13
Atresia, coarction (aorta, pulmonary)	25

*Rarely examined.

Data from published data on 154 individually reported cases.[4]

III. MAGNESIUM AND CARDIOMYOPATHY: CONGENITAL AND GENETIC

A. FETAL AND INFANT CARDIOVASCULAR DEATHS; MATERNAL MAGNESIUM DEFICIENCY

Neonatal Cardiovascular Damage—The early arterial lesions, that were reported in 154 individual case reports and in more than 500 infants reported in pathology summaries of abnormalities found in stillborn infants and in those dying in the first month of life, (Table 1[4,8]) include small vessel intimal fibroblastic proliferation, elastica degeneration, and calcification, changes such as have been produced by experimental magnesium deficiency in laboratory animals.[4–6] The coronary microcirculation was rarely examined in the fetuses and infants tabulated in the 1980 report, but the myocardial and endocardial lesions indicate probable damage to the small arteries of the heart. (Table 2[4,8]) To correlate such changes with magnesium deficiency of the fetus, which must reflect that of the mother, has to be inferential, few data having been reported on the mothers of the cited infants born with these or gross cardiovascular anomalies. In the few reports indicating the maternal condition, spontaneous abortions, toxemias, multiple births, frequent and/or multiple pregnancies and diabetes, were all associated with poor outcomes of pregnancy; in all, magnesium inadequacy is likely.[4,8,11,12,70–87] Also, prenatal magnesium supplements have shown benefit; fewer toxemic pregnancies and low birth weight infants among supplemented vs. control mothers have been reported fron extensive European retrospective,[88,89] epidemiologic,[90] and prospective double-blind studies.[91,92] Since preterm infants have low

TABLE 2
Cardiac Abnormalities Suggestive of Myocardial Ischemia in Stillborn Infants or in Those Dead in First Month of Life*

Location of Lesions	
Mural	36
Subendocardial, papillary muscle	34
Multifocal, disseminated	26
Pathologic changes	
Myocardial necrosis, cell infiltration	41
Myocardial fibrosis	37
Myocardial calcification	28
Myocardial lipid infiltration	2
Endocardial fibroelastosis	80
Conduction System Abnormality*	7

*Rarely examined.

Data from published data on 154 individually reported cases.[4]

TABLE 3
Increase in Fetal Body Magnesium and Magnesium Uptake with Fetal Age

Age (lunar months)	Total Mg Content (Average mg/body)	Average Mg Uptake (mg/day)	(Number of Specimens*)
3	15	0.5	(7)
4	58	1.5	(3)
7	173	2.6	(6)
8	306	4.7	(8)
9	512	7.4	(7)
10	703	9.0	(10)

Note: No data on normal fetuses, which might have higher Mg contents.
*From spontaneous abortions, miscarriages, stillbirths.

Data derived from Widdowson, E.M. and Dickerson, J.W.T., *Mineral Metabolism*, Academic Press, New York, 1962.

magnesium levels, much of the total fetal magnesium being accumulated in the last two lunar months[4,8,93] (Table 3), fetal inadequacy of magnesium seems to be a plausible contributory factor to the higher incidence of cardiovascular abnormalities in such infants, than in term singleton infants born to mature normal non-multiparous mothers.

Maternal and Infant Cardiomyopathy—Microangiopathy/CMP has been reported both in infants,[94–98] and peripartally in mothers,[99–112] particularly with gestational conditions associated with magnesium deficiency. A 1971 study indicated that toxemic pregnancies were complicated by peripartum CMP about six times as frequently as were normal pregnancies, and that 7% of peripartal CMP occurred with twin pregnancies.[103] The much greater vulnerability of marginally malnourished multiparous and toxemic pregnant women to CMP, than of affluent mothers, who usually are better nourished, was editorially commented upon in 1968;[102] the condition is much rarer in the developed world[108,112] than in Africa.[107,110]

Among the primary "idiopathic" cardiomyopathies listed in a 1970[99] paper systematizing the many conditions associated with CMP were fibrotic CMP of infancy (suggested as possibly attributable to hypercalcemia), familial myocardial fibrosis of later childhood and adults, and CMP of pregnancy and the puerperium, in each of which magnesium inadequacy may participate. Magnesium deficiency or tissue loss also occurs in several of the conditions that were listed as contributory to secondary CMP: alcoholism, severe diarrhea, potassium deficiency or excess, abnormal calcium deposition including that of hyperparathyroidism, protein calorie malnutrition, beriberi (both thiamin responsive and resistant), or catecholamine-toxicity. Cited also were magnesium deficiency itself and several dysrhythmias.

Magnesium Deficiency and the Heart: There is growing evidence that magnesium deficiency can contribute to a variety of arrhythmias and that magnesium treatment is increasingly being used for their management worldwide.[4,7,113–124] Although its therapeutic use is based predominantly on its pharmacologic effect, it may also be restoring a deficit.[117–121] Increased intravascular coagulation, a probable factor in CMP,[99] might be contributed to by a low ratio of magnesium to calcium.[12] And finally, the myocardial lesions of experimental magnesium deficiency in rats,[125–128] dogs,[129] and golden Syrian hamsters[130–132] resemble those of CMP.

B. CARDIOMYOPATHIES IN DISEASES WITH GENETIC COMPONENTS INVOLVING MAGNESIUM

The familial occurrence of CMP was commented on in 1957, when the term was first used to indicate myocardial disease without major coronary arterial involvement.[133] Familial myocardial fibrosis was listed as the commonest form of CMP in 1970;[99] that year 13 families with hypertrophic CMP were reported,[134] and the following year 11 more such families were described, with findings compatible with an autosomal dominant gene with high penetrance.[135] The 1964[133] and 1970[99] reviews of genetic diseases associated with cardiovascular abnormalities did not include diabetes mellitus. However, in 1964 and 1967,[137,138] the minute focal areas of myocardial degeneration and fibrosis of these diseases were attributed to medial necrosis of the small intramural arteries and their occlusion by platelet aggregation, which was then proposed as being partially explanatory of not only the rare genetic cardiomyopathies but that of alcoholics and of juvenile diabetics; their similarity to that of experimental magnesium deficiency[129] was noted.[138]

Diabetic Cardiomyopathy—Proliferative lesions of the endothelium that narrow or obliterate the lumina were described in diabetics' intramural small coronary arteries in 1960.[139] Fibrotic CMP, stemming from microangiopathy of diabetes, was described a decade later.[140] In 1972, it was shown that 16 of 73 pts with diabetes mellitus had CMP; autopsies in several showed mural coronary microangiopathy and focal perivascular and interstital fibrosis of the endocardium and subendocardium.[141]

Relationship with Magnesium Loss: It has been known for almost half a century that poorly controlled insulin-dependent diabetes (Type I) causes magnesium loss.[142] An early report also showed that those with juvenile diabetes are likely to have the lowest serum magnesium values.[143] A study with diabetic children, under strict medical and dietary control, found that a third had hypomagnesemia (<1.4 mEq/L of serum) as well as comparably low levels in red cells.[144] It thus appears that diabetic children are more likely to be magnesium deficient than are adult diabetics. As with adults, in a study of 95 Type I children, those not optimally controlled had the lowest serum magnesium values.[145] This was verified in a subsequent study of 63 such children, which further demonstrated the deficit by retention of >40% of a parenteral load of magnesium.[146] In both of those clinical studies, increased urinary magnesium output was noted versus controls. Diabetic gastroenteropathy was suggested as a cause of magnesium malabsorption that contributes to magnesium deficiency. Elevated serum levels of low-density lipids and depressed levels of high-density lipids, of poorly controlled diabetic children were shifted to an improved pattern with better control that was associated with higher serum magnesium.[144] This is in accord with the demonstration that magnesium deficiency causes substantial elevation of low-density and very low-density lipids and decreased high-density lipids, accompanied by increased platelet aggregation and thrombus formation, in rats fed a low-fat diet.[147,148] Since increased magnesium intake protected against these effects, the authors suggested that the dyslipidemia of diabetes mellitus might be mediated, at least in part, by magnesium loss.[147,148]

Possible Relationship with Chromium: The dietary intake of chromium is suboptimal in the American diet, its loss is increased during pregnancy, its deficiency has been linked to maturity-onset diabetes and arterial disease, and its supplementation has increased high-density lipid levels.[150–152] It has been suggested that, although the major mineral abnormality in diabetes is that of magnesium, these effects of chromium justify trial of its supplementation, as well as that of magnesium, in juvenile diabetes.[153]

Congenital Deafness, Syncope, Arrhythmias, and Sudden Death (Jervell-Lange-Nielsen Syndrome)—First reported in 1957, as familial deaf-mutism with prolonged Q-T interval, CMP and attacks of syncope, no abnormal blood electrolytes (calcium, potassium and phosphorus) were then encountered.[154] In a 1964 paper on the obscure cardiomyopathies, involvement of the nutrient arteries of the sinus node and atrioventricular node was suggested to explain the high incidence of arrhythmias and conduction abnormalities in this condition,[137] by the investigator who later noted the similarity of the lesions of human CMP to that produced by experimental magnesium deficiency in dogs.[138] Among nine cases in six sibships, autopsies disclosed focal hemorrhages near the atrioventricular and sinoatrial nodes, with involvement of the left anterior descending coronary artery (from which the nodal arteries arise), and nodal fibrotic lesions.[155] Cochlear abnormalities (absent stria vascularis and spiral ganglion of the ear) were speculated to have been caused by fetal vascular damage. Since then, there have been additional reports of the complete familial syndrome,[156,157] of several members of a family with only sinus node involvement,[158] and of a family in which arrhythmias occurred alone or with deafness.[159]

Deafness Caused or Intensified by Experimental Magnesium Deficiency: Weanling rats, that survived magnesium deficiency that caused noise-induced convulsions and death in litter mates, were found to have markedly decreased response to sound[160] and to have cochlear damage (personal communication). That the infantile impairment of hearing might have been caused by magnesium deficiency is suggested by the significantly greater hearing loss of magnesium-deficient guinea pigs exposed to very loud noise than did animals fed a magnesium-rich diet, exposed to the same noise.[161] Suggested mechanisms of the combined noise plus magnesium deficiency-induced deafness were increased catecholamine release, with constriction of the cochlear artery, and decreased magnesium content of the fluid around the hair cells, which allowed for their increased permeability, with increased calcium and sodium influx and energy depletion in the hair cells.[162,163]

Progressive Muscular Dystrophy, Marfan's Syndrome, and Friedrich's Ataxia—Patients with the rare genetic disorders Marfan's syndrome and Friedrichs ataxia, or with the more common Duchenne's muscular dystrophy, are prone to CMP.[99,136–138] Almost all of those with muscular dystrophy, which is inherited as an X-linked trait, develop cardiac lesions,[164] which resemble those of skeletal muscle.[165] It is of interest that a patient with Duchenne's muscular dystrophy, who died suddenly, had medial degeneration of the nutrient artery to the sinus node and of the node, such as is seen in more generalized CMP, in addition to scattered sites of myocardial necrosis.[165] Electrocardiographic study of 169 pts with this disease disclosed abnormalities in 75%.[166]

Magnesium and Calcium in Duchenne's Muscular Dystrophy: Markedly lower serum magnesium and higher serum calcium levels in dystrophic patients than control values were reported by one group of investigators,[167,168] but not by another, who found that preadolescent patients had significantly higher red blood cell magnesium than did postpubertal patients, which was the reverse of findings in controls.[169] Fetuses at risk of this disease, and a premature infant who later developed typical Duchenne's disease, had 3- to 6-fold increased muscle calcium and a lesser increase of muscle magnesium (18 to 57% above normal); no necrotic fibers were detected.[170] Comparably increased muscle calcium was seen in full-blown dystrophy.[171] The increase in their muscle calcium was considered a non-specific part of the final common pathway leading toward cellular degeneration and death.[170,171]

Among the data on the myocardium from studies with patients with Duchenne's muscular dystrophy, although not directly illustrative of the magnesium status, are a few that can be correlated with animal models of hereditary muscular dystrophy that provide information on magnesium (below). Abnormal myocardial metabolism, suggestive of uncoupling of oxidative phosphorylation (indicated by high inorganic phosphate, increased glycolysis, and the high arteriovenous redox potential of anaerobic metabolism) were detected in 11 Duchenne patients whose blood samples were obtained by coronary sinus catherization.[172] Six patients' biopsied muscle, not yet irreparably damaged by the dystrophic disease, had excess catecholamine accumulation around arterioles.[173]

Magnesium and Calcium, Myocardial Metabolism, and Catecholamines in Models of Genetic Muscular Dystrophy: From a strain of golden Syrian hamsters that had hereditary muscular dystrophy, an inbred strain (BIO 14.6) was developed that consistently developed spontaneous CMP.[174] The authors commented that the lesions resembled those produced by excessive catecholamines in rats and other species: augmentation of oxygen consumption beyond requirements for cardiac work and uncoupling of oxidative phosphorylation. The earliest cardiac abnormality in the CMP hamster, before myocardial necrosis developed, was markedly decreased myocardial magnesium as compared with normal hamsters (73 vs. 115 mg/100 g dry weight), at 29 days of age.[175] At two months, when the cardiac damage was moderate to severe in the CMP strain, the myocardial Mg was the same (109 mg) as in the normal hamsters of the same age. The increase in myocardial calcium, over that of normals, however, was slight at the prenecrotic phase (17 vs. 12 mg) but marked at two months (215 vs. 15 mg). Fed magnesium-deficient diets, the CMP hamsters and hybrids developed much more myocardial necrosis than did normal hamsters fed the deficient diet. Adding MgCl2 (1.0mM/d) to the low magnesium diet prevented myocardial necrosis of normal and hybrid hamsters, but not of the CMP strain.[175] Comparable myocardial mineral findings were reported in another study of this CMP strain that also provided ultramicroscopic evidence of the progressive mitochondrial damage,[176] and in one that also reported trace mineral findings: high zinc levels, possibly in exchange for calcium.[177] The MDx mouse, which has a gene defect at the locus homologous to the defective one in Duchenne's muscular dystrophy,[178] also exhibited elevated myocardial calcium at 10 days, 30 days, and 254 to 347 days of life,[178] and at 5, 10, and 23 weeks.[179] As in the BIO 14.6 hamsters,[175,176] the magnesium changes were not notable. The energy metabolism studies of skeletal muscle of the CMP hamster[180–182] and of myocardium of the MDx mouse[179] yielded findings to those from Duchenne patients.[172, 173] Uncoupling of oxidative phosphorylation occurred in 50 to 80-day-old CMP hamsters, the time of most necrosis. Addition of 3mM MgCl2 to abnormal mitochondria at the beginning of the experiment produced doubling of initial rate of respiration and restored phosphorylative coupling to near normal. This might bear on the lack of response to magnesium of the hamsters with most advanced disease. Mitochondria from MDx mice have decreased respiratory control, an observation also in the dystrophic hamsters and in Duchenne's muscular dystrophy.

Cardiac norepinephrine studies,[183] prior to and during development of congestive heart failure of the CMP hamster, showed that its formation was most above normal in the prenecrotic phase, probably due to increased cardiac sympathetic nerve activity. Similar but lesser increases in the myocardial catecholamine were seen at the intermediate phase of cardiac damage. During the final phase, when congestive heart failure had developed, there was a markedly lower content—possibly from a "dilution" effect of hypertrophy and focal destruction of adrenergic nerve terminals. Directly germane to the low myocardial magnesium/calcium ratio in the prenecrotic myocardium of the CMP hamster is the *in vitro* evidence that a low magnesium/calcium ratio in adrenals[184] and peripheral nerves[185] increases catecholamine secretion.

Vitamin E, Selenium, Zinc, and Magnesium in Cardiomyopathy of Muscular Dystrophy: Among the abnormalities produced by experimentally induced vitamin E deficiency in several species is necrotizing myopathy that has pathologic changes resembling those of human muscular dystrophy (which is not responsive to vitamin E). In swine, the manifestations of tocopherol deficiency include myocardial damage associated with fibrinoid necrosis of the arteries.[186] Apart from the histopathologic similarities between vitamin E deficiency-induced lesions and those of hereditary muscular dystrophy, the deficiency-altered muscles also exhibit excessive oxygen utilization, possibly as a result of uncoupling of oxidative phosphorylation, even before the lesions are detectable.[186] Still another similarity is in the lowered muscle magnesium content of both vitamin E-deficient dystrophic animals[187–189] and those with hereditary muscular dystrophy before development of the histologic lesions.[174–176]

Interrelationships between magnesium and vitamin E are indicated also by precipitation of signs and lesions of magnesium deficiency in vitamin E-deficient normal rats[190] and by prevention of respiratory decline of (hepatic) mitochondria from vitamin E-deficient rats by administration of magnesium.[191] Additionally, lipid peroxidation, which has long been known to be counteracted by the antioxidant effect of vitamin E,[192] is increased by magnesium deficiency in rat liver, muscle, heart, and other organs,[193,194] an effect suggesting that the magnesium deficiency-induced myocardial damage might be mediated by free radicals.[195] This premise has been substantiated by studies with magnesium-deficient golden Syrian hamsters (not the CMP strain). Magnesium deficiency alone caused myocardial injury that was protected against by vitamin E.[132] Intensification of the myocardial damage by injection of the beta-catecholamine analog, isoproterenol, was interpreted as showing impairment of tolerance of oxidative stress by magnesium deficiency.[131,196] Direct evidence that free radicals participate in the myocardial damage caused by magnesium deficiency was provided by a study showing that vitamin E deficiency increased the number and extent of the lesions so induced and that its supplementation was protective.[132] Combined vitamin E and magnesium protected against (erythrocyte) membrane lipid peroxidation in magnesium-deficient Syrian hamsters.[196]

Selenium also protects against peroxidation of lipids, and its deficiency (in China) causes CMP of Keshan disease.[197–199] Its interactions with vitamin E are being considered, as they affect muscle disease.[198–201] It has been found to spare vitamin E, decreasing the amount required by the Syrian golden hamster (after 120 days of E depletion), but it did not prevent the myopathy.[202] A study of the muscle damage (determined by increased release of creatine kinase) of vitamin E and selenium-low calves when transferred from enclosures to open pasture[203] recalls the intensification of neuromuscular signs of magnesium deficiency of ruminants shifted from barns to pasture,[204] which was attributed in part to the change from warmth of the enclosure to stress of exposure to cold, with catecholamine release.[205] The meaning of the increase in intracellular zinc that is associated with increased calcium in the affected heart of dystrophic hamsters[177] is not clear. The investigators hypothesized that zinc might be co-transported with calcium across the cell membrane or substituted for calcium in pathways affected by the high-energy ATP-pump. The role of zinc, not as an antioxidant, but possibly through its ability to stabilize membranes exposed to oxidative stress,[206] might have a protective function. Possibly pertinent zinc, magnesium, and pyridoxine interrelationships are considered below under homocystinuria.

Cystic Fibrosis and Cardiomyopathy; Interaction of Nutrient Deficiencies?—Cystic fibrosis, the most common lethal or semilethal genetic disease in the white population, is inherited in an autosomal recessive manner.[207] Loss of pancreatic enzyme activity is characteristic; complete loss occurs in 80 to 85% of patients.[207,208] Malabsorption, the degree of which depends on the extent of loss of pancreatic function, leads to nutritional deficiencies of vitamin E and other fat-soluble vitamins, and deficiencies of magnesium and selenium—also implicated in myopathies—can also develop. In about 10% of those with cystic fibrosis, CMP, characterized by patchy focal areas of necrosis and fibrosis, complicates the disease.[209] The lesions resemble those of ''idiopathic'' CMP, human muscular dystrophy, and the CMP induced by experimental vitamin E deficiency or magnesium deficiency.

Vitamin E, Magnesium, Calcium, and Selenium in Cardiomyopathy of Cystic Fibrosis: The shortened red blood cell survival of this disease responds to high-dosage vitamin E (10-fold higher than the normal recommended dietary allowance,[209,210] but the muscle weakness (myopathy?) of cystic fibrosis is not responsive even to vitamin E. In an early report[211] (additional to those reviewed in 1988[209]) of two brothers with cystic fibrosis, who had steatorrhea that developed at six months of age, one who died of pneumonia was found at autopsy to have not only pancreatic damage but also endomyocardial fibroelastosis. In their discussion, the authors referred to eight additional reports of myocardial damage in infants and young children with pancreatic insufficiency of cystic fibrosis. Autopsy examination of two additional patients, who had had hypercalcemia, disclosed generalized arterial intimal proliferation and calcification and renal calcification; one had mild rickets.[212] These manifestations resemble those described above in patients with renal wasting of magnesium, with and without magnesium malabsorption.

Few data have been found on the magnesium status of cystic fibrosis patients. A study of erythrocyte levels of magnesium, calcium, zinc, and sodium in eight children with cystic fibrosis, and in four of the parents, showed that the patients had the lowest magnesium, zinc, and sodium values and the highest calcium levels; the parents had higher magnesium levels than did the affected children, but lower than control adult values.[213] There were higher magnesium and lower zinc contents of hair, nails, and duodenal fluid from children with cystic fibrosis;[214] the significance is unknown. They had very high calcium content of their duodenal fluid. In this study the magnesium level in sweat was slightly elevated; in another[215] there was no difference between patients and normal children. Neonatal hair from most of the 13 infants with cystic fibrosis contained water-soluble calcium versus less than 30% of controls and over ten times as much insoluble calcium as controls.[216] There were similar findings with hair magnesium, but of lesser magnitude. It was speculated that inability (of patients' hair) to bind calcium and magnesium might be related to the basic defect.

Tremor, nervousness, weakness, and anorexia (symptoms of latent tetany of magnesium deficiency[60]) developed in a young man with cystic fibrosis complicated by cor pulmonale, who had recently received furosemide (a loop diuretic that causes magnesium loss) for heart failure, but only for several days.[217] When his serum magnesium was found to be 1.4 mEq/L, and a magnesium load of 367 mg Mg disclosed urinary excretion of 612 mg in 48 hours, he was treated with 4 g/d of magnesium for five days, with disappearance of signs and symptoms of the deficiency. The renal excretion of so much magnesium in the face of hypomagnesemia suggests renal wasting. Acutely lowered serum magnesium developed in seven cystic fibrosis patients with bowel obstruction from meconium ileus that had been treated with oral and rectal administration of the mucolytic agent, N-acetylcysteine, and a hypertonic solution of sodium diatrizoate.[218]

Considered above, under muscular dystrophy, are the data on the CMP of selenium deficiency and its interactions with vitamin E.[197–203] A 1982 review of clinical data failed to support a premise that selenium deficiency might be implicated in the pathogenesis of cystic fibrosis.[198] However, since then, in a discussion of the CMP complication, it was pointed out that malabsorption with and without cystic fibrosis has caused low plasma selenium levels.[201] A study with the golden Syrian hamster showed that selenium prevented pancreatic atrophy induced by vitamin E depletion.[219]

Homocystinuria and Cardiomyopathy; Interaction of Nutrient Deficiencies?—*Pyridoxine-Dependence of Homocystinurics:* One of the conditions listed as being associated with CMP is homocystinuria,[99] predominantly a vitamin B_6-dependent disorder.[220–222] There are several metabolic abnormalities that give rise to homocystinuria, the most common of which is a Mendelian recessive trait that causes deficient activity of cystathionine beta-synthase, an enzyme that contains pyridoxal phosphate.[220] Almost half of the patients respond to very high dosage pyridoxine (up to 300 times more pyridoxine than is needed for correction of a simple deficiency; they may have slight (residual) activity of this enzyme.[220,221] Those with a more complete deficiency of the enzyme, or with a metabolic block after formation of cystathionine, also require additional dietary modification and/or supplementation.

Among the pathologic changes in homocystinuric patients are several germane to development of CMP: the occurrence of thrombi, not only in large arteries but in the microvasculature, with fraying of the elastica and premature arteriosclerosis.[223] CMP is not frequently reported, but Marfan's syndrome, another condition associated with CMP,[99,137] has occurred in homocystinuric patients.[220,223]

Additional Nutritional Treatment of Homocystinuria: Folate B_{12}*, Amino Acid Intake Modification and Possibly Magnesium and Zinc:* Patients with deficiency of cystathionine synthase, accompanied by abnormally low serum folate levels have the folate level further lowered by pyridoxine treatment, and folate repletion is necessary for the chemical response to pyridoxine.[220,224] Vitamin B_{12} supplements are necessary for those whose metabolic block is after formation of cystathionine. Those with methionine accumulation also require methionine restriction and cysteine or choline supplements. Perhaps vitamin B_6 treatment from infancy might be effective.[222]

Pyridoxal phosphate, which is necessary for activity of cystathionine synthase, requires magnesium as a cofactor for this and for many other of its enzymatic reactions.[225] Studies in the 1960s[226–229] and more recent investigations[230–233] have shown that vitamin B_6 is necessary for the maintenance of magnesium and zinc tissue levels and that the pyridoxine-dependent enzymes also require these minerals. Thus, adding magnesium and zinc to vitamin B_6 supplementation of patients with unduly high vitamin B_6 requirements seems worthy of trial.

IV. CONCLUDING COMMENTS

Presented here is a postulate that magnesium deficiency caused by genetic variations in magnesium metabolism, in conjunction with marginal magnesium intake, is a contributory factor to gestational complications, including perinatal and neonatal CMP. Severe forms of familial magnesium deficiency, isolated malabsorption, and renal wasting of magnesium have been identified. The possibility that the underlying inherited abnormality in the renal magnesium wasting syndromes is that of magnesium malabsorption is presented. It is suggested that renal magnesium wasting may be caused by damage at the major site of tubular magnesium reabsorption, when infants with hypomagnesemic hypocalcemia are provided calcemic treatment without magnesium repletion. It is proposed that there may be degrees of magnesium malabsorption and that there are genetic differences in plasma and cellular magnesium levels in different ethnic groups; low values have been associated with specific HLA groups. These differences might be part of the metabolic basis of other inherited diseases.

The heritable diseases, that are complicated by cardiovascular damage, especially microangiopathy leading to generalized or nodal CMP, are of particular interest as regards the possibility of contributory magnesium inadequacy, because experimental magnesium deficiency causes similar arterial and myocardial lesions. There is direct evidence of abnormalities in magnesium retention and/or tissue levels in diabetes mellitus, especially in the juvenile form, in which microangiopathy causes serious early complications. Since chromium, like magnesium, participates in carbohydrate and lipid metabolism, and it seems to be linked to maturity-onset diabetes, its role, with and without magnesium in pregnant diabetic women and in infants born to such women, might be worth exploring.

The familial syndrome of congenital deafness, syncope, arrhythmias, and sudden death has not been correlated with magnesium abnormality in the literature, but it bears similarity to manifestations produced by magnesium deficiency. The myocardial lesions involving the conducting tissue and nodes differ from those of the more commonly described cardiomyopathies and those of magnesium deficiency only in their location. The deafness has been presumed to be caused by cochlear damage, caused by damage of the nutrient arteries to the inner ear; it can be correlated with the hearing loss of magnesium-deficient weanling rats and of noise-exposed magnesium-deficient guinea pigs. Might magnesium supplementation during pregnancy of diabetics or of members of families with risk of the Jervell-Nielsen-Lange syndrome protect the infant, and might it slow or limit progression of the disease manifestations in those afflicted?

The skeletal muscle and myocardial lesions of Duchenne's progressive muscular dystrophy have been compared with those produced by magnesium deficiency and are similar to those seen in models of genetic muscular dystrophy, which also have comparable metabolic findings. CMP hamsters were shown to lose skeletal and myocardial muscle magnesium and to gain large amounts of calcium before the necrosis developed, even when not fed a magnesium-deficient diet. They also exhibited high myocardial catecholamine content and uncoupling of oxidative phosphorylation, abnormalities also seen in the human disease. Magnesium supplementation delayed but did not prevent the damage in the CMP hamster, but prevented it in hybrid and normal magnesium-deficient hamsters. The lesions of the human disease also resemble those of vitamin E-deficient animals, which also have low muscle magnesium content. The CMP induced by magnesium deficiency in golden Syrian hamsters has been shown to be intensified by vitamin E deficiency and reduced by its supplementation. The capacity of magnesium-deficient Syrian hamsters to withstand the oxidative stress of catecholamine challenge was diminished without prior supplementation with vitamin E, indicating free radical participation in magnesium deficiency-induced CMP. Treatment with high-dosage vitamin E has not influenced Duchenne's muscular dystrophy. Perhaps treatment with both vitamin E and magnesium might delay its progression.

The shortened erythrocyte survival time of mucoviscidosis or cystic fibrosis requires ten times the normal intake of vitamin E for correction; the CMP that develops in about 10% of the cases is unresponsive. The vitamin E deficiency is attributed to malabsorption, from steatorrhea resulting from loss of pancreatic enzymes that can be complete in up to 85% of the patients with cystic fibrosis. It can be presumed that this condition also interferes with absorption of other nutrients, including magnesium. Few data have been found on the magnesium status, other than magnesium deficiency in several patients under short-term diuretic treatment or other therapy not usually associated with magnesium depletion for complications of mucoviscidosis. However, these few data suggest that these patients may be unduly vulnerable to magnesium loss. The interrelationships of magnesium with vitamin E, additional to those discussed in relation to Duchenne's disease, are indicated by evidence that vitamin E deficiency has precipitated magnesium deficiency in rats, and magnesium has prevented vitamin E-induced respiratory decline and lipid peroxidation. Might vitamin E plus magnesium supplements prove helpful in management of cystic fibrosis? The observation that selenium prevented the pancreatic atrophy induced by vitamin E depletion in the golden hamster might be worthy of exploration in humans.

Almost half of homocystinuric patients (those with some cystathionine synthase activity) require very high-dosage pyridoxine (up to 300 times normal) for correction of the biochemical abnormality. The supplements or dietary restriction required by patients who have a more complete deficiency of the enzyme, or with a metabolic block after formation of cystathionine, can include vitamin B_{12}, folate, methionine restriction, and cysteine supplements. Progression of the disease despite the several nutritional approaches has suggested that (pyridoxine) supplementation be instituted in infancy. CMP has not been reported in homocystinuria, but patients are subject to microvascular thrombosis, which can contribute to CMP, as well as to large artery thromboses. Homocystinuria has been reported in patients with Marfan's syndrome (associated with CMP), and premature arteriosclerosis with frayed elastica and medial degeneration is common. Both because the arteriopathy (and hypercoagulability) resemble that produced by magnesium deficiency, and because pyridoxal phosphate, which is necessary for activity of cystathionine synthase, requires magnesium as a cofactor and is necessary for maintenance of tissue levels of magnesium, addition of magnesium to the therapeutic regimen deserves trial. Since tissue levels of zinc are also dependent on adequate pyridoxine, the effect of addition of zinc is worth determining.

The major emphasis in this paper has been on genetic disorders that are associated with CMP, to which magnesium deficiency, caused by higher than average magnesium requirements, might be contributory. Magnesium is protective against arterial and myocardial damage and interacts with vitamins E and B_6, dependencies or deficiencies of which have been implicated in the genetic diseases complicated by CMP. A few data have been presented on the trace minerals: selenium, chromium, and zinc that may also bear on these diseases.

REFERENCES

1. Seelig, M. S., The requirement of magnesium by the normal adult, *Am. J. Clin. Nutr.*, 14, 342–390, 1964.
2. Seelig, M. S., Magnesium requirements in human nutrition, *Magnesium Bull.*, 3(1a), 26–47, 1981.
3. Seelig, M. S., Nutritional status and requirements of magnesium, with consideration of individual differences and prevention of cardiovascular disease, *Magnesium Bull.*, 8, 170–185, 1986.
4. Seelig, M. S., *Magnesium Deficiency in the Pathogenesis of Disease. Early Roots of Cardiovascular, Skeletal and Renal Abnormalities,* Avioli, L. V., Ed., Publ Plenum Medical Book Co., New York, N.Y., 1980.
5. Seelig, M. S. and Heggtveit, H. A., Magnesium interrelationships in ischemic heart disease: a review, *Am. J. Clin. Nutr.*, 27, 59–79, 1974.
6. Seelig, M. S. and Haddy, F. J., Magnesium and the Arteries. I. Effects of magnesium deficiency on arteries and on retention of sodium, potassium, and calcium, in *Magnesium in Health and Disease,* Cantin, M. and Seelig, M. S., Eds., Spectrum, N.Y., 1980, 2nd Intl. Mg Sympos., Quebec, 1976, 605–638.
7. Seelig, M. S., Cardiovascular consequences of magnesium deficiency and loss: pathogenesis, prevalence and manifestations—magnesium and chloride loss in refractory potassium repletion, *Am. J. Cardiol.*, 63, 4G–21G, 1989.
8. Seelig, M. S., Early nutritional roots of cardiovascular disease, in *Nutrition and Heart Disease,* Naito, H. K., Ed., Proc. 19th Ann. Mtg. Am. Coll. Nutr., 1978, SP Medical & Sci Books, New York, 31–59, 1982.
9. Seelig, M. S., Prenatal and neonatal mineral deficiencies: magnesium, zinc and chromium, In *Clinical Disorders in Pediatric Nutrition,* Lifshitz, F., Ed., Marcel Dekker, N.Y., 1982, 167–196.
10. Seelig, M. S., Nutritional roots of combined system disorders, In *Clinical Disorders in Pediatric Nutrition,* Lifshitz, F., Ed., Marcel Dekker, New York, 1982, 327–351.
11. Seelig, M. S., Magnesium in pregnancy: special needs for the adolescent mother, *J. Am. Coll. Nutr.*, 10, 566, 1991.
12. Seelig, M. S., Interrelationship of magnesium and estrogen in cardiovascular and bone disorders, eclampsia, migraine and premenstrual syndrome, *J. Am. Coll. Nutr.*, 12, 1993, in press.
13. Paunier, L. and Radde, I. C., Normal and abnormal magnesium metabolism, *Bull. Hosp. Sick Childr.*, Toronto, 14, 16–23, 1965.
14. Salet, J., Polonovski, C., DeGouyon, F., Pean, G., Melekian, B., and Fournet, J. P., Hypocalcemic tetany deriving from congenital hypomagnesemia. A new metabolic disease, *Arch. Franc. Pediat.*, 23, 749–768, 1966, in French.
15. Salet, J., Polonovski, C., Fournet, J. P., DeGouyon, F., Aymard, P., Pean, G., and Taillemite, J. L., Demonstration of the familial nature of hypomagnesemia, *Arch. Franc. Pediat.*, 27, 550–551, 1970, in French.
16. Skyberg, D., Stromme, J. H., Nesbakken, R., and Harnaes, K., Neonatal hypomagnesemia with a selective malabsorption of magnesium—A clinical entity, *Scand. J. Clin. Investig.*, 21, 355–363, 1968.
17. Stromme, J. H., Nesbakken, R., Normann, T., Skjorten, F., and Johannessen, B., Familial hypomagnesemia. Biochemical, histological and hereditary aspects studied in two brothers, *Acta Padiat. Scand.*, 58, 433–444, 1969.
18. Gitelman, H. J., Graham, J. B., and Welt, L. G., A new familial disorder characterized by hypokalemia and hypomagnesemia, *Trans. Assoc. Amer. Physicians,* 79, 221–235, 1966.
19. Freeman, R. M. and Pearson, E., Hypomagnesemia of unknown etiology, *Am. J. Med.*, 41, 645–656, 1966.

20. Sann, L., Moreau, P., Longin, B., Sassard, J., and Francois, R., Bartter's syndrome, associated with hypercorticism, phosphorus and magnesium diabetes and familial tubulopathy, *Arch. Franc. Pediat.*, 32, 349–366, 1975, in French.
21. Henrotte, J. G., Plouin, P. F., Levy-Leboyer, C., Moser, G., Sidoroff-Girault, N., Franck, G., Santarromana, M., and Pineau, M., Blood and urinary magnesium, zinc, calcium, free fatty acids, and catecholamines in type A and type B subjects, *J. Am. Coll. Nutr.*, 4, 165–172, 1985.
22. Henrotte, J. G., Genetic regulation of blood and tissue magnesium content in mammals, *Magnesium*, 7, 306–314, 1988.
23. Henrotte, J. G., Pla, M., and Dausset, J., HLA- and H-2-associated variations of intra- and extracellular magnesium content, *Proc. Natl. Acad. Sci. USA*, 87, 1894–1898, 1990.
24. Wacker, W. E. C., *Magnesium and Man*, Harvard University Press, Cambridge, MA, 1980.
25. Vallee, B. L., Metal and enzyme interactions: correlation of composition, function, and structure, in *The Enzymes*, 3, 225–270, 1960.
26. Booth, C. C., Babouris, M. B., Hanna, S., and MacIntyre, I., Incidence of hypomagnesemia in intestinal malabsorption, *Brit. J. Med.*, 2, 141–143, 1963.
27. Paunier, L., Radde, I. C., Kooh, S. W., Conen, P. E. E., and Fraser, D., Primary hypomagnesemia with secondary hypocalcemia in an infant, *Pediatrics*, 41, 385–402, 1968.
28. Paunier, L., Magnesium malabsorption, *Adv. Intern. Med. Pediatr.*, 42, 113–131, 1979.
29. Friedman, M., Hatcher, G., and Watson, L., Primary hypomagnesemia with secondary hypocalcemia in an infant, *Lancet*, 1, 703–705, 1967.
30. Nordio, S., Donath, A., Macagno, F., and Gatti, R., Chronic hypomagnesemia with magnesium-dependent hypocalcemia. I. A new syndrome with intestinal malabsorption. II. Magnesium, calcium and strontium, *Acta Pediatr. Scand.*, 60, 441–448, 449–455, 1971.
31. Main, N. H., Morgan, R. J., Russell, R. I., Hall, M., Mackenzie, J. F., Shenkin, A., and Fell, G. S., Magnesium deficiency in chronic inflammatory bowel disease and requirements during intravenous nutrition, *J.P.E.N.*, 5, 15–19, 1981.
32. Klingberg, W. G., Idiopathic hypomagnesemia and osteochondritis, *Ped. Res.*, 4, 452, 1970; Death from cardiomyopathy (personal communication).
33. Michelis, M. F., Drash, A. L., Linarelli, L. G., DeRubertis, F. R., and Davis, B. B., Decreased bicarbonate threshold and renal magnesium wasting in a sibship with distal renal tubular acidosis, *Metabolism*, 21, 905–920, 1972.
34. Mace, J. W., Hambridge, K. M., Gotlin, R. W., Dubois, R. S., Solomons, C. S., and Katz, F. H., Magnesium supplementation in Bartter's syndrome, *Arch. Dis. Child.*, 48, 485–487, 1973.
35. Booth B. E. and Johanson, A., Hypomagnesemia due to renal tubular defect in reabsorption of magnesium, *J. Pediat.*, 84, 350–354, 1974.
36. Runeberg, L., Collan, Y., Jokinen, E. J., Lahdevirta, J., and Aro, A., Hypomagnesemia due to renal disease of unknown origin, *Am. J. Med.*, 59, 873–882, 1975.
37. Rapado, A. and Castrillo, J. M., Chondrocalcinosis and hypomagnesemia; Nephrocalcinosis and hypomagnesemia, in *Magnesium in Health and Disease*, 2nd Intl. Sympos. on Magnesium, Quebec, 1976; Cantin, M. and Seelig, M. S., Eds., Spectrum, New York, 1980, 355–364, 485–497.
38. Seelig, M. S., Berger, A. R., and Avioli, L. A., Speculations on renal, hormonal, and metabolic aberrations in a patient with marginal magnesium deficiency, in *Magnesium in Health and Disease*, 2nd Intl. Sympos. on Magnesium, Quebec, 1976; Cantin, M. and Seelig, M. S., Eds., Spectrum, New York, 1980, 459–468.
39. Manz, F., Anders, A., Janka, P., Lombeck, I., and Scharer, K., Renal magnesium wasting, incomplete tubular acidosis, hypercalciuria and nephrocalcinosis in 4 children and adults, *Magnesium Bull.*, 1, 151, 1979, in German.
40. Bauer, F. M., Glasson, P., Valloton, M. B., and Courvoisier, B., Bartter's syndrome, chondrocalcinosis, and hypomagnesemia, *Schweiz. med. Wschr.*, 109, 1251–1256, 1979, in German.
41. Guellner, H. G., Gill, J. R., and Bartter, F. C., Correction of hypokalemia by magnesium repletion in familial hypokalemic alkalosis with tubulopathy, *Am. J. Med.*, 71, 578–582, 1981.

42. Mayoux-Benhamou, M. A., Clerc, D., Ganeval, D., Pertuisset, N., and Massias, P., Articular chondrocalcinosis and hypomagnesemia of renal origin; 2 cases, *Rev. Rhum. Mal. Ostoartic.,* 52, 545–548, 1985.
43. Rodriguez-Soriano, J., Vallo, A., and Garcia-Fuentes, M., Hypomagnesemia of hereditary renal origin, *Pediat. Neprol.,* 1, 465–472, 1987.
44. Bianchetti, M. G., Girardin, E., Sizonenko, P. C. C., and Paunier, L., Metabolic studies on a new case of hypomagnesaemia-hypokalaemia, *Magnesium Res.,* 1, 116, 1988.
45. Pronicka, E. and Gruszczynska, B., Familial hypomagnesaemia with secondary hypocalcaemia—autosomal or X-linked inheritance? *J. Inhert. Metab. Dis.,* 14, 397–399, 1991.
46. de-Filippi, J. P., Diderich, P. P., and Wouters, J. M., Hypomagnesemia and chondrocalcinosis, *Ned. Tijdscgr. Geneesk.,* 136, 139–141, 1992, in Dutch.
47. Sutherland, L. E., Hartroft, P., Balis, J. W., Bailey, J. D., and Lynch, M. J., Bartter's syndrome. A report of four cases, including three in one sibship, with comparative histologic evaluation of the juxtaglomerular apparatuses and glomeruli, *Acta Pediat. Scand. Suppl.,* 201, 1–25, 1970.
48. Riggs, J. E., Klingberg, W. G., Flink, E. B., Schochet Jr., S. S., Balian, A. A., and Jenkins, J. J., 3d, Cardioskeletal mitochondrial myopathy associated with chronic magnesium deficiency, *Neurology,* 42, 128–130, 1992.
49. Elin, R. J., Assessment of magnesium status, In *Magnesium in Health and Disease,* Itokawa, Y., Durlach, J., Eds., Libbey, J., London, 1989, Fifth Intl. Mg. Sympos., Kyoto, Japan, 1988, 137–146.
50. Hess, R., MacIntyre, I., Alcock N., and Pease, A. G. E., Histochemical changes in rat kidney in magnesium deprivation, *Brit. J. Exp. Path.,* 40, 80–86, 1959.
51. Welt, L. G., Experimental magnesium depletion, *Yale J. Biol. Med.,* 36, 325–349, 1964.
52. Oliver, J., MacDowell, M., and Whang, R., The renal lesion of electrolyte imbalance. IV. The intranephric calculosis of experimental magnesium depletion, *J. Exp. Med.,* 124, 263–299, 1966.
53. Quamme, G. A., Renal handling of magnesium: drug and hormone interactions, *Magnesium* 5, 248–272, 1986.
54. Seelig, M. S., Calcium and magnesium deposits in disease, In *Handbook on Metal-Ligand Interactions in Biological Fluids,* Berthon, G., Ed., Marcel Dekker, Inc., New York, 1993, in press.
55. Miller, J. F., Tetany due to deficiency in magnesium. Its occurrence in a child of six years with associated osteochondrosis of capital epiphysis of femur, *J. Dis. Child.,* 67, 117–119, 1944.
56. Vainsel, M., Vandervelde, G., Smulders, J., Vosters, M., Nubain, P., and Loeb, H., Tetany due to hypomagnesemia with secondary hypocalcemia, *Arch. Dis. Childh.,* 45, 254–258, 1970.
57. Paunier L. and Sizonenko, P. C., Asymptomatic chronic hypomagnesemia and hypokalemia in a child: cell membrane disease, *J. Pediatr.,* 88, 51–55, 1976.
58. Darlu, P. and Henrotte, J. G., The importance of genetic and constitutional factors in human red blood cell magnesium control, In *Magnesium in Health and Disease,* Cantin, M. and Seelig, M. S., Eds., Spectrum, N.Y., 1980, 2nd Intl. Mg Sympos., Quebec, 1976, 921–927.
59. Henrotte, J. G., Relationship between red blood cell magnesium and HLA antigens, *Tissue Antigens,* 15, 419–439, 1980.
60. Durlach, J., *Magnesium in Clinical Practice,* Translated by Wilson, D., John Libbey & Co., Ltd., London, 1985.
61. Maertens de Noordhout, B., Henrotte, J. G., and Franchimont, P., Latent tetany, magnesium and HLA tissue antigens, *Magnesium Bull.,* 9, 118–121, 1987.
62. Megard, M., Andre-Fouet, E., Guisti, E., Betuel, H., and Gebhurer, L., Articular chondrocalcinosis, Association with antigen B15, *Presse Med.,* 13, 1727, 1984, in French.
63. Darlu, P., Defrise-Gussenhoven, E., Michotte, Y., Susanne, C., and Henrotte, J. G., Possible linkage relationship between genetic markers and blood magnesium and zinc. A twin study, *Acta Genet. Med. Gemell. Roma,* 34, 109–112, 1985.
64. Henrotte, J. G., Genetic regulation of red blood cell magnesium content and major histocompatibility complex, *Magnesium,* 1, 69–80, 1982.

65. Durlach, J., Henrotte, J. G., Lepage, V., Elchidial, A., and Degos, L., HLA-Bw35 antigen, mitral valve prolapse and blood magnesium level, In *Magnesium Deficiency. Physiopathology and Treatment Implications,* Halpern, M. J. and Durlach, J., Eds., First Europ. Mg Congr., Lisbon, 1983, Karger, Basel, 1985, 95–101.
66. Galland, L. D., Baker, S. M., and McLellan, R. K., Magnesium deficiency in the pathogenesis of mitral valve prolapse, *Magnesium,* 5, 165–174, 1986.
67. Durlach, J. and Durlach, V., Idiopathic mitral valve prolapse and magnesium, State of the art, *Magnesium Bull.,* 8, 156–169, 1986.
68. Henrotte, J. G., Hannoun, C., Benech, A., and Dausset, J., Relationship between postvaccinal anti-influenza antibodies, blood magnesium levels, and HLA antigens, *Hum. Immunol.,* 12, 1–8, 1985.
69. Henrotte, J. G., Recent advances on genetic factors regulating blood and tissue magnesium concentrations. Relationships with stress and immunity, In *Magnesium in Health and Disease,* Itokawa, Y. and Durlach, J., Eds., Libbey, J., London, 1989, Fifth Intl. Mg Sympos., Kyoto, 1988, 285–289.
70. Kontopoulos, V., Seelig, M. S., Dolan, J., Berger, A. R., and Ross, R. S., Influence of parenteral administration of magnesium sulfate to normal pregnant and to pre-eclamptic women, In *Magnesium in Health and Disease,* Cantin, M. and Seelig, M. S., Eds., Spectrum, N.Y., 1980, 2nd Intl. Mg Sympos., Quebec, 1976, 839–848.
71. Weaver, K., A possible anticoagulant effect of magnesium in pre-eclampsia, In *Magnesium in Health and Disease,* Cantin, M. and Seelig, M. S., Eds., Spectrum, N.Y, 1980, 2nd Intl. Mg Sympos., Quebec, 1976, 833–838.
72. Franz, K. B., Correlation of urinary magnesium excretion with blood pressure of pregnancy, *Magnesium Bull.,* 4, 73–78, 1982.
73. Weaver, K., Pregnancy-induced hypertension and low birth weight in magnesium deficient ewes, *Magnesium,* 5, 191–200, 1986.
74. Ajayi, G., Serum magnesium concentration in premenopausal, menopausal women, during normal and EPH-gestosis pregnancy and the effect of diuretic therapy in EPH-gestosis, *Magnesium Bull.,* 10, 72–76, 1988.
75. Palla, G. P., Giaquinto, P., Moro, P. R., Maniccia, E., Carelli, G., and Mancuso, S., Magnesium load test in pregnancy hypertension, *Clin. Exp. Hypertens. Pregn.,* B7, 159–163, 1988.
76. Sjogren, A., Gennser, G., and Rymark, P., Reduced concentrations of magnesium, potassium and zinc in skeletal muscle from women during normal pregnancy or eclampsia, *J. Am. Coll. Nutr.,* 7, 408, 1988.
77. Wynn, A. and Wynn, M., Magnesium and other nutrient deficiencies as possible causes of hypertension and low birthweight, *Nutr. Health,* 6, 69–88, 1988.
78. Dawson, E. B. and Kelly, R., Calcium, magnesium and lead interrelationships in preeclampsia, *J. Am. Clin. Nutr.,* 51, 512, 1990.
79. Wibell, L., Gebre-Medhin, M., and Lindmark, G., Magnesium and zinc in diabetic pregnancy, *Acta Pediatr. Scand.,* Suppl. 320, 100–106, 1985.
80. Mimouni, F., Miodovnik, M., Tsang, R. C., Holroyde, J., Dignan, P. S., and Siddiqi, T. A., Decreased maternal serum magnesium concentration and adverse fetal outcome in insulin-dependent diabetic women, *Obstet. Gynecol.,* 70, 85–88, 1987.
81. Miodovnik, M., Mimouni, F., Siddiqi, T. A., and Tsang, R. C., Periconceptional metabolic status and risk for spontaneous abortion in insulin-dependent diabetic pregnancies, *Am. J. Perinatol.,* 5, 368–373, 1988.
82. Palla, G. P., Castaldo, F., Moro, P. R., Giaquinto, P., Carelli, G., Caruso, A., Lanzone, A., and Mancuso S., Intravenous magnesium load test in normal and diabetic pregnant women. *Magnesium Res..* 2, 91–92, 1989, Proc. Fifth Intl. Mg. Sympos., Kyoto, Japan, August 8–12, 1988.
83. Greene, M. F., Hare, J. W., Krache, M., Phillippe, M., Barss, V. A., Saltzman, D. H., Nadel, A., Younger, M. D., Heffner, L., and Scherl, J. E., Prematurity among insulin-requiring diabetic gravid women, *Am. J. Obstet. Gynecol.,* 161, 106–111, 1989.

84. Garner, P. R., D'Alton, M. E., Dudley, D. K., Huard, P., and Hardie, M., Preeclampsia in diabetic pregnancies, *Am. J. Obstet. Gynecol.*, 163, 505–508, 1990.
85. Lin, C. K., Kuo, P. L., Liu, H. C., Yau, K. I., Chang, H. S., Wang, T. R., and Chen, S. H., Clinical analysis of infants of diabetic mothers, *Acta Paediatr. Sin.*, 30, 233–239, 1989.
86. Ranade, A. Y., Merchant, R. H., Bajaj, R. T., and Joshi, N. C., Infants of diabetic mothers—an analysis of 50 cases, *Indian Pediatr.*, 26, 366–370, 1989.
87. Djurhuus, M. S., Klitgaard, N. A., and Beck-Nielsen, H., Magnesium deficiency and development of late diabetic complications, *Ugeskr. Laeger.*, 153, 2108–2110, 1991, in Danish.
88. Conradt, A. and Weidinger, H., The central position of magnesium in the management of fetal hypotrophy—a contribution to the pathomechanism of utero-placental insufficiency, prematurity and poor intrauterine fetal growth as well as pre-eclampsia, *Magnesium Bull.*, 4, 103–124, 1982.
89. Conradt, A., Weidinger, H., and Algayer, H., Magnesium therapy decreased the rate of intrauterine fetal retardation, premature rupture of membranes and premature delivery in risk pregnancies treated with betamimetics, *Magnesium*, 4, 20–28, 1985.
90. Kuti, V., Balazs, M., Morvay, F., Varenka, Z., Szekly, A., and Szucs, M., Effect of maternal magnesium supply on spontaneous abortion and premature birth and on intrauterine foetal development: experimental epidemiological study, *Magnesium Bull.*, 3, 73–79, 1981.
91. Spaetling, L. and Spaetling, G., Magnesium supplementation in pregnancy. A double-blind study, *Brit. J. Obstet. Gynec.*, 95, 120–125, 1988.
92. Kovacs, L., Molnar, B. G., Huhn, E., and Bodis, L., Magnesium substitution in pregnancy. A prospective, randomized double-blind study, *Geburtsch. Frauenheil.*, 48, 595–600, 1988, in German.
93. Widdowson, E. M. and Dickerson, J. W. T., Chemical composition of the body, In *Mineral Metabolism*, 2: Part A, Comar, C. L. and Bronner, F., Eds., Academic Press, New York, 1962, 2–247.
94. Haese, W. H., Maron, B. J., Mirowski, M., Rowe, R. D., and Hutchins, G. M., Peculiar focal myocardial degeneration and fatal ventricular arrhythmias in a child, *New Engl. J. Med.*, 287, 180–181, 1972.
95. Ferguson, II, J. E., Harney, K. S., and Bachicha, J. A., Peripartum maternal cardiomyopathy with idiopathic cardiomyopathy in the offspring. A case report, *J. Reprod. Med.*, 31, 1109–1112, 1986.
96. Weintraub, R. G., Swinburn, M. J., and Lee, L., Neonatal hypertrophic cardiomyopathy: a case report and family study, *Austral. Pediatr. J.*, 23, 249–251, 1987.
97. Reller, M. D. and Kaplan, S., Hypertrophic cardiomyopathy in infants of diabetic mothers: an update, *Am. J. Perinatol.*, 5, 353–358, 1988.
98. McMahon, J. N., Berry, P. J., and Joffe, H. S., Fatal hypertrophic cardiomyopathy in an infant of a diabetic mother, *Pediatr. Cardiol.*, 11, 211–212, 1990.
99. Hudson, R. E. B., The cardiomyopathies: order from chaos, *Am. J. Cardiol.*, 25, 70–77, 1970.
100. Johnson, J. B., Mir, G. H., Flores, P., and Mann, M., Idiopathic heart disease associated with pregnancy and the puerperium, *Am. Heart J.*, 72, 809–816, 1966.
101. Govan, T. A. D., Myocardial lesions in fatal eclampsia, *Scot. Med. J.*, 2, 187–192, 1966.
102. Unsigned Editorial, Cardiomyopathy and pregnancy, *Brit. J. Med.*, 4, 269–270, 1968.
103. Demakis, J. G. and Rahimtoola, S. H., Peripartum cardiomyopathy, *Circulation*, 44, 964–968, 1971.
104. Julian, D. G. and Szekely, P., Peripartum cardiomyopathy, *Prog. Cardiovasc. Dis.*, 27, 223–240, 1985.
105. Adler, A. K. and Davis, M. R., Peripartum cardiomyopathy: two case reports and a review, *Obstet. Gynecol. Surv.*, 41, 675–682, 1986.
106. Homans, D. C., Peripartum cardiomyopathy, *New Engl. J. Med.*, 312, 1432–1437, 1985.
107. Falase, A. O., Peripartum heart disease, *Heart Vessels*, Suppl 1, 232–235, 1985.

108. Cunningham, F. G., Pritchard, J. A., Hankins, G. D., Anderson, P. L., Lucas, M. J., and Armstrong, K. F., Peripartum heart failure: idiopathic cardiomyopathy or compounding cardiovascular events? *Obstet. Gtnecil.*, 67, 157–168, 1986.
109. O'Connell, J. B., Costanzo-Nordin, M. R., Subramanian, R., Robinson, J. A., Wallis, D. E., Scanlon, P. J., and Gunnar, R. M., Peripartum cardiomyopathy: clinical, hemodynamic, histologic and prognostic characteristics, *J. Am. Coll. Cardiol.*, 8, 52–56, 1986.
110. Cenac, A., Gaultier, Y., Soumana, I., Harouna, Y., and Develoux, M., Postpartum cardiomyopathy in the Sudanese-Sahelian area. Clinical and epidemiologic studies of 66 cases, *Arch. Mal. Coeur.*, 82, 553–558, 1989, in French.
111. Lee, W. and Cotton, D. B., Peripartum cardiomyopathy: current concepts and clinical management, *Clin. Obstet. Gynecol.*, 32, 54–67, 1989.
112. Ferriere, M., Sacrez, A., Bouhour, J. B., Cassagnes, J., Geslin, P., Dubourg, O., Komajda, M., and Degeorges, M., Cardiomyopathy in the peripartum period: current aspects. A multicenter study. 11 cases, *Arch. Mal. Coeur.*, 83, 1563–1569, 1990, in French.
113. Iseri, L. T., Chung, P., and Tobis, J., Magnesium therapy for intractable ventricular tachyarrhythmias in normomagnesemic patients, *West. J. Med.*, 138, 823–828, 1983.
114. Dyckner, Th. and Wester, P. O., Magnesium-electrophysiological effects, *Magnesium Bull.*, 8, 219–222, 1986.
115. Iseri, L. T., Magnesium and cardiac arrhythmias, *Magnesium*, 6, 266–267, 1987.
116. Cohen, L., Kitzes, R., and Shnaider, H., Multifocal atrial techycardia responsive to parenteral magnesium, *Magnesium Res.*, 1, 239–242, 1988.
117. Rasmussen, H. S., Justification for intravenous magnesium therapy in acute myocardial infarction, *Magnesium Res.*, 1, 59–73, 1988.
118. Charbon, G. A., Magnesium treatment of arrhythmia: drug or nutritional replenishment? Pitfalls for the experimental design, In *Magnesium in Health and Disease*, Itokawa, Y. and Durlach, J., Eds., Libbey, J., London, 1989, Fifth Intl. Mg. Sympos, Kyoto, 1988, 223–228.
119. Sjogren, A., Edvinsson, L., and Fallgren, B., Magnesium deficiency in coronary artery disease and cardiac arrhythmias, *J. Intern. Med.*, 226, 213–222, 1989.
120. Antoni, D., Engel, M., and Gumpel, N., Magnesium therapy of supraventricular and ventricular arrhythmias, *Magnesium Bull.*, 11, 125–129, 1989, in German.
121. Iseri, L. T., Role of magnesium in cardiac tachyarrhythmias, *Am. J. Cardiol.*, 65, 47K–50K, 1990.
122. Keren, A. and Tzivoni, D., Magnesium therapy in ventricular arrhythmias, *P.A.C.E.*, 13, 937–945, 1990.
123. Perticone, F., Borelli, D., Ceravolo, R., and Mattioli, P. L., Antiarrhythmic short-term protective magnesium treatment in ischemic dilated cardiomyopathy, *J. Am. Coll. Nutr.*, 9, 492–499, 1990.
124. Sager, P. T., Widerhorn, J., Petersen, R., Leon, C., Ryzen, E., Rude, R., Rahimtoola, S. H., and Bhandari, A. K., Prospective evaluation of parenteral magnesium sulfate in the treatment of patients with reentrant AV supraventricular tachycardia, *Am. Heart J.*, 119, 2, Pt 1, 308–316, 1990.
125. Mishra, R. K., Studies on experimental magnesium deficiency in the albino rat. 1. Functional and morphological changes associated with low intake of magnesium, *Rev. Canad. Biol.*, 19, 122–135, 1960.
126. Heggtveit, H. A., The cardiomyopathy of Mg-deficiency, In *Electrolytes and Cardiovascular Diseases*, Bajusz, E., Ed., Karger, S., Basel/NY, 1965, 204–220.
127. Heggtveit, H. A., Herman, L., and Mishra, R. K., Cardiac necrosis and calcification in experimental magnesium deficiency. A Light and electron microscopic study, *Am. J. Pathol.*, 45, 757–782, 1964.
128. Lehr, D., The role of certain electrolytes and hormones in disseminated myocardial necrosis, In *Electrolytes and Cardiovascular Diseases*, Bajusz, E., Ed., Karger, S., Basel, Switzerland, New York, 1965, 248–273.

129. Wener, J., Pintar, K., Simon, M. A., Motola, R., Friedman, R., Mayman, A., and Schucher, R., The effects of prolonged hypomagnesemia on the cardiovascular system in young dogs, *Am. Heart J.*, 67, 221–231, 1964.
130. Bloom, S. and Ahmad, A., Ca channel blockade, inhibition of (Na,K)-ATPase, and myocardial necrosis associated with Mg deficiency, *FASEB J.*, 2, A824, 1988.
131. Bloom, S., Magnesium deficiency cardiomyopathy, *Am. J. Cardiovasc. Path.*, 2, 7–17, 1988.
132. Freedman, A. M., Atrakchi, A. H., Cassidy, M. M., and Weglicki, W. B., Magnesium deficiency-induced cardiomyopathy: protection by vitamin E, *Biochem. Biophys. Res. Commun.*, 170, 1102–1106, 1990.
133. Brigden, W., Uncommon myocardial disease: the non-coronary cardiomyopathies, *Lancet*, 2, 1179–1184, 1957.
134. Goodwin, J. J., Congestive and hypertrophic cardiomyopathies, *Lancet*, 1, 73l–739, 1970.
135. Kariv, I., Kreisler, B., Sherf, L., Feldman, S., and Rosenthal, T., Familial cardiomyopathy. A review of 11 families, *Am. J. Cardiol.*, 28, 693–707, 1971.
136. McKusick, V. A., A genetical view of cardiovascular disease, *Circulation*, 30, 326–357, 1964.
137. James, T. N., An etiologic concept concerning the obscure myocardiopathies, *Progr. Cardiovasc. Dis.*, 7, 43–64, 1964.
138. James, T. N., Pathology of small coronary arteries, *Am. J. Cardiol.*, 20, 679–691, 1967.
139. Blumenthal, H. T., Alex, M., and Goldenberg, A., A study of lesions of intramural coronary artery branches in diabetes mellitus, *Arch. Path.*, 70, 27–42, 1960.
140. Rubler, S., Dlugash, H., Yuceoglu, Y. Z., Kumral, T., Branwood, A. W., and Grishman, A., A new type of cardiomyopathy associated with diabetic glomerulosclerosis, *Am. J. Cardiol.*, 30, 595–602, 1972.
141. Hamby, R. I., Zoneraich, S., and Sherman, L., Diabetic cardiomyopathy, *J. Am. Med. Assoc.*, 229, 1749–1754, 1974.
142. Martin, H. E. and Wertman, M., Serum potassium, magnesium, and calcium levels in diabetic acidosis, *J. Clin. Investig.*, 26, 217–228, 1947.
143. Beckett, A. G. and Lewis, J. G., Serum magnesium in diabetes mellitus, *Clin. Sci.*, 18, 597–604, 1959.
144. Bachem M. G., Strobel, B., Jastram, U., Janssen, E.-G., and Paschen, K., Magnesium and diabetes, *Magnesium Bull.*, 2, 35–39, 1980, in German.
145. Fort, P., Magnesium and diabetes mellitus, In *Clinical Disorders in Pediatric Nutrition*, Lifshitz, F., Ed., Marcel Dekker, New York, 1982, 223–240.
146. Fort, P. and Lifshitz, F., Magnesium status in children with insulin-dependent diabetes mellitus, *J. Am. Coll. Nutr.*, 5, 69–78, 1986.
147. Rayssiguier, Y., Lipoprotein metabolism: Importance of magnesium, *Magnesium Bull.*, 8, 86–193, 1986.
148. Rayssiguier, Y. and Gueux, E., Magnesium and lipids in cardiovascular disease, *J. Am. Coll. Nutr.*, 5, 507–519, 1986.
149. Rayssiguier, Y., Mazur, A., Cardot, P., and Gueux, E., Effects of magnesium on lipid metabolism and cardiovascular disease, In *Magnesium in Health and Disease*, Itokawa, Y. and Durlach, J., Eds., Libbey, J., London, 1989, Fifth Intl. Mg Sympos., Kyoto, 1988, 199–207.
150. Anderson, R. A., Chromium, In *Modern Nutrition in Health and Disease*, Shils, M. E. and Young, V. R., Eds., Lea & Febiger, Philadelphia, PA, 1988, 268–273.
151. Anderson, R. A., Recent advances in the role of chromium in human health and diseases, In *Essential and Toxic Trace Elements in Human Health and Disease*, Prasad, A. S., Ed., Publ. Alan R. Liss, Inc., 1988, 189–197.
152. Anderson, R. A., Chromium metabolism and its role in disease processes in man, *Clin. Physiol. Biochem.*, 4, 31–41, 1986.
153. Tuvemo, T. and Gebre-Medhin, M., The role of trace elements in juvenile diabetes mellitus, *Pediatrician*, 12, 213–219, 1983–1985.
154. Jervell, A. and Lange-Nielsen, F., Congenital deaf-mutism, functional heart disease with prolongation of the Q-T interval, and sudden death, *Am. Heart J.*, 4, 59–68, 1957.

155. Fraser, G. R., Froggatt, P., and James, T. N., Congenital deafness associated with electrocardiographic abnormalities. A recessive syndrome, *Quart. J. Med.*, 33, 361–384, 1964.
156. James, T. N., Congenital deafness and cardiac arrhythmias, *Am. J. Cardiol.*, 19, 627–643, 1967.
157. Koroxenidis, G. T., Webb, Jr., N. C., Moschos, B. B., and Lehan, P. H., Congenital heart disease, deaf-mutism, and associated somatic malformations in several members of one family, *Am. J. Med.*, 40, 149–155, 1966.
158. Spellberg, R. D., Familial sinus node disease, *Chest*, 60, 246–251, 1971.
159. Mathews, Jr., E. C., Blount, Jr., A. W., and Townsend, J. I., Q-T prolongation and ventricular arrhythmias, with and without deafness in the same family, *Am. J. Cardiol.*, 29, 702–711, 1972.
160. Franz, K. B., Hearing thresholds of rats fed different levels of magnesium for two weeks, *J. Am. Coll. Nutr.*, 11, 612, 1992.
161. Ising, H., Handrock, M., Guenther, T., Fischer, R., and Combrowski, Increased noise trauma in guinea pigs through magnesium deficiency, *Acta Otorhino-laryngol.*, 236, 139–146, 1982.
162. Joachim, Z., Ising, H., and Guenther, T., Biochemical mechanisms affecting susceptibility to noise-induced hearing loss, In *Noise as a Public Health Problem*, Rossi, G., Ed., Proc. of 4th Intl. Congr., Vol. 1, Techn. Centro Ricerde Studi Amplifer, Milano, 1983, 243–255.
163. Guenther, T., Ising, H., and Joachims, Z., Biochemical mechanisms affecting susceptibility to noise-induced hearing loss, *Am. J. Otol.*, 10, 36–41, 1989.
164. Appel, S. H. and Roses, A. D., The muscular dystrophies, In *The Metabolic Basis of Inherited Disease*, Stanbury, J. B., Wyngaarden, J. B., Fredrickson, D. S., Goldstein, J. L., and Brown, M. S., Eds., McGraw-Hill Book Co., New York, 1983, 1470–1495.
165. James, T. N., Observations on the cardiovascular involvement (including the cardiac conduction system) in progressive muscular dystrophy, *Am. Heart J.*, 63, 48–56, 1962.
166. Wahi, P. L., Manchanda, S. S., Thind, M. S., and Akhtar, M., Cardiopathy in muscular dystrophy, *Dis. Chest*, 53, 79–84, 1968.
167. Smith, H. L., Fischer, R. L., and Etteldorf, J. N., Studies of serum calcium and magnesium in muscular dystrophy, *Am. J. Dis. Child.*, 100, 714–715, 1960.
168. Smith, H. L., Fischer, R. L., and Etteldorf, J. N., Magnesium and calcium in muscular dystrophy, *Am. J. Dis. Child.*, 103, 771–776, 1962.
169. Boellner, S. W., Olson, E. J., Fredrickson, D., and Hughes, E. R., Plasma and erythrocyte magnesium in muscular dystrophy, *Am. J. Dis. Child.*, 110, 172–175, 1965.
170. Bertorini, T. E., Cornelio, F., Bhattacharya, S. K., Palmieri, G. M., Dones, I., Dworzak, F., and Brambati, B., Calcium and magnesium content in fetuses at risk and prenecrotic Duchenne muscular dystrophy, *Neurology*, 34, 1436–1440, 1984.
171. Shalev, R. S., Shalev, O., Amir, N., Porat, S., and Fawlewski-De-Leon, G., Erythrocyte (Ca2+ + Mg2+)-ATPase activity and calcium homeostasis in Duchenne muscular dystrophy, *J. Neurol. Sci.*, 63, 325–330, 1984.
172. Sundermeyer, J. F., Gudbjarnson, S., Wendt, V. E., denBakker, P. B., and Bing, R. J., Myocardial metabolism in progressive muscular dystrophy, *Circulation*, 24, 348–1355, 1961.
173. Wright, T. L., O'Neill, J. A., and Olson, W. H., Abnormal intrafibrillar monoamines in sex-linked muscular dystrophy, *Neurology*, 23, 510–517, 1973.
174. Bajusz, E., Homburger, F., Baker, J. R., and Opie, L. H., The heart muscle in muscular dystrophy with special reference to involvement of the cardiovascular system in the hereditary myopathy of the hamster, *Ann. N.Y. Acad. Sci.*, 138, 213–231, 1966.
175. Bajusz, E. and Lossnitzer, K., A new disease model of chronic congestive heart failure: studies on its pathogenesis, *Trans. N.Y. Acad. Sci.*, Ser. II, 30, 939–948, 1968.
176. Nadkarni, B. B., Hunt, B., and Heggtveit, H. A., Early ultrastructural and biochemical changes in the myopathic hamster heart, In *Myocardiology. Recent Advances in Studies on Cardiac Metabolism*, Bajusz, E. and Rona, G., Eds., University Park Press, Baltimore, MD, 1972, 251–261.
177. Crawford, A. J. and Bhattacharya, S. K., Excessive intracellular zinc accumulation in cardiac and skeletal muscles of dystrophic hamsters, *Exp. Neurol.*, 95, 265–276, 1987.

178. Dunn, J. F. and Radda, G. K., Total ion content of skeletal and cardiac muscle in the mdx mouse dystrophy: Ca2+ is elevated at all ages, *J. Neurol. Sci.,* 103, 226–231, 1991.
179. Glesby, M. J., Rosenmann, E., Nylen, E. G., and Wrogemann, K., Serum CK, calcium, magnesium, and oxidative phosphorylation in mdx mouse muscular dystrophy, *Muscle Nerve,* 11, 852–856, 1988.
180. Jacobson, B. E., Lundquist, C. G., and Griffith, T. G., Defective respiration and oxidative phosphorylation in muscle mitochondria of hamsters in the late stages of hereditary muscular dystrophy, *Canad. J. Biochem.,* 48, 1037–1042, 1970.
181. Wrogemann, K., Blanchaer, M. C., and Jacobson, B. E., A magnesium-responsive defect of respiration and oxidative phosphorylation in skeletal muscle mitochondria of dystrophic hamsters, *Canad. J. Biochem.,* 48, 1332–1338, 1970.
182. Wrogeman, K., Blanchaer, M. C., and Jacobson, B. E., Abnormal oxidative phosphorylation in skeletal muscle mitochondria of the BIO14.6 dystrophic Syrian hamster, in *Myocardiology. Recent Advances in Studies on Cardiac Metabolism,* Bajusz, E. and Rona, G., Eds., University Park Press, Baltimore, MD, 1972, 289–293.
183. Angelakos, E. T., Carrollo, L. C., Daniels, J. B., King, M. P., and Bajusz, E., Adrenergic neurohumors in the heart of hamsters with hereditary myopathy during cardiac hypertrophy and failure, in *Myocardiology. Recent Advances in Studies on Cardiac Metabolism,* Bajusz, E. and Rona, G., Eds., University Park Press, Baltimore, MD, 1972, 263–278.
184. Douglas, W. W. and Rubin, R. P., The effects of alkaline earths and other divalent cations on adrenal medullary secretion, *J. Physiol.,* 175, 231–241, 1964.
185. Bouillin, D. J., The action of extracellular cations on the release of the sympathetic transmitter from peripheral nerves, *J. Physiol.,* 189, 85–99, 1967.
186. Nelson, J. S., Pathology of vitamin E deficiency, In *Vitamin E. A Comprehensive Treatise,* Machlin, L. J., Ed., Marcel Dekker, Inc., N.Y., 1980, 397–428.
187. Fenn, W. O. and Goettsch, M., Electrolytes in nutritional dystrophy in rabbits, *J. Biol. Chem.,* 120, 41–59, 1937.
188. Blaxter, K. L. and Wood, W. A., The nutrition of the young Ayrshire calf. Composition of the tissues of normal and dystrophic calves, *Brit. J. Nutr.,* 6, 144–163, 1952.
189. Zuckerman, L. and Marquardt, G. H., Muscle, erythrocyte and plasma electrolytes and some other muscle constitutents of rabbits with nutritional muscular dystrophy, *Proc. Soc. Exp. Biol. Med.,* 112, 609–610, 1963.
190. Goldsmith, L. A., Relative magnesium deficiency in the rat, *J. Nutr.,* 98, 87–102, 1967.
191. Schwarz, K., Vitamin E, trace elements and sulfydryl groups in respiratory decline, *Vitamins and Hormones,* 20, 463–484, 1962.
192. Tappel, A. L., Vitamin E as the biological lipid antioxidant, *Vitamins and Hormones,* 20, 493–510, 1962.
193. Guenther, T. and Hoellriegl, V., Increased lipi peroxidation in liver mitochondria from Mg-deficient rats, *J. Trace Elem. Electrol. Hlth. Dis.,* 3, 213–216, 1989.
194. Guenther, T., Vormann, J., Hoellriegl, V., and Gossrau, R., Effect of Mg deficiency and salicylate on lipid peroxidation in vivo, *Magnesium Bull.,* 13, 26–29, 1991.
195. Guenther, T., Vormann, J., Hoellriegl, V., Disch, G., and Classen, H. G., Role of lipid peroxidation and vitamin E in magnesium deficiency, *Magnesium Bull.,* 14, 57–66, 1992.
196. Freedman, A. M., Cassidy, M. M., and Weglicki, W. B., Magnesium deficient myocardium demonstrates an increased susceptibility to an in vivo oxidative stress, *Magnesium Res.,* 4, 185–189, 1991.
197. Draper, H. H., Nutrient interrelationships, in *Vitamin E. A Comprehensive Treatise,* Machlin, L. J., Ed., Marcel Dekker, Inc., N.Y., 1980, 272–288.
198. Levander, O. A., Clinical consequences of low selenium intake and its relationship to vitamin E, *Ann. N.Y. Acad. Sci.,* 394, 70–82, 1982.
199. Levander, O. A., Selenium, In *Modern Nutrition in Health and Disease,* Shils, M. E. and Young, V. R., Eds., Lea & Febiger, Philadelphia, PA, 1988, 263–267.

200. Jackson, M. J. and Edwards, R. H. T., Selenium, vitamin E, free radicals and muscle disease, *Curr. Top. Nutr. Dis.,* 18, 431–439, 1986.
201. Lockitch, G., Selenium: clinical significance and analytical concepts, *Crit. Rev. Clin. Lab. Sci.,* 27, 483–541, 1989.
202. Banks, M. A., Martin, W. G., and Hinton, D. E., Vitamin E deficiency in the adult Syrian golden hamster, *J. Nutr. Growth Cancer,* 4, 91–108, 1987.
203. Arthur, J. R., Effects of selenium and vitamin E status on plasma creatine kinase activity in calves, *J. Nutr.,* 118, 747–755, 1988.
204. Larvor, P., Relations between plasma composition and symptoms of grass tetany in bovines, *Ann. Zootech.,* 11, 135–149, 1962, in French.
205. Rayssiguier, Y. and Larvor, P., Hypomagnesemia following stimulation of lipolysis in ewes: effects of cold exposure and fasting, In *Magnesium in Health and Disease,* Cantin, M., Seelig, M. S., Eds., Spectrum, N.Y, 1980, 2nd Intl. Mg Sympos., Quebec, 1976, 67–72.
206. Machlin, L. J. and Gabriel, E., Interactions of vitamin E with vitamin C, vitamin B_{12} and zinc, In *Micronutrient Interactions: Vitamins, Minerals and Hazardous Elements, Ann. N.Y. Acad. Sci.,* 355, 98–108, 1980.
207. Talamo, R. C., Rosenstein, B. J., and Berninger, R. W., Cystic fibrosis, in *The Metabolic Basis of Inherited Disease,* Stanbury, J. B., Wyngaarden, J. B., Fredrickson, D. S., Goldstein, J. L., and Brown, M. S., Eds., McGraw-Hill, New York, 1983, 1889–1917.
208. Shwachman, H., Kowalski, M., and Khaw, K. T., Cystic fibrosis: a new outlook. 70 patients above 25 years of age, *Medicine,* 56, 129–149, 1977.
209. Farrell, P. M., Deficiency states, pharmacologic effects, and nutrient requirements, In *Vitamin E. A Comprehensive Treatise,* Machlin, L. J., Ed., Marcel Dekker, Inc., N.Y., 1980, 520–620.
210. Farrell, P. M., Vitamin E, in *Modern Nutrition in Health and Disease,* Shils, M. E. and Young, V. R., Eds., Lea & Febiger, Philadelphia, PA, 1988, 340–354.
211. Sacrez, R., Klein, F., Hoffmann, B., Levy, J. M., Geisert, J., and Korn, R., Hypoplasia of exocrine pancreas, associated with endomyocardial fibrosis in one of two brothers, *Ann. Pediatrie,* 16, 43–48, 1969, in French.
212. Kuhn, J., Rosestein, B. J., and Oppenheimer, E. H., Metastatic calcification in cystic fibrosis. A report of two cases, *Radiology,* 97, 59–64, 1970.
213. Foucard, T., Gebre-Medhin, M., Gustavson, K. H., and Lindh, U., Low concentrations of sodium and magnesium in erythrocytes from cystic fibrosis heterozygotes, *Acta Padiat. Scand.,* 80, 57–61, 1991.
214. Kopito, L. and Schwachman, H., Spectroscopic analysis of tissues from patients with cystic fibrosis and controls, *Nature,* 202, 501–502, 1964.
215. Paunier, L., Girardin, E., Sizonenko, P. G., Wyss, M., and Megevand, A., Calcium and magnesium concentration in sweat of normal children and patients with cystic fibrosis, *Pediatrics,* 52, 446–448, 1973.
216. Kopito, L., Elian, E., and Shwachman, H., Sodium, potassium, calcium, and magnesium in hair from neonates with cystic fibrosis and in amniotic fluid from mothers of such children, *Pediatrics,* 49, 620–624, 1972.
217. Orenstein, S. R. and Orenstein, D. M., Magnesium deficiency in cystic fibrosis, *South. Med. J.,* 76, 1586, 1983.
218. Godson, C., Ryan, M. P., Brady, H. R., Bourke, S., and FitzGerald, M. X., Acute hypomagnesaemia complicating the treatment of meconium ileus equivalent in cystic fibrosis, *Scand. J. Gastroenterol.,* Suppl 143, 148–150, 1988.
219. Banks, M. A., Martin, W. G., and Hinton, D. E., Long-term histological observations in the liver and pancreas of vitamin E and selenium-deficient adult Syrian golden hamster, *J. Nutr. Growth Cancer,* 4, 109–128, 1987.
220. Mudd, S. H. and Levy, H. L., Disorders of transsulfuration, in *The Metabolic Basis of Inherited Disease,* Stanbury, J. B., Wyngaarden, J. B., Fredrickson, D. S., Goldstein, J. L., and Brown, M. S., Eds., McGraw-Hill, New York, 1983, 522–559.

221. Laster, L., Spaeth, G. L., Mudd, S. H., and Finkelstein, J. D., Homocystinuria due to cystathionine synthase deficiency, Combined Staff Conf., NIH, *Ann. Intern. Med.,* 63, 1117–1142, 1965.
222. Frimpter, G. W., Andelman, R. J., and George, W. F., Vitamin B_6—dependency syndromes. New horizons in nutrition, *Am. J. Clin. Nutr.,* 22, 794–805, 1969.
223. Gibson, J. B., Carson, N. A. J., and Neill, D. W., Pathological findings in homocystinuria, *J. Clin. Path.,* 17, 427–437, 1964.
224. Wilcken, B. and Turner, B., Homocystinuria. Reduced folate levels during pyridoxine treatment, *Arch. Dis. Childh.,* 48, 58–62, 1973.
225. White, A., Handler, P., Smith, E. L., Hill, R. L., and Lehman, I. R. *Principles of Biochemistry,* 6th ed., McGraw-Hill, N.Y., 1978.
226. Aikawa, J. K., Effects of pyridoxine and desoxypyridoxine on magnesium metabolism in the rabbit, *Proc. Soc. Exp. Biol. Med.,* 104, 461–463, 1960.
227. McCormick, D. B., Gregory, M. E., and Snell, E. E., Pyridoxal phosphokinases. I. Assay, distribution and properties, *J. Biol. Chem.,* 236, 2076–2084, 1961.
228. Hsu, J. M., Zinc content in tissues of pyridoxine deficient rats, *Proc. Soc. Exp. Biol. Med.,* 119, 177–180, 1965.
229. Durlach, J., Studies on synergistic mechanisms between vitamin B_6 and magnesium, *J. Med. Besancon,* 5, 349–359, 1969, in French.
230. Kubena, K. S., Edgar, S. E., and Veltman, J. R., Growth and development in rats and deficiency of magnesium and pyridoxine, *J. Am. Coll. Nutr.,* 7, 317–324, 1988.
231. Majumdar, P., Boylan, M., and Driskell, J. A., Effect of pyridoxine hydrochloride on tissue magnesium levels in the rat, *FASEB J.,* 2, A441, 1988.
232. Harriman, A. E., Dietary pyridoxine loadings affect incidence of "spontaneous" seizures among magnesium-deprived Mongolian gerbils (Meriones unguiculatus), *Percept. Motor Skills,* 66, 327–337, 1988.
233. Boylan, L. M. and Spallholz, J. E., Ultraviolet and fluorometric spectral evidence for the in vitro binding of zinc and magnesium to vitamin B-6, *FASEB J.,* 3, 669, 1989.

Chapter 18

NONNUTRITIVE DIETARY SUPPLEMENTS IN PEDIATRICS

Michael B. Zimmermann and Norman Kretchmer

TABLE OF CONTENTS

I. Introduction 225

II. Fat Substitutes 225

III. Sugar Substitutes 227
  A. Currently Approved Sweeteners 228
  B. Other Sweeteners 230

IV. Potential Pediatric Applications of Nonnutritive Substances 232
  A. Diabetes 232
  B. Dental Caries 232
  C. Hyperlipidemia 233
  D. Obesity 233

V. Summary 234

References 235

I. INTRODUCTION

Changing perceptions about diet and the role of food selection in promoting health have prompted a proliferation of foods containing reduced calories, fat, and sugar. For the past several decades, the use of nonnutritive supplements has evolved from limited use of a few nonnutritive sweeteners in foods for diabetics to the present widespread use of a variety of nonnutritive sweeteners and fat substitutes in the general population. This essay will discuss available nonnutritive sweeteners and fat substitutes and their potential application in pediatrics, including use in diabetes, hyperlipidemia, dental caries, and obesity.

II. FAT SUBSTITUTES

Efforts by the food industry to develop dietary fat substitutes have generated substantial interest among both the public and the medical profession. Intense

0-8493-2764-4/95

TABLE 1
Current and Proposed Fat Substitutes

Carbohydrate-based materials
Modified glucose polymers
Modified tapioca, potato, rice, and corn starches
Gums and algins
Cellulose derivatives
Protein-based materials
Microparticulated proteins
Lipid-based materials
Fatty acid esters of sugars and sugar alcohols
Alkyl glyceryl ethers
Polycarboxylic acid and propoxylated glyceryl esters
Substituted siloxane polymers
Branched (sterically hindered) triglyceride esters

scrutiny by industry and government has focused on the toxicity and functionality of these additives, while few data are available evaluating their nutritional impact on the intake of energy, fat, and cholesterol when incorporated into human diets. Currently available and proposed fat substitutes comprise a heterogeneous group whose potential nutritional benefits remain largely unproven. Unlike many food additives, fat substitutes are macronutrients and may constitute a substantial portion of the future diet. For this reason, there are novel issues relating to the safety assessment of these substances in both children and adults.[1,15] Fat substitutes that are presently available or in development for food use are listed in Table 1. These substitutes can be carbohydrate-, protein-, or lipid-based materials. Carbohydrate-based substances function as fat substitutes by stabilizing large quantities of water into gel-like structures. These gels have flow and lubricant properties that mimic the texture and mouth-feel of fat. In contrast, protein-based materials simulate fat texture due to their unique structure and flow properties. A new microparticulated protein food ingredient (Simplesse®), produced from egg white or whey protein, consists of spheroidal hydrated protein particles 0.1 to 3 microns in diameter.[2] These small particles are not perceived by the tongue as individual grains but rather as a creamy or fatty fluid. Children who are allergic to whey or egg white should avoid foods that include microparticulated proteins produced from these products. The carbohydrate- and protein-based ingredients cannot be used in most baking applications, cannot be used to fry foods, and because of their high water content may have a reduced shelf life. Carbohydrate- and protein-based fat substitutes are partially or fully digested and absorbed, but because of their low energy density have reduced caloric value. Most of the materials have little influence on other food components and micronutrients. Partially digested material, such as cellulose derivatives and gums, may reduce absorption of certain micronutrients and minerals, and may act as bulking agents, reducing intestinal transit time and producing laxative effects.[1] Concerns about widespread use of microparticulated protein products have focused on the question of increased dietary protein and renal dysfunction. The majority of the adult and pediatric population in the United States have daily intakes of protein in excess of that considered sufficient to meet physiological requirements.[3] Because increased dietary protein increases glomerular filtration rate,[4] and because low protein diets slow the rate of progression of certain types of renal disease,[6] it has been suggested the hyperfiltration induced by protein loads may lead to accelerated glomerular damage.[5,6] However, this relationship in normal adults and children remains largely unproven. It has been suggested that the estimated increase in protein intake that might occur with the use of microparticulated proteins as fat substitutes, even considering a profound switch to microparticulated protein-based foods, would be very modest.[7]

The other major category of fat substitutes are the lipid-based materials. Because they have functional properties similiar to the fats they replace and are generally heat-stable, applications in food and food-processing may be very broad.[8] They are resistant to digestive enzymes and have little or no caloric value. Passing through the gut undigested, they may sequester lipophilic substances and influence the absorption of fats and fat-soluble compounds.[8,59] Decreased plasma values of vitamin A[9–11] and vitamin E[9–12] have been observed in studies of olestra (Figure 1), a lipid-based substitute. In addition, these materials may influence the absorption and metabolism of cholesterol[9–12] and bile acids.[59] The bioavailability of several lipid-soluble drugs does not appear to be affected by substitution of olestra for dietary fat.[13] Other considerations in the use of these substances are potential adverse effects of the unabsorbed material on gut epithelium, regional lymphatic tissue, and colonic microbial populations.[15] It appears unlikely that suggested applications for use of these materials would increase the likelihood of essential fatty acid deficiency.[1] However, there is concern that supplementation may affect the ratio of dietary saturated to unsaturated fat,[1] as most fats targeted for application of these substitutes are unsaturated. In addition, there is concern that the increased use of fat substitutes may lead to nutritional imbalances. Consumption of classes of foods that are presently avoided because of high fat and calorie content might increase at the expense of other, more nutrient-dense foods.[14,15]

III. SUGAR SUBSTITUTES

Newborns and children demonstrate a positive response to sweetness that appears to be a reflexive, innate reaction rather than a learned response.[16,20] In school-age children, the percent of total calories from sucrose and other simple carbohydrates is estimated to be around 15%.[17] Current recommendations are for reducing refined and processed sugars to 10% or less of total calories.[18] Sucrose and high fructose corn syrups are the most frequently used nutritive sweeteners in foods and beverages. Fructose and polyols (sorbitol, mannitol, xylitol, and hydrogenated starch hydrolysate) are other nutritive sweeteners that can replace sucrose in foods and produce equivalent sweetness with lower caloric content. Recent reviews of nutritive sweeteners are available.[34,35] Nonnutritive sweeteners

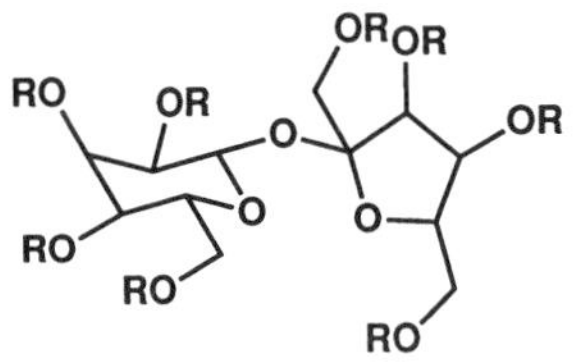

Figure 1. Olestra (6, 7, or 8 fatty acid groups). R = fatty acid.

TABLE 2
Relative Sweetness of Various Nutritive and Nonnutritive Sweeteners

	Factors vs. Sucrose
Sucrose	1.0
High fructose corn syrup (90% extraction)	1.5
Sodium saccharin	300
Aspartame	180
Acesulfame K	200
Sodium cyclamate	30
Alitame	2000
Sucralose	600

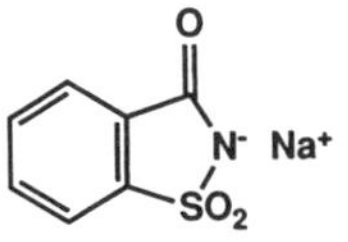

Figure 2. Sodium saccharin.

are defined as substances having less than 2% of the caloric value of sucrose per unit of sweetening capacity.[19] Nonnutritive sweeteners are the fastest growing segment of the food additive industry. Because of changing dietary habits and improving product quality, consumption of nonnutritive sweeteners is expected to exceed current levels by more than 50% by the late 1990s.[37] Nonnutritive sweeteners include a heterogeneous group of natural and synthetic substances (Table 2). Currently approved nonnutritive sweeteners used in the United States include saccharin, aspartame, and acesulfame K. Cyclamates, alitame, and sucralose are nonnutritive sweeteners currently awaiting FDA approval.[19]

A. CURRENTLY APPROVED SWEETENERS

Saccharin—Saccharin (1,2-benzisothiazol-3-one-1,1-dioxide) (Figure 2), introduced commercially in 1901, is currently used in foods, drugs, and tabletop sweeteners at an estimated total annual consumption of 6 million pounds.[21] It is approximately 300× sweeter than sucrose. Saccharin is stable at physiologic pH and temperature and is well absorbed. It does not accumulate in body tissues and is excreted unchanged in the urine and feces; there are no known metabolites in

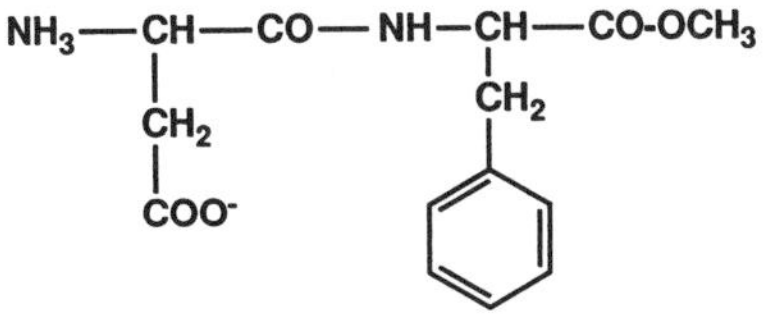
NH3—CH—CO—NH—CH—CO-OCH3
CH2
COO-
CH2

Figure 3. Aspartame.

man.[22] Saccharin does cross the placenta during pregnancy.[22] It has been extensively studied for mutagenicity and carcinogenicity. The cumulative evidence from 1955 to 1978 was critically reviewed by the National Academy of Sciences and the National Research Council in 1978,[23] and it was concluded that in single-generation studies saccharin did not induce cancer in any organ. However, numerous chronic toxicity tests extended over several generations have consistently indicated that saccharin causes a dose-related increase in the number of bladder tumors in rats, particularly when the animal is exposed *in utero* and exposure continues after birth.[22,23] A proposed ban by the FDA in 1977 produced a professional and public outcry which prompted Congress to adopt a moratorium on the proposed ban, which has been extended repeatedly. There remain significant questions regarding the validity of the numerous saccharin studies and their conclusions as regards to evidence of health hazards in humans under normal or exaggerated conditions of use. Generally recognized as safe (GRAS) recommendations from 1955 suggested limiting saccharin use to 500mg/day in children. Because of the availability of alternative sweeteners, current intake in the pediatric population is generally considered to be much lower than the GRAS recommendations.[24] Continued research is needed to identify safe levels for children. Heavy use during pregnancy and early childhood should probably be avoided.

Aspartame—Aspartame is a dipeptide methyl ester (L-aspartyl-L-phenylalanine methylester) (Figure 3) that is approximately 180× as sweet as sucrose. Aspartame is the most widely consumed nonnutritive sweetener and is used in a variety of foods. Because it is labile to heat and $pH > 7.0$, it cannot be used in thermally processed foods.[37] Its shelf life has been estimated to be six months to a year and is converted with time to diketopiperazine with loss of sweetness.[32] Aspartame is rapidly hydrolyzed in the intestine to aspartate, phenylalanine, and methanol. Methanol is converted by the liver into formate and carbon dioxide. Although there have been concerns about the potential ocular toxicity of the metabolites of methanol, the amount of methanol produced by the complete hydrolysis of the acceptable daily intake of aspartame is negligible and does not appear to be hazardous.[25] Both of the amino acids are considered GRAS. Dietary phenylalanine is contraindicated in children with phenylketonuria,[33] and foods and beverages containing aspartame must carry a warning label. Phenylalanine enters the plasma free amino acid pool after partial conversion to tyrosine in the liver, and variable amounts of aspartate are transaminated in the enterocyte before entering the plasma free amino acid pool.[26] Safety issues regarding aspartame have centered

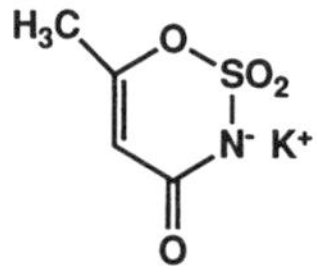

Figure 4. Acesulfame potassium.

around possible central nervous system toxicity of its constituents. Although phenylalanine and aspartate given separately in very large amounts can produce chemical and functional effects in the CNS, it is generally accepted that current levels of consumption pose no risk for CNS toxicity in adult populations.[36] There have been many alleged adverse effects of aspartame,[28,29] including seizures, behavior changes, memory loss, nausea, allergic reactions, and, most commonly, headaches. Careful independent scrutiny by the FDA and advisory groups has generally not supported these claims. Aspartame appears to have no clinically significant effect on activity in children.[27] The FDA has recommended an acceptable daily intake for aspartame to be 50 mg/kg body weight in adults. Young children, because of their small body weight, may conceivably exceed this level of intake. Aspartame appears to be safe in lactating women at levels recommended for healthy adults.[31] At intake levels 3× the 99th percentile of the projected daily intake of aspartame, there is no evidence of risk to the fetus from maternal ingestion.[30] However, further research has been recommended to determine safety of aspartame in children and in pregnant or lactating women.[25]

Acesulfam Potassium—Acesulfam K (6-methyl-3,4-dihydro-1,2,3-oxathiazin-4-one 2,2-dioxide) (Figure 4), a complex derivative of acetoacetic acid, was discovered in 1967. It received limited FDA approval in 1988 for use in tabletop sweeteners, chewing gums, powdered beverages, and other applications.[19] Acesulfam K is approximately 200× sweeter than sucrose. It is water soluble and stable at temperatures up to 225°C and moderate pH.[37] It is well absorbed, not metabolized, and excreted unchanged in urine and feces. Because of its structural resemblance to saccharin, concerns about its possible carcinogenic potential have been expressed.[37] However, the FDA, after extensive analysis of toxicologic studies including teratology, reproduction, and genotoxicity, granted approval for its use,[19] establishing an acceptable daily intake of 0 to 15 mg/kg body weight. The World Health Organization has established an acceptable daily intake of 0 to 9 mg/kg body weight.

B. OTHER SWEETENERS

Cyclamate—Cyclamate (cyclohexanesulfamate) (Figure 5) was discovered in 1937, approved for use as a sweetener in 1949, and was placed on the original GRAS list. It was banned by the FDA in 1970 because of evidence suggesting it was a carcinogen. In Canada and continental Europe, cyclamate has not been banned, and the acceptable daily intake is set at 11 mg/kg body weight. Cyclamate is usually found in foods as the calcium or sodium salt, often in a 10:1 mixture

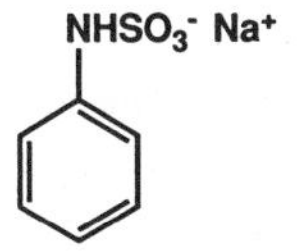

Figure 5. Sodium cyclamate.

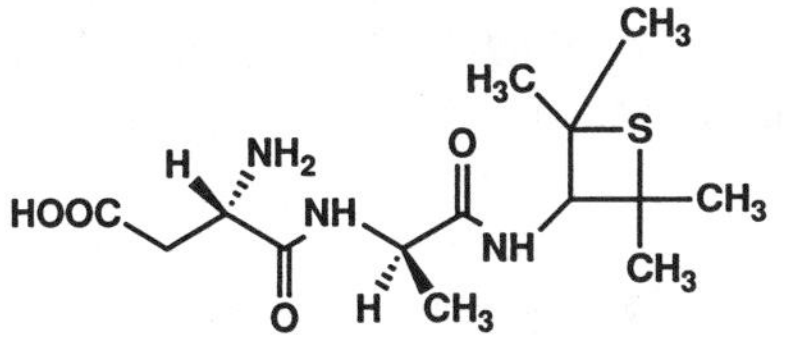

Figure 6. Alitame.

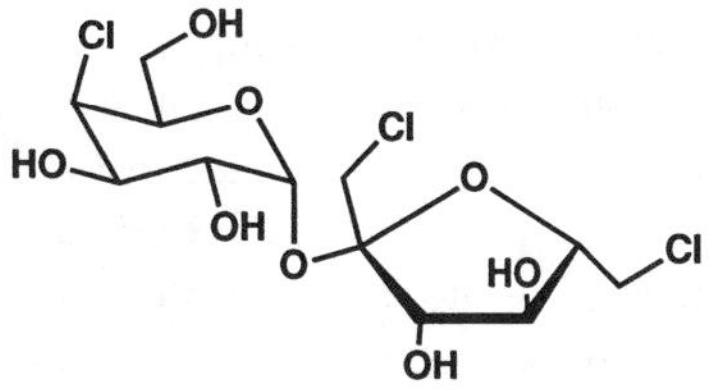

Figure 7. Sucralose.

with saccharin. It is 30× sweeter than glucose, is not heat labile, and has a long shelf life.[37] It is slowly and only partially absorbed and is excreted largely unchanged in the urine.[38] A metabolite, cyclohexylamine, has been identified in human urine following cyclamate administration.[39] It is considered more toxic than cyclamate, and there is evidence that large doses can produce testicular atrophy in animals.[38] The FDA ban in 1970 was precipitated by a long-term study where rats who were fed mixtures of cyclamate, saccharin, and cyclohexylamine showed an increased incidence of bladder tumors.[41] There has been continuing debate over the potential carcinogenicity of cyclamate. The National Academy of Science/National Research Council reviewed the evidence in 1985,[40] and concluded that cyclamate alone is not a carcinogen. However, it was also concluded that cyclamate appears to have cancer-promoting or co-carcinogenic activity. A petition is currently pending before the FDA to reinstate cyclamate as a nonnutritive sweetener.[19]

Alitame—Alitame (1-α-aspartyl-N-(2,2,4-tetramethyl-3-thietanyl)-d-alanimide) (Figure 6) is a dipeptide-based amide derived from aspartate and alanine that is 2000× sweeter than sucrose.[42] No safety issues have been raised to date, and a food additive petition is currently pending before the FDA.[19]

Sucralose—Sucralose (1,4,6-trideoxy-trichlorogalactosucrose) (Figure 7) is a derivative of sucrose in which three chlorine groups have been substituted for three hydroxyl groups.[43] It is approximately 600× sweeter than sucrose. Sucralose is soluble in water and stable within a large range of temperature and pH.[43,34] It is poorly absorbed and passes through the gut mostly undigested. In man, less than 5% of an oral dose is biotransformed into a glucoronide conjugate and is excreted in the urine.[43] A food additive petition was filed with the FDA in 1987 and is pending.[19]

IV. POTENTIAL PEDIATRIC APPLICATIONS OF NONNUTRITIVE SUBSTANCES

A. DIABETES

Dietary modification is one of the cornerstones of any comprehensive therapeutic program for the treatment of diabetes in children. The issue of dietary sucrose has been extensively debated in discussion of the dietary management of diabetes. Statements from both the American Diabetes Association (ADA)[45] and the FDA[46] have suggested that there may be no adverse effects associated with moderate sucrose consumption in individuals with diabetes. On the other hand, multiple prospective studies[44,47,48] have suggested that consumption of moderate amounts of sucrose may result in metabolic effects that may be deleterious and are associated with an increased risk for coronary artery disease, an issue of special concern in diabetic patients. ADA dietary recommendations[49] state that it appears there are no contraindications to the moderate use of nonnutritive sweeteners in diets for children. Nonnutritive sweeteners do not affect concentrations of blood glucose in diabetics,[50] but the question of whether they are necessary or useful in the diabetic diet is unresolved. In many diabetics, sucrose does not elevate postprandial blood glucose significantly more than readily digestible starches,[51,52] so substitution of nonnutritive sweeteners for sucrose in foods may not be of value in reducing postprandial blood glucose fluctuations. However, the use of nonnutritive sweeteners may allow for more flexibility in planning meals. For example, substitution of a nonnutritive sweetener may allow for additional potato, rice, or fruit to be included in the meal. Nonnutritive sweeteners in soft drinks and between-meal snacks may be helpful when these items are not included in the meal plan. However, if childhood is a time when habits and preferences for the intense sweetness of additives is learned and established, it may be preferable to direct the pediatric patient toward the sweetness of whole foods. Teaching a child the importance of nutrition in diabetes and emphasizing foods which constitute a healthy diet may establish habits that will be carried into adulthood.

B. DENTAL CARIES

A potential area in pediatric health where nonnutritive sweeteners could potentially be useful is in the reduction or prevention of dental caries. The perceived importance of restricting dietary sucrose to prevent caries has waned, largely because of the significant reduction in caries incidence due to fluoridation and other preventive methods in dentistry.[19] Evidence does overwhelmingly implicate exposure to dietary sucrose as the major factor in the complex etiology of both coronal and root caries.[53] Because strict restriction of sucrose intake in children is often difficult to achieve, recommendations have been made for substituting nonnutritive sweeteners for sucrose in the diets of children prone to rampant caries.[53,54] This may be of special value in food and beverage products consumed by these children between meals. The nonnutritive sweeteners are not metabolized by oral microflora, are nonacidogenic in plaque pH measurements, and are noncariogenic in animals.[53] In addition, in *in vitro* systems, acesulfam K and saccharin have demonstrated inhibitory effects on the growth of *Streptococcus mutans* and other streptococci implicated in caries production.[55] In children with rampant caries, moderate use of nonnutritive sweeteners may be of value when dietary avoidance of sucrose is unachievable.

C. HYPERLIPIDEMIA

Children with certain familial hyperlipidemic syndromes who are placed on highly fat-restricted diets may benefit from the increased variety and palatability of foods made from fat substitutes. Studies in adults[8–12] have demonstrated that ingestion of olestra can reduce serum triglycerides, total cholesterol, and low-density lipoprotein levels without affecting high-density lipoprotein levels. Effects were most evident in hyperlipidemic individuals. The magnitude of the changes in plasma lipids appears to be greater than those expected from simple replacement of fat and cholesterol by olestra substitution. In preliminary trials, addition of foods containing 50% olestra shortening to the diets of homozygotic children with Type 1 hyperchylomicronemia and heterozygotic relatives appears not to adversely affect serum lipid profiles.[59] Further studies are needed to determine the potential value of fat substitutes in the management of pediatric hyperlipidemic syndromes.

D. OBESITY

Children have unique nutritional and energy needs for normal development and growth. Although it appears most children are capable of self-regulating their food intake to meet energy needs, there is growing concern over the incidence of childhood obesity and questions about appropriate cholesterol and fat intake in childhood.[56,58] A recent Department of Agriculture survey of children ages one through four revealed total dietary fat intake of 34% of calories, with saturated fat intake of 13% of calories.[57] Studies of school-age children have shown total fat intake to be 38 to 41% of calories.[58] Current dietary guidelines for adults recommend limiting consumption of total fat to 30% or less of calories and saturated fat to less than 10% of calories.[61,18] There appears to be a growing consensus that similiar recommendations are appropriate for children over the age of two.[56] In the U.S., about 87% of ten-year-olds exceed these requirements for fat intake.[62] High levels of cholesterol in children may be predictive of high levels in adulthood and increased risk of coronary heart disease.[56,58] Studies show that around 20% of school-age children have serum cholesterol values greater than 185 mg/dl (90th percentile);[63,64] skin-fold thicknesses above the 70th percentile in children are associated with elevated serum cholesterol.[59] The incidence of childhood obesity appears to be increasing, and it is estimated that 70% of obese children will become obese adults.[59]

Effective strategies for the management of childhood and adolescent obesity are clearly needed. It has been suggested that fat substitutes may potentially be of use in preventing weight gain in children and in the treatment of pediatric obesity.[59] Incorporating fat substitutes and nonnutritive sweeteners into specific food products can substantially reduce caloric content. However, the effect of fat substitutes on total dietary fat intake in childhood may be modest, due to proposed regulatory limits on the uses of these substances. Meat and meat products, milk, cheese, eggs, and whole grains contribute the majority of the fat in U.S. diets,[60] and fat substitutes will have little effect on these food categories.[1] In addition, the effects of long-term consumption of foods containing nonnutritive substances on macronutrient and energy intake during attempts at weight loss are uncertain. Multiple studies[65–67] have used a variety of fat substitutes, alone or in combination with nonnutritive sweeteners, to manipulate the energy density of foods, and several studies have examined compensatory energy intake with the use of nonnutritive sweeteners.[68,69] Although the methodology of many of these studies has been criticized, it appears that healthy human subjects usually compensate well for dilutions in food energy by increasing consumption to maintain energy intake. Based on limited data from studies in man and rat, nonnutritive sweeteners may be of little practical use in the management of pediatric obesity. Although further studies are needed to evaluate the impact of fat substitutes on long-term energy intake in humans, the potential nutritional benefits appear, at this time, largely unproven. Although fat-substituted foods may prove to be of some value as adjuncts in individuals in weight loss programs, modification of behavior continues to be fundamental in determining the success of these programs.

V. SUMMARY

There is currently widespread and increasing use of nonnutritive sweeteners and fat substitutes in the general population. An expanding list of new products may be available in the near future, as the food industry responds to consumer demand. Questions concerning the use of these substances in childhood remain unresolved. The potential usefulness of nonnutritive substances in pediatrics is limited by insufficient knowledge of long-term safety and efficacy of these substances in children. Moderate use of nonnutritive sweeteners in the dietary management of childhood diabetes may be of some value in flexible meal planning. As part of a preventive program in children prone to dental caries, moderate use of nonnutritive sweeteners may be useful if dietary avoidance of sucrose is unachievable. Fat substitutes may prove to be of value in the dietary management of children with hyperlipidemic syndromes. Nonnutritive fats and sweeteners may be of only limited value in the management of childhood and adolescent obesity.

Recommendations concerning the use of nonnutritive substances in pediatrics must be somewhat guarded awaiting further long-term studies of these substances in children.

REFERENCES

1. Mela, D. J., Nutritional implications of fat substitutes, *J. Am. Diet. Assoc.*, 92, 472–476, 1992.
2. Gaull, G. E., Role of microparticulated protein fat substitutes in food and nutrition, *Ann. NY Acad. Sci.*, 623, 350–355, 1991.
3. FAO, WHO Technical Report, *Energy and protein requirements*, Series No. 724, Geneva, World Health Organization, 1985, 1–206.
4. Levine, M. M., Kirschenbaum, M. A., Chaudhari, A., Wong, M. W., and Bricker, N. S., Effect of protein on glomerular filtration rate and prostanoid synthesis in normal and uremic rats, *Am. J. Physiol.*, 251, 635–641, 1986.
5. Brenner, B. M., Meyer, T. W., and Hostetter, T. H., Dietary protein intake and the progressive nature of kidney disease: the role of hemodynamically mediated glomerular injury in the pathogenesis of progressive glomerular sclerosis in aging, renal ablation and intrinsic renal disease, *New Eng. J. Med.*, 307, 652–659, 1982.
6. Klahr, S., Effects of protein intake on the progression of renal disease, *Ann. Rev. Nutr.*, 9, 87–108, 1989.
7. Young, V. R., Fukagawa, M. D., and Pellet, P. L., Nutritional implications of microparticulated protein, *J. Am. Coll. Nut.*, 9, 418–426, 1990.
8. Jandacek, R. J., Studies with sucrose polyester, *Int. J. Obes.*, 8, 13–21, 1984.
9. Fallat, R. W., Glueck, C. J., Lutmer, R., and Mattson, F. H., Short-term study of sucrose polyester: a nonabsorbable fat-like material as a dietary agent for lowering cholesterol, *Am. J. Clin. Nutr.*, 29, 1204–1215, 1976.
10. Glueck, C. J., Mattson, F. H., and Jandacek, R., The lowering of plasma cholesterol by sucrose polyester in subjects consuming diets with 800, 300, or less than 50 mg of cholesterol per day, *Am. J. Clin. Nutr.*, 37, 347–354, 1979.
11. Glueck, C. J., Jandacek, R., Hogg, E., Allen, C., Baehler, L., and Tewksbury, M., Sucrose polyester: substitution for dietary fats in hypocaloric diets in the treatment of familial hypercholesterolemia, *Am. J. Clin. Nutr.*, 37, 347–354, 1983.
12. Mellies, M. J., Jandacek, R. J., Taulbee, J. D., Tewksbury, M. B., Lamkin, G., Baehler, L., King, P., Boggs, D., Goldman, S., Gouge, A., Tsang, R., and Glueck, C. J., A double-blind placebo-controlled study of sucrose polyester in hypocholesterolemic outpatients, *Am. J. Clin. Nutr.*, 37, 339–346, 1983.
13. Roberts, R. J. and Leff, R. D., The influence of absorbable and nonabsorbable lipids and lipid-like substances on drug bioavailability, *Clin. Pharmacol. Ther.*, 45, 299–304, 1989.
14. Owen, A. L., The impact of future foods on nutrition and health, *J. Am. Diet. Assoc.*, 90, 1217–1222, 1990.
15. Munro, I. C., Issues to be considered in the safety evaluation of fat substitutes, *Food Chem. Toxicol.*, 28, 751–753, 1990.
16. Beauchamp, G. K. and Cowart, B. J., Development of sweet taste, in *Sweetness*, Dobbing, J., Ed., London, Springer Verlag, 127–140, 1987.
17. Frank, G. C., Webber, L. S., Farris, R. P., and Berenson, G. S., Eds., *Dietary data book: quantifying dietary intakes of infants, children, and adolescents—the Bogalusa heart study, 1973–1983*, From the Nutrition Core and Planning and Analysis Core Componenets of the National Research and Demonstration Center—Artheriosclerosis (NRDC-A). New Orleans, LA, Louisiana State University Medical Center, 1986.

18. USDHHS, *The Surgeon General's Report on Nutrition and Health,* Washington, D.C., Government Printing Office, 1988, [DHHS publication (PHS) 88 50210.]
19. Glinsmann, W. H. and Dennis, D. A., Regulation of nonnutritive sweeteners and other sugar substitutes, in *Sugars and sweeteners,* Kretchmer, N. and Hollenbeck, C. B., Eds., Boca Raton, CRC Press, 1991, 257–282.
20. Leiberman, H. R., Sugars and behaviors, *Clin. Nutr.,* 4, 195–199, 1985.
21. Oser, B. L., The saga of man-made sweeteners, in *Sweeteners: Health Effects,* Williams, G. L., Ed., Princeton, Princeton Scientific Publishing Co., 1988, 25–38.
22. Miller, S. A. and Fratali, V. P., Saccharin, in *Sugars and sweeteners,* Kretchmer, N. and Hollenbeck, C. B., Eds., Boca Raton, CRC Press, 1991, 245–255.
23. National Academy of Sciences/National Research Council, *Saccharin: technical assessment of risks and benefits,* 1978, Report No. 1, Committee for a Study on Saccharin and Food Safety Policy.
24. American Dietetic Association, Appropriate use of nutritive and nonnutritive sweeteners, *J. Am. Diet. Assoc.,* 87, 1690–1694, 1987.
25. Leon, A. S., Hunninghake, D. B., Bell, C., Rassin, D. K., and Tephly, T. R., Safety of long-term large doses of aspartame, *Arch. Intern. Med.,* 149, 2318–2324, 1989.
26. Stegnik, L. D. and Filer, J. R., Eds., *Aspartame-Physiology and Biochemistry,* New York, Marcel Dekker, Inc., 1984, 29–160.
27. Kruesi, M. J. P. and Rapoport, J. L., Aspartame and children's behavior, in *Sweeteners: Health Effects,* Williams, G. M., Ed., Princeton, Princeton Scientific Publishing Co., 1988, 173–178.
28. Bradstock, M. K., Serdula, M. K., Marks, J. S., Barnard, R. J., Carne, N. T., Remmington, P. L., and Trowbridge, F. L., Evaluation of reactions to food additives: the aspartame experience, *Am. J. Clin. Nutr.,* 43, 464–469, 1986.
29. Department of Health and Human Services, *Quarterly report on adverse reactions associated with aspartame ingestion,* 1986.
30. Sturtevant, F. M., Use of aspartame in pregnancy, *Intl. J. Fertil.,* 30, 85–87, 1985.
31. Stegink, L. D., Filer, L. J., and Baker, G. L., Plasma erythocyte and human milk levels of free amino acids in lactating women administered aspartame or lactose, *J. Nutr.,* 109, 2173–2179, 1979.
32. Aspartame: ruling on objectives and notice of hearings before a public board of inquiry, *Fed. Reg.,* 44, 317, 1979.
33. Council of Scientific Affairs, Aspartame: review of safety issues, *JAMA,* 254, 400–402, 1985.
34. Guesry, P. R. and Secretin, M. C., Sugars and nonnutritive sweeteners, in *Nestle Nutrition Workshop,* series vol. 25, Kretchmer, N., Gracey, M., and Rossi, E., Eds., New York, Vevey/Raven Press, 1991, 33–54.
35. Kretchmer, N. and Hollenbeck, C. B., Eds., *Sugars and Sweeteners,* Boca Raton, CRC Press, 1991, vii–297.
36. Fernstrom, J. D., Central nervous system effects of aspartame, in *Sugars and Sweeteners,* Kretchmer, N. and Hollenbeck, C. B., Eds., Boca Raton, CRC Press, 1991, 151–173.
37. Bertorelli, A. M., Czarnowski, and Hill, J. V., Review of present and future use of nonnutritive sweeteners, *Diab. Edu.,* 16, 415–420, 1991.
38. Collings, A. J., Metabolism of cyclamate and its conversion to cyclohexylamine, in *Sugars and Sweeteners,* Kretchmer, N. and Hollenbeck, C. B., Eds., Boca Raton, CRC Press, 1991, 217–227.
39. Bopp, B. A., Sonders, R. C., and Kesterton, J. W., Toxicological aspects of cyclamate and cyclohexylamine, *CRC Crit. Rev. Toxicol.,* 16, 213–218, 1986.
40. Committee on the Evaluation of Cyclamate and Carcinogenicity: evaluation of cyclamate for carcinogenicity, Washington, D.C., National Academy of Science, National Resource Council, 1985.
41. Price, J. M., Blava, B., Oster, B. L., Steinfeld, K., and Ley, H. L., Bladder tumors in rats fed cyclohexylamine or high doses of a mixture of cyclamate and saccharin, *Science,* 1967, 1131–1139, 1970.

42. *Alitame: a New High-Intensity Sweetener,* Groton, Pfizer Central Research, 1987.
43. *Sucralose: Fact Sheet.,* Skillman, McNeil Specialty Products Co., 1990.
44. Reiser, S., Bickard, M. C., Hallfrisch, J., Michaelis, O. E., and Prather, E. D., Blood lipids and their distribution in lipoproteins in hyperinsulinemic subjects fed three different levels of sucrose, *J. Nutr.,* 111, 1045–1052, 1981.
45. American Diabetes Association, *The physician's guide to type II diabetes (NIDDM): diagnosis and treatment,* New York, American Diabetes Association, 1984.
46. Glinsmann, W. H., Irausquin, H., and Park, Y. K., Evaluation of health aspects of sugars contained in carbohydrate sweeteners, Reports of Sugars Task Force, Washington, D.C., U.S. Food and Drug Administration, 1986.
47. Coulston, A. M., Hollenbeck, C. B., Donner, C. C., Williams, R., Choiu, Y.-A. M., and Reaven, G. M., Metabolic effects of added dietary sucrose in individuals with non-insulin-dependent diabetes mellitus (NIDDM), *Met,* 34, 962–969, 1985.
48. Coulston, A. M., Hollenbeck, C. B., Swislocki, A. L. M., Chen, Y.-D. I., and Reaven, G. M., Deleterious metabolic effects of high carbohydrate sucrose containing diets in patients with NIDDM, *Am. J. Med.,* 84, 213–221, 1987.
49. American Diabetes Assocation, Nutritional recommendations and principles for individual with diabetes mellitus, *Diab. Care,* 10, 126–132, 1987.
50. Niewoehner, C. B., Use of sweeteners in the diabetes diet, in *Sweeteners: health effects,* Williams, G. M., Ed., Princeton, Princeton Scientific Publishing Co., Inc., 1988, 95–106.
51. Bantle, J. P., Laine, D. C., Castle, G. W., Thomas, J. W., Honogwerf, B. J., and Geotz, P. C., Postprandial glucose and insulin responses to meals containing different carbohydrates in normal and diabetic subjects, *New Engl. J. Med.,* 309, 7–12, 1983.
52. Slama, G., Jean-Joseph, P., Gicolea, I., Elgrable, F., Haardt, M. J., Costagliola, D., Bornet, F., and Tchobroutsky, G., Sucrose taken during a mixed meal has no additional hyperglycemic action over isocaloric amounts of starch in well controlled diabetes, *Lancet,* 2, 122–127, 1984.
53. Newbrun, E., Dental effects of sugars and sweeteners, in *Sugars and sweeteners,* Kretchmer, N. and Hollenbeck, C. B., Eds., Boca Raton, CRC Press, 1991, 175–202.
54. Ikeda, T., Sugar substitutes: reasons and indications for their use, *Int. Dent. J.,* 32, 33–41, 1982.
55. Brown, A. T., Breeding, L. C., and Grantham, W. C., Interaction of saccharin and acesulfame with Streptococcus mutans, *J. Dent. Res.,* 61, 191–198, 1982.
56. Coronary artery disease prevention: cholesterol, a pediatric perspective, in *Preventive Medicine,* Wynder, E. L., Ed., 18, 323–409, 1989.
57. U.S. Department of Agriculture, Nationwide food consumption survey, continuing survey of food intake by individuals, U.S. Department of Agriculture Report 85–1, Hyattsville, Md., Human Nutrition Information Service, 1985.
58. Frank, G. C., Farris, R. P., Cresanta, J. L., and Nicklas, T. A., Perspectives on cardiovascular risk in early life, in *Causation of Cardiovascular Risk Factors in Children,* Berenson, G. S., Ed., New York, Raven Press, 1986, 254–291.
59. Bergholz, C. M., Olestra and the potential role of a nonabsorbable lipid in the diets of children, *Ann. N.Y. Acad. Sci.,* 623, 356–368, 1991.
60. Block, G., Dresser, C. M., Hartman, A. M., and Carroll, M. D., Nutrient sources in the American diet: quantitative data from the NHANES II survey. II. Macronutrients and fats, *Am. J. Epidemiol.,* 122, 27–40, 1985.
61. Committee on Diet and Health, Food Nutrition Board, Commission on Life Sciences, National Research Council, *Diet and health, implications for reducing chronic disease risk,* Washington, D.C., National Academy Press, 1989.
62. Cresanta, J. L., Farris, R. P., Croft, J. B., Webber, L. S., Frank, G. C., and Berenson, G. S., Trends in fatty acid intakes of 10-year-old children, *J. Am. Diet. Assoc.,* 88, 1973–1982, 1988.
63. Garcia, R. E. and Moodie, D. S., Routine cholesterol surveillance in childhood, *Pediatrics,* 88, 751–755, 1989.
64. Lauer, R. M., Lee, J., and Clarke, W. R., Factors affecting the relationship between childhood and adult cholesterol levels: the Muscatine study, *Pediatrics,* 82, 309–318, 1988.

65. Foltin, R. W., Fischman, M. W., Moran, T. H., Rolls, B. J., and Kelly, T. H., Caloric compensation for lunches varying in fat and carbohydrate content by humans in a residential laboratory, *Am. J. Clin. Nutr.,* 52, 969–980, 1990.
66. Rolls, B. J., Pirraglia, P., Jones, M., and Peters, J., Effects of covert fat replacement with olestera on 24-hour food intake in lean adults, *FASEB,* 5, 2, A1077, 1991.
67. Mellies, M. J., Vitale, C., Jandacek, R. J., Lamkin, G. E., and Glueck, C. J., The substitution of sucrose polyester for dietary fat in obese, hypercholesterolemic outpatients, *Am. J. Clin. Nutr.,* 41, 1–12, 1985.
68. Rogers, P. J. and Blundell, J. E., Evaluation of the influence of intense sweeteners on the short-term control of appetite and caloric intake: a psychobiological approach, in *Progress in Sweeteners,* Grenby, T. H., Ed., London, England, Elsevier Applied Science, 1989, 267–289.
69. Rolls, B. J., Effects of intense sweeteners on hunger, food intake, and body weight: a review, *Am. J. Clin. Nutr.,* 53, 872–878, 1991.

Chapter 19

THE ECOLOGY OF POVERTY, UNDERNUTRITION, AND LEARNING FAILURE

Robert J. Karp

TABLE OF CONTENTS

I. Introduction 239

II. Eugenics and "Hereditary Feeble-Mindedness" 240

III. Eugenics and Racist Theories 241

IV. Malnutrition and Mental Development 242

V. Lead Poisoning 245

VI. Iron Deficiency 245

VII. Alcohol 246

VIII. Making the Connections 246

IX. Summary 248

References 248

I. INTRODUCTION

Why is it that poverty and ignorance in one generation recur in the next and in succeeding generations? The hypothesis presented here is that nutrition-related disorders including iron deficiency (Lozoff, 1990; Walter et al., 1989), lead poisoning (Needleman, 1988), growth retardation (Winick, 1993; Karp et al., 1992) and, of most recent concern, *in utero* exposure to alcohol (Streissguth et al., 1991) have contributed to trans-generational deprivation and antisocial behavior within the observed families (Cravioto & Delicardie, 1972; Hallberg, 1989; Sewell et al., 1993).

The problem, as noted by Hallberg (1989), is that malnutrition among the poor in developed countries occurs in social environments which affect nutritional status and learning ability synergistically. Rather than continue the "nature *vs.*

0-8493-2764-4/95

Figure 1. Pauline "Kallikak." (From Karp, R. J., *Malnourished Children in the United States: Caught in the Cycle of Poverty*, Karp, R. J., Ed., Springer Publishing Co., Inc., New York 10012, 1993. With permission.)

nurture" debates of recent years (Gould, 1984; Kevles, 1985; Myers et al., 1991), we know how disorders with developmental consequences interact with the social environment to produce transgenerational poverty—a "nature *cross* nurture" model.

II. EUGENICS AND "HEREDITARY FEEBLE-MINDEDNESS"

Though now universally rejected, until recently, eugenic theories were taken quite seriously (Rosenberg, 1976; Kevles, 1985). Into the late 1950s, in Philadelphia, children with mild mental retardation were placed in classes for the "Orthogenetically Backward (O.B.)" and classified as "morons, imbeciles, and idiots" according to their developmental age. Segregation rather than education was the goal of the educational system.

At the height of the eugenics movement in the United States, at the turn of the 20th century—two sets of families, the Jukes and the Kallikaks, were used to "document" the hereditary nature of mental retardation, poverty, and antisocial behavior (Dugdale, 1877; Goddard, 1912 & 1914; Kevles, 1985). Although they are very different studies leading to very different conclusions, the Jukes and the Kallikaks are always cited together.

In 1877, Richard Dugdale published his studies of "crime, pauperism, disease, and heredity" among the inhabitants of New York state prisons, many of whom were related and to whom he gave the name "Jukes." ". . . [I]n idiocy and insanity . . . ," Dugdale notes (1877), "heredity is the preponderating factor in determining career; but it is, even then, capable of marked modification for better or worse by the character of the environment (65) . . . The essential characteristics of [the Jukes] are great vitality, ignorance, and poverty. They have never had a training which would bring into activity the aesthetic tastes, the habits of reasoning, or indeed a desire for the ordinary comforts of a well-ordered home" (66).

By contrast, Henry Goddard, director of the Vineland Training School for Feeble-minded Boys and Girls in New Jersey, reported, ". . . no amount of education or of good environment can change a feeble-minded individual into a normal one (1913)" and "[S]ince feeble-mindedness is in all probability transmitted in accordance with the Mendelian Law of heredity, the way is open for eugenic procedure which will mean much for the future welfare of the race" (1914, 590).

Though the eugenicists were bigoted in their social outlook and biased in their science, it is likely that they made accurate observations: feeble-minded Kallikak parents produced feeble-minded Kallikak children. It is well recognized now that Goddard was seeing the cumulative effects of poverty and problems with parent-child interaction (Kevles, 1985). To these can be added the consequences of one or more nutrition-related disorders on the development of the children.

A review of Goddard's writings and investigation of his original data at Vineland confirm this hypothesis. Data are available to show that alcohol exposure *in utero*—the Fetal Alcohol Syndrome (FAS)—plus the consequence of living in the home of alcoholics (defined by Goddard as "drunkards") were associated with the syndrome of "hereditary feeble-mindedness" (Karp et al., 1993; Karp, 1993).

These data were presented by Goddard in 1914 with the explanation ". . . one might say without fear of dispute that more people are alcoholic because they are feeble-minded than vice-versa" (1914, 493). Such reasoning provided a satisfactory explanation for the association between alcoholism and learning failure for several generations of investigators and, in part, for why the fetal alcohol syndrome went unrecognized for so long.

III. EUGENICS AND RACIST THEORIES

Though the Jukes and Kallikaks were of European ancestry, the same causal theories are maintained currently to explain black/white differences in performance on Stanford-Binet and other tests used to assess learning ability (Sewell et al., 1993). It may be comforting, in a peculiar way, to maintain discredited racial theories, because they explain and justify societal or personal inaction. However, the general success of interventions that are given time to work provides the best

TABLE 1[a]

Cumulative Data from the Vineland School for Feeble-minded Boys and Girls as Assembled in 1911 and Reanalyzed (Karp et al., 1993)

	OFFSPRING	
PARENTS	**Feebleminded**	**Normal**
"Not Drunkard"	1085	1269
"Drunkard"	388	127
Chi square = 151; p<0.001		

Note: Our data show that mentally retarded offspring of non-drunkard parents were more likely to have normal siblings than the mentally retarded children of drunkard parents; note the few normal siblings in families of alcoholics.

	OFFSPRING	
PARENTS	**Feebleminded**	**Died in Infancy**
"Not Drunkard"	1085	393
"Drunkard"	388	216
Chi square = 29.5; p<0.001		

Note: Our data show that mentally retarded offspring of non-drunkard parents were less likely to have siblings who died in infancy than the mentally retarded children of drunkard parents; note the many deaths in infancy in families of alcoholics.

[a] Data derived from Goddard, H. H., *Feeblemindedness: Its Causes and Consequences,* Macmillan, New York, 1914, 474–492.

challenge to the validity of these theories (Gould, 1984; Kevles, 1985; Sewell et al., 1993). Moreover, in their writings the eugenicists make an artificial distinction between theory (recognizing differences in aptitude or performance which coincide with current social class and racial hierarchies) and practice (assigning opportunity on the basis of social class or race/ethnicity). Once eugenic theories are accepted, intent becomes irrelevant. The consequences—institutionalization, sterilization and extermination—are inevitable.

IV. MALNUTRITION AND MENTAL DEVELOPMENT

In the United States, children from the poorest and least educated families consistently perform poorly on measures of cognitive ability. Many of these children are malnourished. The nutritionist's problem has been an over-simplified definition of causation with the expectation that poverty leads directly to learning failure by altering "cell growth" or "CNS function" of the child (Dobbing, 1990; Johnston & Markowitz, 1993). The data are there to construct the argument that malnutrition causes learning failure, but the construct is far more complex than these simple theories allow.

Another concern has been the influence of racism (as opposed to race, per se). Why, asks Jonathan Kozol (1991), are twice as many white children than black

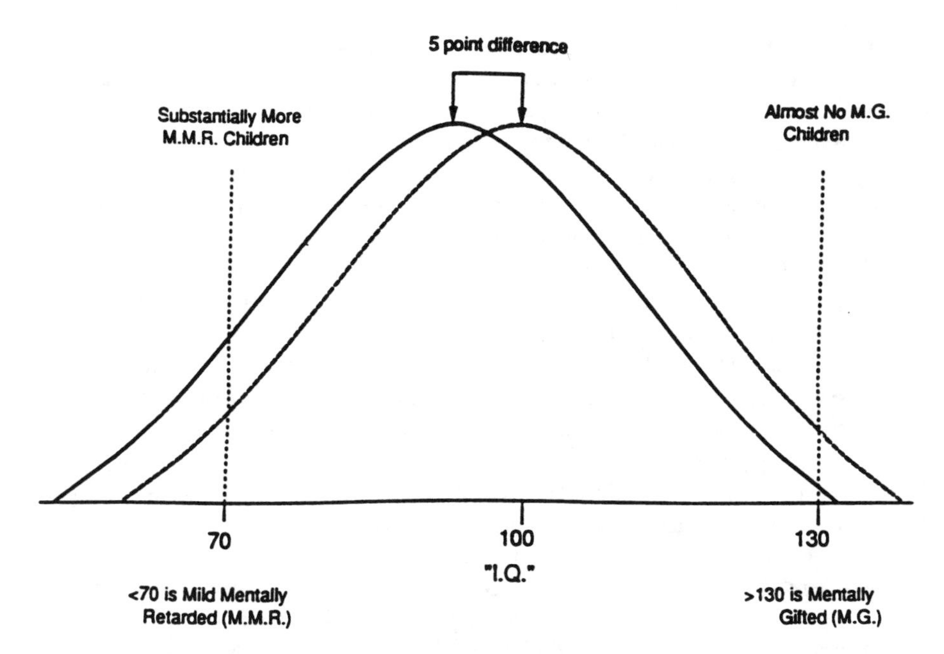

Figure 2. Effects of low-level lead exposure on the distribution of IQ scores in a community (From Needleman, et al. 1990). A downward shift in mean IQ score of 5 points may have little effect on the life of an individual child, but the effect on the community is essentially to empty the Mentally Gifted classes (M.G.) where children have an IQ two standard deviations above the mean (>130). The small change at the mean causes the classes for children who are mildly Mentally Retarded (M.R.)—IQ two standard deviations below the mean (<70)—to become full to capacity. (From Harris, P., Clarke, M., and Karp, R. J., *Malnourished Children in the United States: Caught in the Cycle of Poverty,* Karp, R. J., Ed., Springer Publishing Co., Inc., New York 10012, 1993. With permission.)

in "mentally gifted" classes and why does the reverse hold true for assignment to classes for children with mild mental retardation? One answer to Kozol's question is that the nutritional factors noted above—lead poisoning, iron deficiency, undernutrition at critical ages, and the influence of alcohol on the fetus—would have exactly this effect. Race then is a shorthand notation for socioeconomic differences (Williams, 1987) and an additional contribution to the "cluster effects of poverty" (Sameroff et al., 1987; Hertzig, 1992). It is my belief that racism shapes public policy (e.g., the failure to address the housing, nutrition, and education needs of the poor) and thus affects school achievement and future success in life for the poor of all racial groups.

It is tempting to conclude that retarded mental development is caused by the permanent cellular changes induced by early undernutrition. But, as noted above, malnutrition is not an isolated condition in the lives of affected children. Outside cycles of disadvantage, the "brain growth theory" has proved wanting (Winick, 1993). The concept is not without merit, however, as two recent studies (Galler et al., 1984; Lucas et al., 1990) have shown that there are residual effects if early malnutrition is not addressed aggressively. Temperament and activity level are affected by caloric undernutrition, and prevention of growth retardation by careful screening and the provision of supplements can prevent associated learning failure.

With respect to school-age children, no correlation has been found between anthropometric measures (the nutritional variable) and development, aptitude, and school achievement among well-nourished population of middle-class children (Pollit et al., 1982). These observations do not carry over to disadvantaged populations. Among the inner-city poor, nutritional status, including anthropometrics, correlates closely with borderline or deficient cognitive and behavioral development (Hepner & Maiden, 1971; Meyers et al., 1991).

In Philadelphia, school-age children with decreased height for age were found to be older than their classmates, reflecting delayed entry into school (Karp et al., 1976). These undernourished children appeared to be of "normal" size and performance by grade (but not by age) when compared to their younger classmates. Moreover, the parents displayed ineffective parenting behavior likely to diminish both school performance and nutritional status (Karp et al., 1984).

Karp et al. (1992) looked at five measures of nutritional status (including height and weight), three measures of cognitive function and development (including a standardized test of Visual Motor Integration (VMI)), and a composite measure of school achievement in the children attending this same Philadelphia school. Non-verbal skills were chosen for testing as an attempt to limit the effect of cultural experiences on the measurement of neurodevelopment.

The data demonstrated that all anthropometric measures were significantly related to achievement scores, but the anthropometric variables correlated better with academic achievement than with measures of cognitive ability or visual-motor-perceptual ability. This suggests that common antecedents of undernutrition and learning failure affect school achievement in disadvantaged children (Hallberg, 1989). Problems in the micro-environment of the child precede both growth and learning failure (see Hallberg, 1989; Wachs, 1993; Sameroff et al., 1987).

This study suggests that for undernourished children, growth, nonverbal aspects of neurodevelopment, and learning are all on a "slow track", reflecting a response to an environment that may be stressful, nonstimulating (Sameroff et al., 1987), or frankly abusive (Karp et al., 1989).

It is noteworthy that the anthropometric variables did not relate significantly to non-verbal measures of cognitive ability. This seems to indicate that the protein-energy nutritional status of children who were at the low end of the anthropometric scale was not significant enough to cause damage to basic cognitive processes. This provides explanation, in part, as to why undernourished children adopted into homes where they are well nourished and nurtured achieve in school even when growth is slow (Winick et al., 1975) and why stimulation programs aid in the development of low and normal weight infants in families with limited parental education (Brooks-Dunn, 1992). Growth was related, however, to the measure of visual-motor-perceptual ability which does require some enrichment of the environment, suggesting that the performance ability of these children may have been affected.

V. LEAD POISONING

Lead poisoning and iron deficiency each contribute independently to learning failure among children (Needleman, 1988). However, when these conditions coexist, children are four times more likely to show symptoms usually attributed to lead poisoning alone (Clark et al., 1988). The two curves shown in Figure 2 represent the distribution of "IQs" of groups of children with or without appreciable lead exposure. IQ in this example could represent any of a variety of measures of psychosocial or motor development. As shown by Herbert Needleman, low lead exposure with a serum lead between 25 and 40 pg/dl is associated with a 5 point drop in the mean IQ, accounting for numerous confounding social variables. It is likely that the association of malnutrition and lead poisoning magnifies the problems of lead poisoning alone (Harris et al., 1993; Sewell et al., 1993).

VI. IRON DEFICIENCY

An identical set of "IQ" curves shown in Figure 2 could be drawn for modest reductions in hemoglobin level. In the past, learning failure was ascribed to deficiency of storage iron in the extrapyramidal regions of the brain (Lozoff, 1990). It is known from rat studies that "brain iron accumulates from gestation to early adulthood, and brain iron levels are more seriously affected by iron deficiency in the very young animal than in the adult . . . rats who sustained early iron deficiency also have persisting behavioral and learning deficits . . . although these lasting changes have yet to be explained, it is possible that they relate to altered CNS neurotransmission. Thus far, in iron-deficient animals, alterations have been found in the dopamine, serotonin, and GABA (gamma-aminobuteryric acid) systems" (Lozoff, written communication).

But the focus on neurotransmitter activity in the central nervous system, however elegant it may be, seems less critical than the effect of iron deficiency on hemoglobin levels. For as the work of Lozoff (1990) and Walter (1989) have shown, during the phases of iron depletion (decreased ferritin level) and iron deficiency without anemia (decreased serum iron and percent saturation of transferrin associated with elevated red cell porphyrins), no effects on learning have been found. In this last phase of iron deficiency, with the presence of anemia, the ability of the infant and child to sustain motor activity seems to be limited. Since children learn through exploration and repetition, learning is affected. The two important points here are (1) temperament and activity level are affected by iron deficiency, and (2) prevention of iron deficiency anemia by careful screening and the provision of supplements can prevent the associated learning failure.

VII. ALCOHOL

The effects of alcohol exposure *in utero* vary from the grotesque (25% of children born to Plains Indian Tribes have the full Fetal Alcohol Syndrome (FAS)) (May et al., 1988) through the modest (about 10% of white children in a poor industrial suburb of Philadelphia show Fetal Alcohol Effects (FAE)) (Marino et al., 1987). These syndromes are just as likely to occur in affluent communities with a disproportionate availability of special schools for dyslectic children whose families simply drink too much. As May et al. (1988) shows, alcohol use crosses all social class and racial lines.

With respect to neurodevelopmental outcome, children exposed to alcohol show multiple long-term disabilities. The more extreme the dysmorphology, the greater the disability. These children have impaired cognitive functioning. Generally, those with FAS are mentally retarded in the mild to moderate range (Streissguth et al, 1991), with an average IQ of 68. They also show concomitant motor and language deficits. These children often suffer from attentional difficulties with hyperactivity and have serious behavioral problems. Their attentional problems make learning all the more difficult, further impairing the acquisition of new information.

The long-term development of exposed children is of great concern. Shaywitz and Shaywitz write, "No one yet can determine the effect of prenatal alcohol consumption on the occurrence of attention deficit disorder and other conditions affecting learning" (Shaywitz & Shaywitz, 1983). This description of uncoordinated children matches observations by Marino et al., (1987) who found that 33% of the children in the special education classes of an American industrial town (approximately 10% of the children in the school) had the traits of FAS and associated clumsiness.

FAS affects temperament, gross and fine motor development, and perception of self (Kylermann et al., 1985). Swedish investigators discovered varying degrees of motor dysfunction during testing of FAS children who experienced tremor, balance difficulties, and low muscle tone, all indicative of motor pathology rather than developmental delay. Furthermore, these impairments appeared more pronounced in those with the morphological stigmata of FAS. These findings taken as a whole "support the view that prenatal damage is the major etiologic factor for growth restriction and lower motor skills in . . . [children of alcoholic mothers]" (Kyllerman, 1985).

VIII. MAKING THE CONNECTIONS

The connections which link POVERTY → LEARNING FAILURE → and continued POVERTY require attention. In the current era, no one would make a

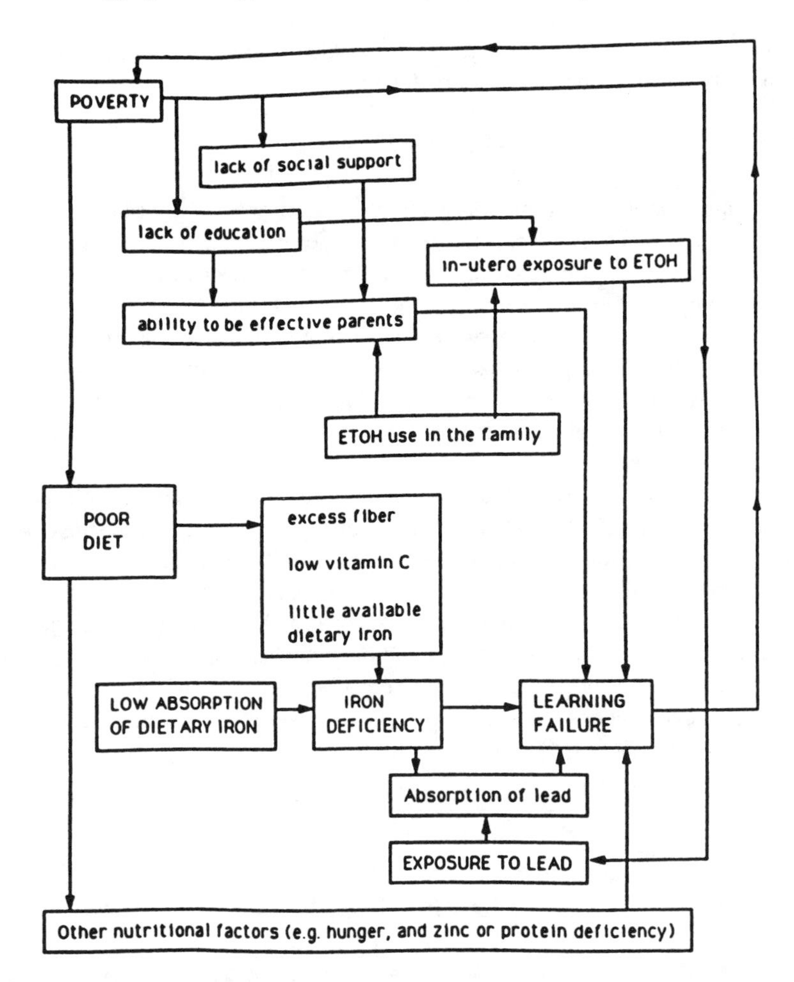

Figure 3. The ecology of poverty, undernutrition, and learning failure. (Derived in part from the works of Cravioto, J. and De Licardie, G., 1972, and Hallberg, L., 1989. From Sewell, T., Price, V., and Karp, R. J., *Malnourished Children in the United States: Caught in the Cycle of Poverty,* Karp, R. J., Ed., Springer Publishing Co., Inc., New York 10012, 1993. With permission.)

direct connection between poverty and learning failure. This would reflect badly on the many highly accomplished people from poverty backgrounds and would be based on the discredited eugenic theories described above.

However, it is possible to make the connection using social variables which interfere with parent-child interaction. This theme dominates lead industry testimony on lead poisoning. Note alcohol consumption is separated from poverty. Alcoholism is a ubiquitous problem in our society which crosses all social and economic lines.

It is equally possible to make the connection using biologic variables. Lead poisoning, a direct consequence of poverty, is not under the control of the poor parent.

One must put these pieces together to consider the complexity of the ecology of poverty, malnutrition, and learning failure.

IX. SUMMARY

Malnutrition in its various forms is not a single or even a leading cause of these "cycles of disadvantage" (Rutter & Madge, 1987; Karp, 1990). The prevalence of undernutrition in young children and the consequences are related so closely to chronic poverty that one can only rarely distinguish the consequences of these disorders from the consequence of living in the milieu in which undernutrition occurs. No single "cause" for learning failure among the poor—genetic potential, various forms of undernutrition or malnutrition, or laying responsibility on the shoulders of mothers for their parenting—provides an adequate explanation when taken alone.

The origins of cycles of disadvantage have no simple or single explanation, nor would feeding children, alone, suffice to disrupt them. There is no single path out of the cycles. But as Schorr writes, solutions are "within our reach" (1988). The results of policies that nurture and nourish children include remarkable improvements in health and social stability. If society and the individual can produce the cycle of poverty, then society and the individual can succeed in terminating it. The involvement of pediatricians and other health care providers is critical to successful interventions to disrupt the cycle of poverty (Chamberlin, 1992; Sia, 1992; Karp, 1993).

REFERENCES

Aronson, M., Kyllerman, M., Sabel, K. G., et al., Children of alcoholic mothers: Developmental, perceptual, and behavioral characteristics as compared to matched controls, *Acta. Pediatr. Scand.*, 74, 27–35, 1985.

Chamberlin, R. W., Preventing low birth weight, child abuse, and school failure: The need for comprehensive, community-wide approaches, *Pediatr. Rev.*, 13, 64–71, 1992.

Clark, M., Royal, J., and Seeler, R., Interaction of iron deficiency and lead and the hematologic findings in children with severe lead poisoning, *Pediatrics*, 81, 247–254, 1988.

Cravioto, J. and DeLicardie, G., Environmental characteristics of severe clinical malnutrition and language development in survivors from kwashiorkor and marasmus, in *Nutrition, The Nervous System and Behavior*, PAHO Scientific Publication No. 251, Washington, D.C., 1972.

Dugdale, R., *The Jukes: A Study in Crime, Pauperism, Disease, and Heredity*, GP Putnam and Co., New York, 1876, 1st Ed., and 1910, 4th Ed.

Galler, J. R., et al., The influence of early malnutrition on subsequent behavioral development. III. Learning disabilities as a sequel to malnutrition, *Pediatr. Res.*, 18, 309–313, 1984.

Goddard, H. H., *The Kallikak Family: A Study in Feeble-mindedness*, Macmillan, New York, 1912.

Goddard, H. H., One man's wild oats, *Scientific Temperance Journal*, January, 1913, 51–52.

Goddard, H. H., *Feeble-mindedness: Its Causes and Consequences*, Macmillan, New York, 1914, 474–492.

Gould, S. J., Human equality is a contingent fact of history, *Natural History,* 93, 26–28, 1984.

Hallberg, L., Search for nutritional confounding factors in the relationship between iron deficiency and brain function, *Am. J. Clin. Nutr.,* 50, 598–606, 1989.

Harris, P., Clarke, M., and Karp., R. J., The prevention of lead poisoning, in *Malnourished Children in the United States: Caught in the Cycle of Poverty,* Karp, R. J., Ed., Springer Publishing Co., New York, 1993.

Hepner, R. and Maiden, N. C., Growth rate, nutrient intake and "mothering" as determinants of malnutrition in disadvantaged children, *Nutr. Rev.,* 29, 219–223, 1971.

Hertzig, M. E., Mental health and developmental problems of children in poverty, *Bull. NY Acad. Med.,* 68, 25–31, 1992.

Johnston, F. E. and Markowitz, D., Do poverty and malnutrition affect children's growth and development: Are the data there? in *Malnourished Children in the United States: Caught in the Cycle of Poverty,* Karp, R. J., Ed., Springer Publishing Co., New York, 1993.

Karp, R. J., Nuchpakdee, M., Fairorth, J., and Gorman, J. M., The school health service as a means of entry into the inner-city family for the identification of malnourished children, *Am. J. Clin. Nutr.* 29, 216–218, 1976.

Karp, R. J., Snider, E., Fairorth, J. W., et al., Parental behavior and the availability of foods among undernourished inner-city school children, *J. Fam. Prac.,* 18, 731–735, 1984.

Karp, R. J., Scholl, T. O., Decker, E., and Ebert, E., Growth of abused children contrasted with the non-abused in an urban poor community, *Clin. Pediatr.,* 28, 317–320, 1989.

Karp, R. J., Martin, R., Sewell, T., et al., The relationship of height, weight, cognitive ability and neurodevelopmental level to academic achievement in inner-city kindergarten children, *Clin. Pediatr.,* 32, 336–340, 1992.

Karp, R. J., Introduction: An overview of malnourished children in the United States, in *Malnourished Children in the United States: Caught in the Cycle of Poverty,* Karp, R. J., Ed., Springer Publishing Co., New York, 1993.

Karp, R. J., What the health care provider can do to break the cycle of poverty, in *Malnourished Children in the United States: Caught in the Cycle of Poverty,* Karp, R. J., Ed., Springer Publishing Co., New York, 1993.

Kevles, D., *In the Name of Eugenics: Genetics and the Uses of Human Heredity,* Alfred A. Knopf, New York, 1985.

Kozol, J., *Savage Inequalities: Children in American Schools,* Crown Publishers, New York, 1991.

Kyllerman, M., Aronson, M., Sabel, K. G., et al., Children of alcoholic mothers: Growth and motor performance compared to matched controls, *Acta Pediatr. Scand.,* 74, 20–26, 1985.

Lozoff, B., Has iron deficiency been shown to cause altered behavior in infants? in *Brain, Behavior, Iron in the Infant Diet,* Dobbing, J., Ed., London, Springer Verlag, 1990, 107–131.

Lucas, A., Morley, R., Cole, T. J., et al., Early diet in preterm babies and developmental status at 18 months, *Lancet,* 335, 1477–1481, 1990.

Marino, R., Scholl, T. O., and Karp, R. J., Minor physical anomalies and learning disabilities: Is there a prenatal component? *J. Nat. Med. Assoc.,* 79, 37–39, 1987.

May, P. A., Hymbaugh, K. J., Aase, J. M., et al., Epidemiology of fetal alcohol syndrome among American Indians of the Southwest, *Soc. Biol.,* 30, 374–385, 1988.

Myers, A. F., Sampson, A. E., and Weitzman, M., Nutrition and academic performance in school, *Clin. Appl. Nutr.* 1, 13–25, 1991.

Needleman, H., The persistent threat of lead: the medical and sociological issues, *Current Problems in Pediatrics,* 1988, December, 702–744.

Pollitt, E., Mueller, W., and Leibel, R. L., The relation of growth to cognition in a well-nourished population, *Child. Dev.,* 53, 1157–1163, 1982.

Richards, G. E., Marshall, R. N., and Kreuser, I. L., Effect of stature on school performance, *J. Pediatr.,* 106, 841–842, 1985.

Rosenberg, C., The bitter fruit: heredity, disease and social thought in the 19th century, *Perspectives in American History,* 8, 189–235, 1974.

Rutter, M. and Madge, N., *Cycles of Disadvantage—a review of the literature,* Heinemann, London, 1976.

Sameroff, A., Seiffer, R., Barocas, R., et al., Intelligence quotient scores of 4-year-old children, *Pediatrics,* 79, 343–350, 1987.

Schorr, L. B., *Within Our Reach: Breaking the Cycle of Disadvantage,* Anchor Press/Doubleday, New York, 1990.

Sewell, T., Price, V., and Karp, R. J., The ecology of poverty, undernutrition, and learning failure, in *Malnourished Children in the United States: Caught in the Cycle of Poverty,* Karp, R. J., Ed., Springer Publishing Co., New York, 1993.

Shaywitz, S. E. and Shaywitz, B. A., Biological influences in attentional deficit disorders, in *Developmental-Behavioral Pediatrics,* Levine, M. D., Carey, N. B., Crocker, A. C., et al., Eds., Philadelphia, W. B. Saunders, 1983.

Streissguth, A. P., Fetal alcohol syndrome: an overview and implications for patient management, in *Alcoholism—Development, Consequences, and Interpretations,* 3rd ed., Estes, N. J. and Heinemann, M. E., St. Louis, C. V. Mosby Co., 1986.

Streissguth, A. P., et al., Fetal alcohol syndrome in adolescents and adults, *JAMA,* 265, 1961, 1991.

Susser, M., The challenge of causality: human nutrition, brain development and mental performance, *Bull. N.Y. Acad. Med.,* 65(10), 1032–1049, 1989.

Wachs, T. D., Early experience and the development of disadvantaged children, in *Malnourished Children in the United States: Caught in the Cycle of Poverty,* Karp, R. J., Ed., Springer Publishing Co., New York, 1993.

Walter, T., deAndraca, I., Chadud, P., et al., Adverse effect of iron deficiency anemia on infant psychomotor development, *Pediatrics,* 84, 7–17, 1989.

Wilson, W. J., *The Truly Disadvantaged: The Underclass and the Inner-City,* Chicago, University of Chicago Press, 1987.

Winick, M., Meyer, K. K., and Harris, R. C., Malnutrition and environmental enrichment by early adoption, *Science,* 190, 1173, 1975.

Winick, M., Forward in *Malnourished Children in the United States: Caught in the Cycle of Poverty,* Karp, R. J., Ed., Springer Publishing Co., New York, 1993.

Chapter 20

NUTRITION EDUCATION IN MEDICAL SCHOOLS AND RESIDENCY TRAINING

Myron Winick

Nineteen ninety-four promises to be the year of "health reform" in the United States. Several plans exist, some of which propose radical changes in how we deliver health care, whereas others propose only modest changes in the existing system. Despite these vast differences, there is one thing that seems to have universal support: a new emphasis on health promotion and disease prevention. A major component of this new emphasis is acknowledged to be diet and nutrition. Specifically, nutrition has been mentioned by members of the president's task force, the secretary of Health and Human Services, members of Congress and representatives of both professional and consumer groups, as an area that needs more attention. For those of us who have worked hard to introduce nutrition into the educational process for physicians, this should be good news. And yet I fear that the forces within the medical education establishment resisting meaningful changes in what physicians are taught are so strong that prevention in general and nutrition in particular will not easily be assimilated into our present system. In fact, I am not sure that the present system of educating physicians will ever be able to emphasize prevention and nutrition to the extent necessary.

The subject of nutrition education in medical schools is not new. In 1985 a committee to study the state of nutrition in American medical schools and to recommend any needed improvements was organized by the National Research Council (NRC).[1] The report made the following recommendations:

1. Nutrition should be a *required* course in every medical school. (At that time about 25% of medical schools had such courses.)
2. A minimum of 25 hours would be necessary to include the basic principles of nutrition. (At that time only one or two medical schools had 25 hours or more of material specifically labeled as nutrition.)
3. The National Boards should create a mechanism by which basic nutritional knowledge could be tested. (At that time less than 3% of the questions had anything to do with nutrition. Even more distressing, the balance of questions was badly uneven. For example, 12% of the pediatric portion related to nutrition, whereas there was not a single question in the obstetrics portion relating to nutrition in three consecutive years reviewed. In addition, there were no questions related to nutrition and cancer, nutrition and the elderly, or enteral or parenteral nutrition.)

0-8493-2764-4/95

4. There should be a separate nutrition department in all medical schools. (At that time only one medical school had such a department, and only one or two schools had a free-standing institute or program.)

Since this report was published, a number of other groups, including a committee of the American Society for Clinical Nutrition have re-examined the status of nutrition education in medical schools.[1,2,3] No progress has been made! In 1992, the United States Congress instructed the Department of Health and Human Services to "describe the appropriate federal role to insure that students enrolled in the United States medical schools and physicians practicing in the United States have access to adequate training in the field of nutrition and its relation to human health". A panel of experts was convened in Washington to address this charge.

A background report on the state of nutrition education in United States medical schools was sent to the participants before the meeting. The report indicated that between 20% and 25% of medical schools had a required course in nutrition, only one school had a free-standing nutrition department, and no mechanism had been introduced by the National Boards to ensure the inclusion of questions on nutrition and health—still no progress since the NRC or the ASCN reports. The nutrition community has been talking to itself, and the medical educators have been conducting business as usual.

What is preventing nutrition from being included in the required curriculum for medical students? Recently, I have discussed what I believe to be the core of the problem[5] and the inability of the present hospital-based tertiary-care-oriented system to integrate primary care, preventive medicine, and nutrition into the curriculum. Medical education is no longer the full-time system that emerged after the Flexner report. Medical schools have become dominated by tertiary care and by modern technology. This had come about partly because of the perception that the frontiers of medicine lie in these directions and more recently because medical schools are financed to a larger and larger extent by the funds generated by providing high-tech tertiary care.

The Flexner report changed medical education from essentially an apprentice system employing volunteer faculty to a system in which the full-time faculty member conducting clinical research became dominant. There is no doubt that this change led to a revolution in medical research and better patient care, from the 1950s to the early 1980s, as federal funds poured into medical schools, and clinical research flourished, and thousands of new discoveries were made. The pharmaceutical industry grew from modest size to giant proportions. New companies were developed employing the new medical technologies first explored in medical schools. Some have described this period as the golden age of United States medicine. During this period the hospitals began to exert more and more control over medical education, first by controlling post-graduate training and then by invading the training of undergraduates. Rotating internships were abolished, not because straight internships were educationally superior, but because straight interns soon became more efficient in taking care of patients. As more and more medical schools made most of the fourth year elective, the hospitals put pressure on the students to take electives in the specialty that the student was entering. If you wanted a good residency, then you were expected to prepare in advance by taking special electives. Medical schools were no longer training future physicians; they were training future residents. Just as medical schools dictated pre-medical curricula, hospitals dictated the clinical phase of undergraduate medical education. Thus, the modern medical school trains residents for hospitals who, in turn, provide the bulk of tertiary care and the majority of clinical training for undergraduate medical students. However, it is clearly the wrong system for emphasizing health promotion, disease prevention, and nutrition. These areas must be learned primarily outside the hospital, in the community where residents rarely go.

Thus, the present-day dichotomy: an acknowledged need to train our future doctors in preventive medicine, nutrition, and other ways of promoting wellness and a system of medical education which, if left unchanged, is incapable of achieving these aims. In addition, the incentive to keep the system from changing is very strong because more and more medical education is financed through the private practice of medicine which, in turn, generates most of its funds from high-tech, hospital-based medicine.

How then can we teach our future physicians to incorporate health promotion, disease prevention, nutrition, and the whole concept of wellness into medical practice? I believe the time has come for a new Flexner report. A panel should be convened to examine medical education in light of the new realities. The panel must be made up of people of different interests and backgrounds. It should not be dominated by members of the medical education establishment, although they must be represented. The panel should be charged with finding ways to train medical students (and some residents) outside the hospital setting. And whatever ways are suggested must include a mechanism for financing. Within such a new training environment, the medical schools should be encouraged to develop innovative techniques for teaching nutrition and other aspects of preventive medicine. Only when this is accomplished can we supply incentives for students to enter primary care and expect these incentives to result in the creation of a physician able to integrate prevention, nutrition, and wellness into his or her practice. At the same time, these community-based teaching sites must become the centers for clinical research into prevention of disease, maintenance of good health, nutrition, gerontology, and a host of new areas that will soon become obvious. Only by establishing a firm scientific basis from which practical recommendations can be made will we know whether our recommendations are appropriate. What I am proposing is doing for primary care what we did so well for tertiary care, but doing it in a different setting with different kinds of patients—and most important, finding a stable way to finance such an effort.

Even if we started today, the kinds of changes I am proposing would take years before they are accomplished. What, if anything, can be done in the interim? I believe that while it may be possible to increase the number and extent of didactic nutrition courses required during the first two years, we will not be able to introduce nutrition in a meaningful way in the clinical years. Introducing nutrition and other disciplines related to prevention into the clinical years must await the changes I have outlined above. Therefore, I believe the focus should change to those residencies concerned with primary care: family medicine, pediatrics, general internal medicine, and perhaps obstetrics. Within these disciplines, we must seek ways to expose these residents to nutritional principles directly related to patient care. There are a number of ways in which this can be done, ranging from didactic courses built around patients to nutrition conferences at all outpatient clinics. We must try new and innovative approaches for each of the different specialties. Family medicine has already recognized the importance of nutrition by requiring an experience for all residents. The leaders of this young specialty are seeking ways to do this. We in the nutrition community can help. Perhaps, if we can train young physicians in this specialty in nutrition, it will serve as a catalyst for training physicians in the other primary-care specialties. Imagine an obstetrician who can prescribe a proper diet for a pregnant woman or an internist who can really put together a weight reduction program based on sound nutritional principles. That certainly would be a first step in the process of training every physician in preventative medicine and nutrition.

REFERENCES

1. Committee on Nutrition in Medical Education, Food and Nutrition Board, Council on Life Sciences, National Research Council, Nutrition education in U.S. medical schools, Washington, D.C., National Academy Press, 1985.
2. Weinsier, R. H., Boker, J. R., Brooks, C. M., et al., Priorities for nutrition content in a medical school curriculum: a national consensus of medical educators, *Am. J. Clin. Nutr.*, 50, 707–712, 1989.
3. Nestle, M., Nutrition in medical education: new policies needed for the 1990s, *J. Nutr. Educ.*, 90, 5–35, 1988.
4. Winick, M., Gunzburger, L., and Loesch, R., Meeting the American Society of Clinical Nutrition's standards in the medical school curricula, *Acad. Med.*, 66, 456, 1991.
5. Winick, M., Nutrition education in medical schools, *Am. J. Clin. Nutr.*, 58, 825–829, 1993.

INDEX

A

Absorption
  alteration of, with diarrhea and malnutrition, 119–122
  calcium, low birth weight infant, 35–36
  iron, 55
    and iron deficiency, 55
  magnesium, heritable, 198–202
  phosphorus, low birth weight infant, 36
Acesulfam potassium, nonnutritive dietary supplement, 230
Acidemia, isovaleric, 77
Acidity, gastric, diarrhea, 111
Acquired immunodeficiency, *see* AIDS
Adults, obesity in, 86
AIDS, diarrhea, 112
Alcohol exposure in utero, poverty and, 246
Alitame, sugar substitute, 231
Allergy, to food, 103, *see* Food allergy
Aluminum, 173
Anemia, iron deficiency, 56
Arrhythmia, with magnesium deficiency, 205–206
Arsenic, 175
"Artificial" feeding, of infants, 3
Aspartame, nonnutritive dietary supplement, 229–230

B

Bacterial overgrowth, and diarrhea, 116–117
Balance, of iron, and iron deficiency, 54–55
Bayle Scale of Infant Development, measure, and iron deficiency, 58
Behavioral development, iron deficiency, 58
Biotinidase, 77
Bone disease, parenteral nutrition, 192
Boron, 175
Brain growth, toddler, 49–50
Breast milk jaundice, 29
Breastfeeding, 21–31,
  breast milk jaundice, 29
  Cesearean section, 27
  engorgement, 25
  infection, maternal, 27
  jaundice, 29
  lactational infertility, 23
  let-down reflex, 23
  mastitis, milk duct, 27
  maternal fever, 27
  maternal illness, 28
  maternal nutrition, 28
  medication, maternal, 28
  milk duct, blocked, 27
  milk production, 22
  nipple
    inverted, 26
    sore, 26–27
  "nipple confusion," 24
  nursing technique, 24–26
  prolactin, 22
  prolactin-inhibiting factor, 23
Bronchopulmonary dysplasia, low birth weight infant, 38

C

Calcium, low birth weight infant, 34–37
Cardiomyopathy, magnesium deficiency, 197–224
  absorption, heritable, 198–202
  arrhythmia, 205–206
  congenital, 202–204
  cystic fibrosis, 209–210
  deafness, congenital, 205–206
  diabetic cardiomyopathy, 204–205
  Friedrich's ataxia, 206–209
  genetic, 204–211
  Jervell-Lange-Nielsen syndrome, 205–206
  Marfan's syndrome, 206–209
  maternal, 204–204
  muscular dystrophy, 206–209
  renal excretion, 198–202
  renal tubular magnesium wasting, 199–202
  sudden death, 205–206
  syncope, 205–206
  tissue levels, 198–202

Cardiovascular death, magnesium deficiency, 202–204
Cardiovascular disease
family history, 14
prevention of in adulthood, 13–20
Cesearean section, 27
Children, obesity in, 86–87
Cholecystitis, parenteral nutrition, 191
Cholelithiasis, parenteral nutrition, 191
Cholera epidemic, 1
Cholesterol
cardiovascular disease
family history, 14
prevention of in adulthood, 13–20
growth retardation, nutritional, 15
hypercholesterolemia, 14
lowering, interventions, by diet, 16–17
non-cardiovascular mortality, 15
screening, 13–20
with dietary intervention, 13–20
dietary intervention, cardiovascular disease prevention, 13–20
step 2 diet, 16
Cholesterol-lowering interventions, by diet, 16–17
Chromium, 166–169
Congenital hypertrophic pylori stenosis, gastrointestinal disease, 103
Controversial issues, in nutrition, 5–12
Copper, 164–166
Cyclamate, nonnutritive dietary supplement, 230–231
Cystic fibrosis, and cardiomyopathy, 209–210

D

Darrow, Daniel, 2
Deafness, congenital, magnesium deficiency, 205–206
Dental caries, nonnutritive dietary supplement, 232–233
Diabetes, nonnutritive dietary supplement, 232
Diabetic cardiomyopathy, 204–205
Diarrhea, 99–103
absorption, alteration of, with malnutrition, 119–122
acute, 99–102
AIDS, 112
bacterial overgrowth, 116–117
chronic, 99–102
enteric infection, development of, 108–113
fecalism, 109–110
host factors, 110–111
immune function, 112–113
nutritional practices, 110
gastric acidity, 111
gastrin, 111
gastroenteritis, 108
hemorrhagic colitis, 110–111
infectious, effect on intestine, 114–116
intractable, in infancy, 102
malnutrition, 107–135,
milk, 109
nutrition, interaction, 113–114
poverty, 109
therapy for, 122–125
Dietary intervention, with cholesterol screening, 13–20
cardiovascular disease prevention, 13–20
Dietary protein intolerance, 103
Dietary supplements, nonnutritive, *see* Nonnutritive dietary supplement
Discharge planning, parenteral nutrition, 188
Dwarfing, 144–145

E

E-Ferol, vitamin E supplement, low birth weight infant, 38
Egg yolk, iron deficiency, 53
Energy, importance of, in pediatric nutrition, 2
Engorgement, breastfeeding, 25
Enteric infection, development of, 108–113
fecalism, 109–110
host factors, 110–111
immune function, 112–113
nutritional practices, 110
Enterocolitis, necrotising, low birth weight infant, 38
Environmental causes, of obesity, 83–84
Ethnicity, and obesity, 87–88

Eugenics
  and "hereditary feeble-mindedness," 240–241
  and racist theory, 241–242

F

Family history, cardiovascular disease, 14
Fat substitute, nonnutritive dietary supplement, 225–227
Fecalism, 109–110
"Ferritin model," iron deficiency, 56
Ferrous furnarate, iron deficiency, 54
Ferrous sulfate, iron deficiency, 53
Fluoride, 174
Food allergy, 61–69
  clinical manifestations, 62
  diagnosis, 62–63
  immunologic mechanisms, 62
  management of, 63–67
Formula, iron-fortified, 54
Friedrich's ataxia, and magnesium deficiency, 206–209

G

Galactosemia, 76
Gamble, James, 2
Gastric acidity, diarrhea, 111
Gastrin, diarrhea, 111
Gastroenteritis, diarrhea, 108
Gastrointestinal disease, 99–105
  allergy, food, 103
  diarrhea, 99–103
    acute, 99–102
    chronic, 99–102
    intractable, in infancy, 102
  dietary protein intolerance, 103
  Hirschsprung disease, 103
  hypertrophic pylori stenosis, 103
  inflammatory bowel disease, 103
  malabsorptive disorders, 103
Gender-related differences, iron absorption, 55
Genetic causes, of obesity, 83
Growth, 143–157
  and dietary iron absorption, 54
  dwarfing, 144–154
  nutritional growth retardation, 143–154
  retardation, nutritional, 15

H

Hemorrhage, intraventricular, low birth weight infant, 38
Hemorrhagic colitis, diarrhea, 110–111
"Hereditary feeble-mindedness," eugenics and, 240–241
Hess, Alfred, 2
Hirschsprung disease, 103
HIV
  body composition, 138
  diagnosis of nutritional complication from, 139–140
  malnutrition, pathogenesis of, 139
  treatment of nutritional complications, 140–141
Human immunodeficiency virus, *see* HIV
Human milk, vitamin A, low birth weight infant, 39
Hypercholesterolemia, 14
Hyperlipidemia, nonnutritive dietary supplement, 233
Hypertrophic pylori stenosis, gastrointestinal disease, 103

I

Immune function, and enteric infection, 112–113
Infection
  diarrhea, effect on intestine, 114–116
  iron deficiency, 58
  maternal, 27
  parenteral nutrition, 189–190
Inflammatory bowel disease, 103
Intestine
  bacterial overgrowth, 116–117
  effect of infectious diarrhea on, 114–116
  malnutrition, effect on, 117–119
Intraventricular hemorrhage, low birth weight infant, 38
Iodine, 173–174
Iron deficiency, 53–60, 161–162
  absorption, 55
    gender-related differences, 55

anemia, 56
balance of iron, 54–55
Bayle Scale of Infant Development measure, 58
behavioral development, 58
consequences of, 57–59
egg yolk, 53
"ferritin model," 56
ferrous furnarate, 54
ferrous sulfate, 53
growth, and dietary iron absorption, 54
infection, 58
iron deficiency anemia, contrasted, 56–57
iron-fortified formula, 54
isotope study, 55
magnitude of problem, 55–56
poverty and, 245–246
rice cereal, 54
stable isotope studies, 54–55
studies, Mackay, H.M.M., 53
uptake of iron, 54, 57
Iron deficiency anemia
iron deficiency, contrasted, 56–57
non-hematologic complications, infant, 50–52
Iron-fortified formula, 54
Isotope study, iron, 54–55,
Isovaleric acidemia, 77

J

Jaundice, 29
Jervell-Lange-Nielsen syndrome, and magnesium deficiency, 205–206
Lactational infertility, 23
Latta, Thomas, 1
Lead poisoning, and poverty, 245
Learning failure, poverty, *see* also Mental development
connection, overview, 246–248
Let-down reflex, 23
Liebig, pediatric nutrition studies, 2
Lipids, parenteral nutrition, 190
Liver disease, parenteral nutrition, 191
Low birth weight infant
bronchopulmonary dysplasia, 38
calcium, 34–37
E-Ferol vitamin E supplement, 38
human milk feeding, 34
intraventricular hemorrhage, 38
necrotising enterocolitis, 38
nutrition, 33–42
optimal nutrition, 34
phosphorus, 34–37
PTH secretion, 34
retinol binding protein, 39
retinopathy of prematurity, 38
vitamin A, 35, 38–39
in human milk, 39
parenteral nutrition, 39
vitamin D, 35–37
vitamin E, 35, 37–38
vitamin E supplement, 38

M

Mackay, H.M.M., iron deficiency, studies, 53
Magnesium deficiency
cardiomyopathy, 197–224
absorption, heritable, 198–202
arrhythmia, 205–206
congenital, 202–204
cystic fibrosis, 209–210
deafness, congenital, 205–206
diabetic cardiomyopathy, 204–205
Friedrich's ataxia, 206–209
genetic, 204–211
Jervell-Lange-Nielsen syndrome, 205–206
Marfan's syndrome, 206–209
muscular dystrophy, 206–209
renal excretion, 198–202
renal tubular magnesium wasting, 199–202
sudden death, 205–206
syncope, 205–206
tissue levels, 198–202
cardiovascular death, 202–204
malabsorption, 198
maternal, 202–204
Malabsorption, magnesium, 198
Malnutrition
absorption, alteration of, with diarrhea, 119–122
diarrhea, 107–135

therapy for, 122–125
effect on intestine, 117–119
with HIV
diagnosis of, 139–140
pathogenesis of, 139
treatment, 140–141
and mental development, 242–245
Manganese, 169–170
Maple syrup urine disease, 76–77
Marfan's syndrome, and magnesium deficiency, 206–209
Mastitis, milk duct, 27
Maternal fever, 27
Maternal illness, 28
Maternal magnesium deficiency, 202–204
Maternal nutrition, 28
Medical school education, in pediatric nutrition, 251–254
Megacolon, 103
congenital, 103
Mellanby, pediatric nutrition studies, 2
Mental development, and malnutrition, 242–245
Metabolism
inborn errors of, 71–80
biotinidase, 77
galactosemia, 76
isovaleric acidemia, 77
maple syrup urine disease, 76–77
phenylketonuria, 72–75
maternal, 75–76
urea cycle defects, 78
obesity, 82–83
parenteral nutrition, 190–191
Milk
diarrhea, 109
duct, blocked, 27
breastfeeding, 27
production, 22
Molybdenum, 170–171
Muscular dystrophy, and magnesium deficiency, 206–209

N

Necrotising enterocolitis, low birth weight infant, 38
Nickel, 175
Nipple
inverted, 26
sore, 26–27
"Nipple confusion," 24
Non-hematologic complications, with iron deficiency anemia, infant, 50–52
Nonnutritive dietary supplement
cyclamate, 230–231
fat substitute, 225–227
pediatric applications, 232–234
dental caries, 232–233
diabetes, 232
hyperlipidemia, 233
obesity, 233–234
sugar substitute, 227–232
acesulfam potassium, 230
alitame, 231
aspartame, 229–230
saccharin, 228–229
sucralose, 231–232
Nursing technique, 24–26
Nutritional growth retardation, 143–154
Nutritional science, overview
"artificial" feeding, of infants, 3
cholera epidemic, 1
Darrow, Daniel, 2
Gamble, James, 2
Hess, Alfred, 2
Latta, Thomas, 1
O'Shaughnessy, William Brooke, 1
Park, Edward, 2
Powers, Grover, 2

O

Obesity, 81–98
in adults, 86
causes, 82–86
in children, 86–87
determining, criteria for, 89–90
environmental causes, 83–84
ethnicity and, 87–88
genetic causes, 83
metabolic assessment, 92–95
metabolic causes, 82–83
nonnutritive dietary supplement, 233–234
nutritional assessment, 92–95
nutritional causes, 83–84

treatment of, problems, 90–92
O'Shaughnessy, William Brooke, 1

P

Parenteral nutrition
access, 186–187
assessment, ongoing, 193
bone disease, 192
cholecystitis, 191
cholelithiasis, 191
complications of, 188–190
contraindications, 184–186
discharge planning, 188
formula, 187–188
implementation, 186
indications, 184–186
infection, 189–190
lipids, 190
liver disease, 191
metabolic problems, 190–191
mortality, 192
platelets, 190
renal disease, 192
trace elements, 159–181, 190–191
aluminum, 173
arsenic, 175
boron, 175
chromium, 166–169
copper, 164–166
fluoride, 174
iodine, 173–174
iron, 161–162
manganese, 169–170
molybdenum, 170–171
nickel, 175
rubridium, 175
selenium, 171–173
silicon, 174
vanadium, 175
zinc, 162–164
visual function, 192
vitamin A, low birth weight infant, 39
vitamins, 191
Park, Edward, 2
Pediatric nutrition, *see* also under specific headings noted below
breastfeeding, 21–31
cholesterol screening, 13–20
controversial issues in, 5–12
diarrhea, and malnutrition, 107–135
dietary supplements, nonnutritive, 225–238
food allergy, 61–69
gastrointestinal disease, 99–105
growth, 143–157
historical overview, 1–3
HIV, 137–141
iron deficiency, 53–60
low birth weight infant, 33–42
magnesium deficiency in cardiomyopathy, 197–224
medical school education in, 251–254
metabolism
inborn errors of, 71–80
obesity, 81–98
parenteral, trace elements in, 159–181
parenteral nutrition, 183–196
poverty, influence of, 239–250
residency training in, 251–254
Phenylketonuria, 72–75
maternal, 75–76
Phosphorus, low birth weight infant, 34–37
Platelets, parenteral nutrition, 190
Potassium, acesulfam, nonnutritive dietary supplement, 230
Poverty
alcohol exposure in utero, 246
diarrhea, 109
"hereditary feeble-mindedness," eugenics and, 240–241
iron deficiency, 245–246
lead poisoning, 245
learning failure, connection, overview, 246–248
malnutrition, and mental development, 242–245
racist theory, eugenics and, 241–242
Powers, Grover, 2
Prematurity, retinopathy of, low birth weight infant, 38
Prevention of cardiovascular disease, in adulthood, 13–20
Prolactin, breastfeeding, 22
Prolactin-inhibiting factor, 23
Pylori stenosis, hypertrophic, gastrointestinal disease, 103

R

Racist theory, eugenics and, 241–242
Renal disease, parenteral nutrition, 192
Renal excretion, magnesium deficiency, 198–202
Renal tubular magnesium wasting, 199–202
Residency training, in pediatric nutrition, 251–254
Retardation of growth, *see* Growth
Retinol binding protein, low birth weight infant, 39
Retinopathy of prematurity, low birth weight infant, 38
Rice cereal, iron, 54
Rubridium, 175
Ruebner, pediatric nutrition studies, 2

S

Saccharin, 228–229
Selenium, 171–173
Silicon, 174
Step 1 diet, 17
Step 2 diet, 16
Sucralose, sugar substitute, 231–232
Sudden death, with magnesium deficiency, 205–206
Sugar substitute, nonnutritive dietary supplement, 227–232
  acesulfam potassium, 230
  alitame, 231
  aspartame, 229–230
  cyclamate, 230–231
  saccharin, 228–229
  sucralose, 231–232
Syncope, with magnesium deficiency, and cardiomyopathy, 205–206

T

Toddler
  brain growth, 49–50
  diet of, 46–49
  weaning, food, 44–46
Trace elements, parenteral nutrition, 159–181, 190–191
  aluminum, 173
  arsenic, 175
  boron, 175
  chromium, 166–169
  copper, 164–166
  fluoride, 174
  iodine, 173–174
  iron, 161–162
  manganese, 169–170
  molybdenum, 170–171
  nickel, 175
  rubridium, 175
  selenium, 171–173
  silicon, 174
  vanadium, 175
  zinc, 162–164

U

Uptake, iron, and iron deficiency, 54
Urea cycle defects, 78

V

Vanadium, 175
Visual function, parenteral nutrition, 192
Vitamin A
  human milk, low birth weight infant, 39
  low birth weight infant, 35, 38–39
  maternal status, 38
  parenteral nutrition, low birth weight infant, 39
Vitamin D, low birth weight infant, 35–37
Vitamin E
  low birth weight infant, 35, 37–38
  supplement, E-Ferol, low birth weight infant, 38
Vitamins, parenteral nutrition, 191

W

Weaning, food, 44–46

Z

Zinc, 162–164